Shell Structures

Mitchell Gohnert

Shell Structures

Theory and Application

 Springer

Mitchell Gohnert
University of the Witwatersrand
Johannsesburg, South Africa

ISBN 978-3-030-84809-5   ISBN 978-3-030-84807-1   (eBook)
https://doi.org/10.1007/978-3-030-84807-1

This Springer imprint is published by the registered company Springer Nature Switzerland AG
The registered company address is: Gewerbestrasse 11, 6330 Cham, Switzerland

To my wife and children, whom I adore.

Foreword

From our birth and through life's expectations, we grow and develop into the person we become. The path we follow through life, with its many turns, mountains to climb and crossroads to choose, eventually leads us to our place today, and our final destination. Professor Mitchell Gohnert, my former student at Brigham Young University, has experienced a lifelong desire to be a structural engineer, with a special interest in the design and application of shell structures. This book is the result of his lifelong quest.

Just as we are influenced by the teaching and training of others, we also influence others by the teaching and training we give throughout our lives. Mitch has gained much from the many people he has been involved with, and conversely, he has influenced, assisted and guided many through his mentorship and his expertise in the field of shell structures.

Good ideas, called inspiration by some, come to each of us during our life. When acted upon with persistence and determination, these ideas result in great accomplishments that benefit others.

Emeritus Professor, Brigham Young   Arnold Wilson
University, Provo, UT, USA

Contents

Chapter 1
Introduction

Abstract Shells, and in particular domes, have a symbolic meaning in many of the world's civilizations. Some of the most iconic structures incorporate domes as their architectural center piece. Interestingly, most of these structures are religious gathering places, shrines, or temples and are found in cathedrals, stupas, mausoleums, and mosques. The symbolism of domes is of ancient origin and is associated with deity and the celestial expanse of the heavens. The home of the ancients appeared to be under a heavenly dome, and the circle of the dome, a never ending line, is a symbol of eternity. This symbolism continues to this day, but has evolved over the centuries to include other closely associated meanings. Alexander the Great, and early Persian rulers, adopted domes in their architecture to symbolize their divine authority and power (Smith, The dome: study in the history of ideas. Princeton University Press, New Jersey, 1950; Hiscock, The symbol at your door. s.l.:SP Birkhauser Verlag Basel, 2010). This practice has endured to modern times, and domes are seen in some of the most famed architectures, such as the US Capitol Building, the Vatican, and many other religious and governmental buildings around the world. The intent is to portray strength, influence, legitimacy of power, and piety. For these reasons, religious and centers of government will always include domes in their architecture.

Shells, and in particular domes, have a symbolic meaning in many of the world's civilizations. Some of the most iconic structures incorporate domes as their architectural center piece. Interestingly, most of these structures are religious gathering places, shrines, or temples and are found in cathedrals, stupas, mausoleums, and mosques. The symbolism of domes is of ancient origin and is associated with deity and the celestial expanse of the heavens. The home of the ancients appeared to be under a heavenly dome, and the circle of the dome, a never ending line, is a symbol of eternity. This symbolism continues to this day, but has evolved over the centuries to include other closely associated meanings. Alexander the Great, and early Persian rulers, adopted domes in their architecture to symbolize their divine authority and power (Smith 1950; Hiscock 2010). This practice has endured to modern times, and domes are seen in some of the most famed architectures, such as the US Capitol Building, the Vatican, and many other religious and governmental buildings around

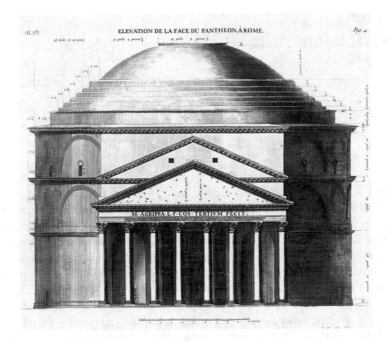

Fig. 1.1 Antoine Desgodetz's 1682 rendition of the Pantheon (Claude-Antoine Jombert)

the world. The intent is to portray strength, influence, legitimacy of power, and piety. For these reasons, religious and centers of government will always include domes in their architecture.

The endurance of this symbolism is fitting and appropriate, considering the seemingly mystical strength and the eternal nature of these structures. If we go back in history more than a thousand years, remnants of antiquity usually amount to nothing more than a pile of rubble, with the exception of a few columns standing here and there. However, of the structures that have survived, the vast majority are domes and arches. The Pantheon in Rome (126 AD) (Fig. 1.1) and Hagia Sophia in Turkey (537 AD) (Fig. 1.2) are prime examples. Both of these structures were built in seismic active regions and have been subjected to several major earthquakes; yet they stand today and are in remarkably good condition.

Despite the religious and political affiliations of domes, the emergence of shell structures is undoubtedly the result of a practical need for a wide-open space, without being inhibited by column supports. Shells are able to span over enormous spaces, and therefore are ideal for places of large gatherings, such as religious assemblies, theaters, and sport arenas.

The key to understanding the strength of shells is to focus on the shape. As will be discussed in later chapters, load flows in a natural and predictable pattern. By matching the shape of the structure with the natural flow of loads, an optimal structure is created. This matching of form and flow results in materially efficient structures that are able to resist a multitude of severe loading events. Unfortunately,

Fig. 1.2 Dome of Hagia Sophia (Wikimedia Commons, Robert Raderschatt)

the majority of modern structures are linear and block-like, which results in a mismatch in structural geometry and the natural flow of stress. Where a mismatch occurs, bending moments and shears are the result. Engineers are able to design buildings to resist bending and shear forces, but at a tremendous cost to structural strength and material economy.

Should nature be our mentor? Interestingly, many shell shapes are found in nature. Biological structures are never square or angular, but are curved, and composed of shells and tubes. Albert Einstein once stated, "Look deep into nature, and then you will understand everything better." It makes perfect sense to pattern our structures after biological forms, which will include the design of shell structures. It should be noted that after cataclysmic events, albeit meteorological or seismic, it is usually the biological structures that survive. For instance, consider Fig. 1.3. The photograph depicts the destruction of the Haiti National Palace after the 2010 earthquake. The palace is in ruins, but the biological structures in the forefront (the palm trees) do not show any discernible damage. The biological structures have survived, but the manmade structures have not. Biological structures have, perhaps, evolved over millions of years into forms that are resistant to local calamitous events; this has led to the survival of the species. Strangely, nature is all around us, yet its connection to structural engineering has gone largely unnoticed; we fail to see the obvious or learn from our surroundings. Interestingly, the National Palace incorporated shells in its roof, which is the only part of the structure that has survived, although displaced from its original position.

Fig. 1.3 Manmade and biological forms in an earthquake zone (Wikimedia Commons file, Logan Abassi)

Most of the shells derived in this book are those of regular and familiar shapes, such as cylinders, spheres, cones, and ellipsoids. However, these shapes are not always found in nature, and therefore certain inefficiencies are found in the shells we design. Fortunately, the effects of the mismatch are usually restricted to the boundaries, or foundations, and have limited impact due to Saint Venant's principle (i.e., diminishing stress with distance). Shells, for the most part, carry the load by membrane action (in-plane compression or tension stresses) and are highly efficient load carrying structures. Because shells carry stresses efficiently, only minimal material is required, and this is the reason why these structures are often referred to as thin shells. The shell does not need to be thick to carry load. As a general rule, the generally accepted classification of a thin shell is where

$$\frac{r}{h} > 20 \tag{1.1}$$

where r is the radius and h is the thickness.

This rule, however, is only a "rule of thumb." Structures have been designed and built with smaller or much greater ratios. For example, St. Peter's cathedral in Rome has a radius to thickness ratio of 14, and the Smithfield Market shell in London has a ratio of 900. High ratios can also be seen in nature; the ratio of an ordinary chicken egg is about 100 (Heyman 1995). Both these structures (i.e., manmade and biological) have performed exceptionally well, especially the chicken egg. The material of an egg shell is composed of calcium carbonate crystals and is semipermeable. The shell material is relatively weak. Yet, the strength of an egg is, without question, the result of its shape. If one takes an egg and squeezes it in hand, membrane stresses are produced in the shell, and it is almost impossible to break it. To crack an egg, a deliberate sharp blow is required. Despite having a large radius to thickness ratio, an egg is able to resist loads that far exceed typical scaled up manmade structures. However, it is not the egg shell material that resists the loads, but its strength is derived from its shape. Thus, the key to structural strength lies primarily in the shape.

Traditionally, all structural designs are based on the consideration of three criteria: strength, stiffness, and stability (Heyman 1995). As will soon be apparent, shell structures are intrinsically strong. The stress levels are usually only a fraction of the capacity of the shell. This gives an enormous factor of safety against failure, and is the reason why shell structures tend to survive cataclysmic events. Stiffness is usually also a noncritical consideration in shell structures. Stiffness, as a design criteria, is related to deflections. Typically, the deflections in domes are relatively small, especially when compared to framed structures. In arches, stability is related to the formation of hinges and collapse mechanisms. In domes, stability is controlled by the buckling capacity of a given shape.

The favorable characteristics of shell structures have not gone unnoticed by our engineering forefathers. They understood the benefit of matching the natural flow of stress with the axial geometry of the structure. Although they did not have access to the mathematical and analytical tools we have today, they had physical models that gave insight into optimal structural forms. Engineer-architects, such as Antoni Gaudi, designed the Sagrada Familia from a series of hanging chains (Fig. 1.4) (Huerta 2006). Chains have very unique structural properties: A hanging chain is a pure tension structure, and void of bending moments and shears. If the shape is inverted, the form becomes an arch, and the stresses reverse to pure compressions (which is ideal for masonry and concrete structures). Although a hanging chain is a two-dimensional form, the shape is often transformed into three dimensions, creating catenary or funicular domes, which was the source of Gaudi's creations.

The use of chains in design seems to have its beginnings in the design of St. Paul's Cathedral in London (Fig. 1.5) (Heyman 1998). Christopher Wen, the architect, lived in the time of Robert Hooke. Hooke is credited as being the first to understand the connection of a hanging chain and its application to structural forms (Heyman 1998; Block et al. 2006); Wen undoubtedly sourced and applied these concepts into the design of St. Paul's Cathedral. The dome of St. Paul is a composite of three domes: The outer and inner domes are merely decorative, but the center conical dome is the structural support. The conical shape is due to the heavy weight of the lantern, which is located on top of the shell. If weights are hung from a hanging chain to represent the self-weight of the shell, and proportionally heavier weights are placed near the bottom of the chain to model the lantern, the geometry of the chain will be conical in shape. History has implied that these models were incorporated into the design of St. Paul's Cathedral and represent a major innovation in shell design.

Hanging chains have an additional valuable feature. A thin chain represents the location of the thrust-line, or the center of the flow of stress. If the geometry of the

Fig. 1.4 Gaudi's Sagrada
Familia (Wikimedia
Commons file)

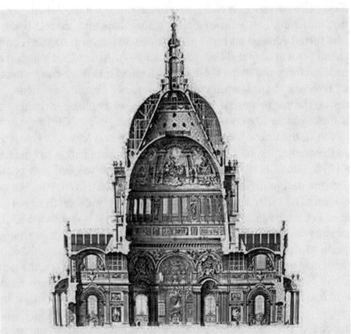

Fig. 1.5 Wen's drawing of the three domes of St. Paul's Cathedral

shell matches the thrust-line, an optimal structure is achieved. Wen's cathedral achieved this objective, using chains as an engineering tool.

This book was written with several purposes in mind: The first purpose is to provide a complete derivation of the theory of shells. A literature review will reveal that most books on shells are not complete, if one wishes to understand the mathematics behind shell theory. Typically, only a summary of key equations are given; this is understandable, considering the complexity and enormity of the derivation. For an academic, however, this leaves a void in understanding shell theory. For the engineer, it is equally important to remove the proverbial "black box" approach to shell design. For this reason, a complete derivation is given. The second purpose is to provide a step-by-step procedure of shell analysis. For the most part, shell solutions require a sequence of analytical steps, which is not always clearly explained in many texts on shell theory. The approach adapted is akin to a "cook book" approach, where an outcome is achieved by following a specified sequence of analytical steps. The final purpose is to introduce concepts and theories on matching the geometry of the shell to the natural flow of stresses, and therefore funicular and catenary shells will be introduced. These shapes are, in fact, the most structurally efficient forms known to humankind.

An additional purpose is to provide a brief description of construction materials, with a focus on how they react to stresses and patterns of stress flow. The objective is to understand the limitations of materials, and how best to apply these materials to achieve the most durable and robust shell. Although a discussion on construction materials is rarely combined with shell theory, the conjugal relationship of this information may radically alter our choice of shell, the materials we use, and its geometric proportions.

References

Block, P., Dejong, M. & Ochsendorf, J., 2006. As Hangs the Flexible Line: Equilibrium of Masonry Structures. *Nexus Network Journal.*

Heyman, J., 1995. *The Stone Skeleton, Structural Engineering of Masonry Architecture.* Cambridge: Cambridge University Press.

Heyman, J., 1998. *Structural Analysis, A Historical Approach.* Cambridge: Cambridge University Press.

Hiscock, N., 2010. *The Symbol at your Door.* s.l.:SP Birkhauser Verlag Basel.

Huerta, S., 2006. Structural Design in the Work of Gaudi. *Architectural Science Review,* 49(4), pp. 324–339.

Smith, E. B., 1950. *The Dome: Study in the History of Ideas.* New Jersey: Princeton University Press.

Chapter 2
Construction Materials and Stress Flow

Abstract This chapter covers background information on stress and strain, shear failures, economy of forces, the flawed nature of construction materials, stress patterns in curved walls, membrane stresses, stress flow, openings in shell walls, and the basic optimal shapes that ensures that stress flows within the plane of the wall, or axis of the member. This information is given to assist the designer in selecting the most suitable construction materials, provide an understanding of how materials react to stress, and the type of shell that is most appropriate for the design.

2.1 Introduction

A basic understanding of stresses and strains seems to elude many practitioners and researchers alike, yet it is fundamental to the understanding of structural materials and the basic principles of design. When clearly understood, one begins to understand why shells are so efficient. We begin to understand why our current alliance with linear box-like structures is highly inefficient, and wasteful of materials. We begin to understand what stresses are preferred and why others should be avoided. We also understand how forces flow in structures and how to match these flows with structural forms to achieve an optimal design.

In this chapter, we briefly explore fundamental concepts of stress and strain, misconceptions related to shear forces, the inefficiencies of designing with bending moments, and the effects of flaws (i.e., microcracks) in construction materials. We also review how forces flow in membrane structures and stress concentrations around openings. All of these concepts are necessary to achieve the most optimal structural design. Most importantly, it enhances our ability to create shell structures that are robust, durable, and time enduring.

M. Gohnert, *Shell Structures*, https://doi.org/10.1007/978-3-030-84807-1_2

2.2   The Basic Characteristics of Stresses and Strains

The basic characteristics of stress and strain in structural members may be broken down into fundamental principles, which are expressed in a series of postulates. The first postulate states:

1. *Only two fundamental stresses exist in structural members, which are tensions and compressions*

   In structural members, there are three types of possible forces: bending moments, shears, and axial loads. These forces are, in fact, resultants of integrated stresses, which may be broken down into fundamental stresses, which are tensions and compressions. As shown in Fig. 2.1, bending moments are merely a force couple, composed of tensions and compressions, characterized by a linear variation of stress across a section. An axial stress is either compressive or tensile. Therefore, without dispute, bending moments and axial forces comply with the postulate.

   The question arises: Is shear composed of tensions and compressions as postulated? The answer is yes. However, shears are components of tensile and compressive stresses. In Fig. 2.2, the left differential element illustrates the stress pattern for a pure shear case, which is approximately true near the ends of a simply supported beam. The shears are oriented vertically and horizontally with equal magnitudes, but oriented in different directions. The shear stresses may be transformed into an

Fig. 2.1 Stress couple of a bending moment

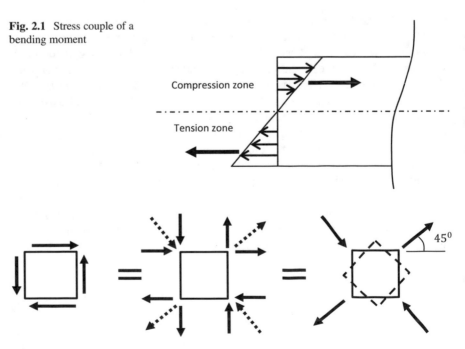

Fig. 2.2 Shear stresses are composed of tension and compression stresses, oriented at 45° from the shear plane

Fig. 2.3 Mohr's circle for
the case of pure shear

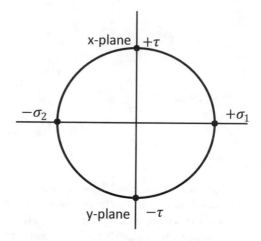

equivalent differential element of tensions and compressions, applied at 45° (the right differential element of Fig. 2.2), by assuming that shears are components of tensions and compressions. These stresses are the principal (maximum) stresses. This assumption is validated by Mohr's circle (Timoshenko 1958).

Mohr's circle is given in Fig. 2.3 for the pure shear case. The shears on the vertical and horizontal planes of the differential element are plotted on the Mohr's circle as $x(0, +\tau)$ and $y(0, -\tau)$. Rotating 90° from these points will give the principal stresses, which are tensions and compressions. As a reminder, the angles in Mohr's circle are double to the actual values. This leads to the second postulate:

2. *Shear stresses are components of tensile and compressive stresses, applied at an angle to the x and y axes*

Therefore, what we call a shear stress is actually composed of tensions and compressions, applied at an angle to the x and y axes. Furthermore, if the shear stress is combined with bending, the principal stresses will be located at some other angle to the x or y axis (not necessarily at 45°). Whatever the case may be, shears are components of tension and compression stresses.

The third postulate states:

3. *In construction materials, two stresses will always exist, and orientated orthogonally to one another*

Poisson's theory states that a strain applied in one direction will cause an opposite strain, perpendicular to the applied load. Thus, a compressive strain applied in one direction will cause a tension strain at 90°, and vice versa (Ohlsen et al. 1976).

Furthermore, Hooke's law gives the relationship of stress, strain, and Poisson ratio, given by Eqs. 2.1 and 2.2 for plane or membrane structures.

$$\varepsilon_x = \frac{1}{E}\left(\sigma_x - v\sigma_y\right) \qquad (2.1)$$

$$\varepsilon_y = \frac{1}{E}\left(\sigma_y - v\sigma_x\right) \qquad (2.2)$$

Rearranged,

$$\varepsilon_x E = \sigma'_x = \sigma_x - v\sigma_y \qquad (2.3)$$

$$\varepsilon_y E = \sigma'_y = \sigma_y - v\sigma_x \qquad (2.4)$$

Thus, a stress applied in one direction will cause an opposite stress in the orthogonal direction, and the magnitude is proportionate to Poisson's ratio.

Stress, however, is the internal resistance to strain. Internal resistance occurs in most construction materials and structural elements, which include membrane structures (i.e., shell structures). Materials such as reinforced concrete, masonry, and structural steel have sufficient depth and width and are bulky or constructed of plate elements that provide internal restraint. The interaction of the web and flanges of I-beams and reinforcement in concrete are sources of internal restrainment. In concrete columns, we call this restrainment confinement steel. In most structural elements, orthogonal stresses are caused by Poisson's effect, as postulated by Eqs. 2.3 and 2.4. Only in a few select cases will this not apply, such as long thin bars and thermal loading in unrestrained members. In shell structures, Poisson's effect is usually ignored in the direction of the thickness, but in-plane, the stiffness of the wall material and internal restraint will cause stresses in two orthogonal directions. This is critical, considering that many construction materials (i.e., concrete and masonry) are adversely affected by tension stresses. Thus, it is typically assumed (i.e., derivation of shell theory) that a stress applied in one direction will cause a stress in the opposite direction (Billington 1982; Chatterjee 1971; Zingoni 1997; Domone and Illston 2010).

The fourth postulate states:

4. *Only two modes of material failure exist: tension and compression*

If only two stresses exist, then it is only rational to conclude that only two material failure mechanisms exist—tension and compression. In thin shells, or plated structures, buckling is the third possible mode of failure, but this is referred to as a geometric failure, rather than a material failure. The question arises: Does the postulate apply to what we call a shear failure?

In Fig. 2.4, a reinforced concrete beam was tested to failure. Cracking in the beam is visible, extending from the support toward the first load point. The observed failure line is typically referred to as a shear crack. Literature has repeatedly described the mechanisms by which shear forces affect structures, and how they are resisted: dowel action, aggregate interlock, interface shear transfer, etc. (Kong and Evans 1987). However, what we call a shear failure is not congruous with shear failure mechanisms.

Fig. 2.4   Shear failure at the end of a beam. (Photo by author)

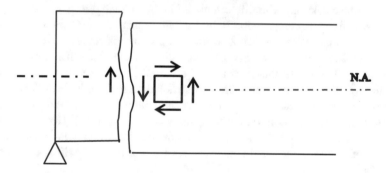

Fig. 2.5   Shear failure mechanism

Near the end of a simply supported beam, where shear is maximum and the bending moment is minimum, the stresses resemble the pure shear case, as illustrated in Fig. 2.5. As observed, the shear stresses are orientated vertically and horizontally, which would precipitate failure patterns in the same direction; a shear failure would be vertical or horizontal. However, vertical failure lines rarely form (if ever). In the horizontal direction, failure sometimes occurs along the interface between composite materials (i.e., between in situ concrete and precast concrete or between a concrete slab and steel beam), but these are usually the only cases where shear failures occur. The diagonal failure line in Fig. 2.4 is not vertical or horizontal, which would imply that the crack is not the result of shear stresses. Furthermore, shear failure would cause material slippage along the failure plane, which is not observed. Typically, what we call a shear failure, a crack commences at the bottom of the beam, propagates at an angle, and terminates near the top of the beam. The movement of the crack is 90° to the direction of the crack (movement of a shear crack would be parallel to the direction of the crack), which would imply that the observed failure is a tension failure. Unfortunately, most shear failures are misinterpreted, or misunderstood. In the majority of cases, shear cracks are actually diagonal tension cracks. Nevertheless, once we understand that shear stresses are components of principal tension stresses (see Fig. 2.2), then the mechanisms of failure become clearer to our understanding. This leads to the fifth postulate:

5. *In construction materials, such as reinforced concrete and masonry, the shear and compressive strength is much greater than the tensile strength, thus tensile failures precipitates*

Shear forces exist, but most construction materials rarely fail in shear. Concrete and masonry are highly anisotropic. In these materials, the shear and compressive strengths are much greater than the tensile strength; thus, failures in these materials are primarily tensile. In fact, almost all the cracking we see in concrete and masonry are from tensile stresses. The old adage, "a chain will break at its weakest link," is true for concrete and masonry; the weakest link is the tensile strength.

When testing concrete cubes for compression strength, the mode of failure is puzzling—the cracks are primarily orientated in the vertical direction, parallel to the compression force. A compression failure would precipitate horizontal failure lines, rather than vertical. What we call a compression failure is actually a tension failure (i.e., the cracks are vertical, and consistent with a tensile mode of failure). Tension strains are orthogonal to the direction of the applied compression force, and the magnitude of the tension strains are proportional to Poisson's ratio. The tension strength is commonly estimated as 10% of the compression strength, but a better estimate is the value of Poisson's ratio (e.g., $v = 0.12$, or 12%). However, we can still refer to a compression test as a compression test, but with an understanding that the point of failure is the maximum compression the material will resist, before a tensile failure occurs.

The fifth postulate does not necessarily apply to structural steel. Structural steel elements (i.e., I-beams, angles, channels, etc.) are composed of an arrangement of thin plates, and the predominant mode of failure is buckling. However, a vast number of shells are constructed of concrete or masonry (almost all shells found in antiquity), and therefore the final postulate is applicable to shells constructed of these materials.

2.3   Economy of Stresses

An important concept to improve the strength of shell structures (or any structure) is to understand that some forces are expensive forces and others are far more economical, in terms of stress magnitudes and the material that is required to resist that stress. To illustrate this concept, a simple example is given in Fig. 2.6. In the figure,

Fig. 2.6 A force applied to a beam and column of the same size

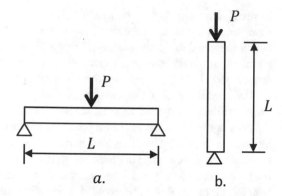

Table 2.1 Stress ratios of various loading and support configurations

Load condition	End condition	L/h	Stress ratio (r)
Point load	Simply supported	16	24
Uniform distributed load	Simply supported	16	12
Point load	Fixed supports	28	21
Uniform distributed load	Fixed supports	28	14

two identical members support the same magnitude of load. However, in Fig. 2.6a, the load is resisted by bending in a beam, and in Fig. 2.6b, the load is resisted by a column.

The maximum bending stress (σ_b) in the beam (Fig. 2.6a.) is expressed by Eq. 2.5 for an elastic distribution of stress.

$$\sigma_b = \frac{3PL}{2bh^2} \tag{2.5}$$

where L is the length, h is the depth of the beam, and b is the width.

If the same load is resisted by a column of the same length and size, the maximum compressive stress (σ_c) (Fig. 2.6b) in the column is given by Eq. 2.6.

$$\sigma_c = \frac{P}{bh} \tag{2.6}$$

If the bending stress (σ_b) is divided by the column stress (σ_c), the ratio of the stresses (r) is given by Eq. 2.7.

$$r = \frac{\sigma_b}{\sigma_c} = \frac{3L}{2h} \tag{2.7}$$

The value (L/h) is the span-to-depth ratio, which usually falls within a typical range of values, depending on the code of practice. Table 2.1 is a list of stress ratios for a variety of loads and support conditions. The stress ratio represents the increase in stress in a beam (subjected to a bending moment) compared to an axially loaded member. As indicated in the first row of Table 2.1, the stress in a simply supported beam, subjected to a point load, is 24 times higher than the stress in an axially loaded member. Similarly, the stress increase in other beam load/support configurations ranges from 12 to 21 times the stress in a column.

This simple example demonstrates that resisting forces by bending is highly uneconomical, compared to axial resistance. In both cases, an identical structural member is used to resist the same point load. The only difference is that in one case the load is carried by bending, and in the other, the load is carried along the axis of the member. By carrying the load along the axis, the stresses are reduced by more than 90%. Therefore, to achieve an optimal structure (where stresses are minimized), bending moments should be avoided. If bending moments are avoided, significantly less construction materials are required; the structure is more materially efficient and the shell is stronger (i.e., the shell has a greater capacity to resist abnormal increases in stress levels). The question may arise: Is it possible to avoid bending moments? Yes, but never entirely. However, Table 2.1 clearly demonstrates that greater efficiency is achieved by directing the flow of stress along the axis of the member. In the example, the flow of stress is directed along the axis of a column. In shells, the same applies—higher structural efficiency is achieved by directing the flow of stress along the axis, or in the plane of the shell. The flow of stress along the axis is referred to as membrane stresses. An inherent characteristic of shells is that the stresses are primarily membrane, which is the source of its structural efficiency. The premise is clear; bending moment should be avoided to achieve an optimal and efficient structure.

For the most part, bending moments in shells are eliminated, except, perhaps, near the base or foundation of the shell. These moments are termed "boundary effects," and usually only influence a small portion of the shell. The axis, however, does not necessarily have to be straight or linear (like a column). Stress will flow along the axis or middle surface of a shell, despite being curved. The geometry of the shell, however, will determine the magnitude of the stress, and whether or not the stress is compressive or tensile.

2.4   The Flawed Nature of Construction Materials

In the past, the majority of shells were constructed of masonry. Today, a large portion of the shells are constructed of reinforced concrete. Concrete as well as most masonry materials are anisotropic and, to a greater or lesser extent, plagued with flaws. As a building material, concrete and masonry are poor structural materials. To clarify this assertion, the compression strength is far greater than the tensile strength, and therefore concrete is not ideal. An ideal material will have equal strengths in tension and compression. It is well known that the tensile strength of concrete is only about a tenth of the compressive strength. Masonry responds similarly to concrete. In an age when so many strong materials are available, we may wonder why we use

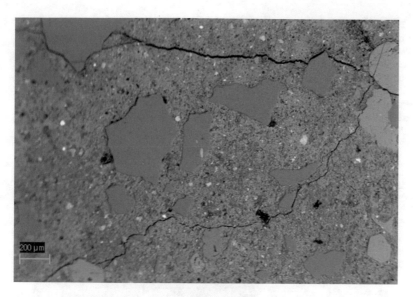

Fig. 2.7 View of a concrete sample using an electron microscope. (Photo by Y Ballim)

such poor materials in construction; the answer is simple: concrete is affordable. Because it is cheap, we are motivated to overlook its flaws and look for ways to compensate for its poor strength in tension (i.e., the inclusion of reinforcing steel). The question may arise: Why are these construction materials so weak in tension? The answer lies in the structure of the matrix (the composition of the material after hardening). In Fig. 2.7, with the use of an electron microscope, a magnified view of the matrix may be seen. An assortment of fine and course aggregates are easily identified, as light and darker shapes. However, on close inspection, numerous microcracks can also be seen. These microcracks are a source of weakness in tension (Popovics 1998).

To illustrate this concept, if we apply an even tension to a piece of paper, it is difficult to cause a rupture in the paper (Fig. 2.8a). However, if we make a small rip on the edge of the paper, only a very small tension stress is required to pull the paper apart (Fig. 2.8b). This simple example is analogous to what occurs in many construction materials—flaws (microcracks) significantly reduce the strength of the material in tension and is the primary reason why the tension strength of concrete is only about 10% of the compressive strength. Unfortunately, all construction materials are plagued with flaws, to a greater or lesser extent. In concrete, the boundaries between the different components of the matrix are weak, and tend to fracture. Inspecting Fig. 2.7 will reveal that crack propagation often follows along the boundaries of the materials.

In Fig. 2.9, an ideal (flawless) material is illustrated on the left, subjected to an even tension stress. In the center and the right-hand side, the same material is given, but with flaws of differing lengths. Flaws, in the form of cracks, obstruct the flow of stress. The stress is required to flow around the crack, which increases in intensity

Fig. 2.8 An illustration
with paper to show the
effects of a microcrack in
construction materials
(photos by author). (**a**)
Tension applied to flawless
paper; (**b**) paper is easily
ripped if a flaw (small rip) is
introduced

a.   Tension applied to flawless paper

b.   Paper is easily ripped if a flaw (small rip) is introduced

Fig. 2.9 Stresses
distribution in a tension
member with cracks

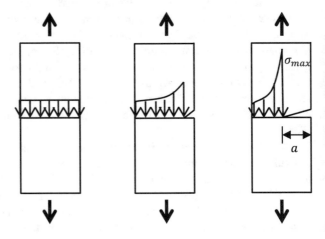

around the crack tip. The flow of stress is very similar to the flow of water through a
channel. If we place an obstruction in a channel, the water will flow around the
obstruction. Close to the obstruction, the velocity of the water increases, similar to
the stress concentrations around the crack tip. The magnitude of the stress concentrations is dependent on the shape, size, and direction of the flaw.

The tension force is the same in all three materials of Fig. 2.9. Therefore, if we
sum the stress, the resultant must also be the same in all cases to satisfy equilibrium
(the internal stress must equal the external stress). Since the width available for the

Table 2.2 Approximate values for the material dissipated energy value (G) and Young's modulus (E)

Construction material	G (kJ/m²)	E (GPa)
High strength steel	107	210
Mild steel	12	210
Concrete	0.5	10–30
Timber	0.5–2	2
Glass	0.01–0.5	70

stress to flow is reduced in the materials with flaws, the stress intensity must increase to achieve the same resultant. Thus, a longer crack will result in higher localized stresses.

The relationship between stress and the crack length is given by Eq. 2.8 (Roylance 2001).

$$\sigma_f = \sqrt{\frac{EG}{\pi a}} \qquad (2.8)$$

where σ_f is the fracture stress, a is the length of the crack, and G is the dissipated energy. The dissipated energy may be broken down into two parts: the surface energy (γ) and the plastic dissipation (G_p).

$$G = 2\gamma + G_p \qquad (2.9)$$

For brittle materials, such as concrete, Eq. 2.9 reduces to the following form:

$$G \approx 2\gamma$$

and for ductile materials,

$$G \approx G_p$$

Table 2.2 gives the approximate G and E values for different construction materials. However, these values may change considerably from those listed, since the properties of materials can vary. The intension of the table is to give the reader an indication of magnitudes, for the different types of construction materials The larger the dissipated energy, the more stress that is required to facture the material, and vice versa. It is interesting to note that for brittle materials, such as concrete, very little tensile stress is required to fracture the material. Thus, the equation indicates what we already know from practical experience.

When the applied stress exceeds the fracturing stress, the crack will "run" and propagation is unstoppable in brittle materials. As the crack length increases, the stress at the crack tip similarly increases, leading to a state of instability and uncontrollable crack growth. Thus, the material will fracture, or fail, almost simultaneously (Allen and Zalewski 2010).

Fig. 2.10 Spring-ball analogy to represent contraction and expansion of molecular bonds

Fig. 2.11 Molecular lattice
representing a solid material

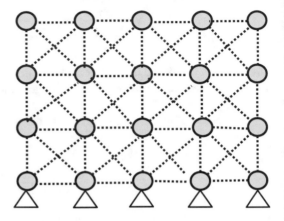

If we consider materials at a molecular level, solid materials are formed by intermolecular bonds. In Fig. 2.10, a ball and spring analogy is used, which is a simplistic but convenient method of demonstrating molecular interaction of bonding forces. The balls represent the molecules and the spring represents the bond between molecules. The bond between molecules responds similarly to a spring, which may be subjected to tensions or compressions. If we apply a tension or compressive force to the molecules, the bonds will expand or contract respectively. If the force is removed, the molecules will be restored back to its original unstressed position (elastic response). However, if the force exceeds the bond strength, breakage will occur and the molecules will separate.

In Fig. 2.11, molecules are arranged in an idealized lattice to represent a solid material. The dashed lines represent the molecular bonds, which are assumed to act like springs. Molecules in the lattice will interact with the surrounding molecules. Thus, the interaction lines (dashed lines) emanating from each molecule are connected to the surrounding molecules. The resulting configuration of interaction represents a truss, which is simplistic and rudimentary, but a convenient tool in understanding how forces flow through solid materials and how materials react to stress.

In Fig. 2.12, we trace the flow of stress for a force applied to the edge of the lattice. The flow will resemble the flow of stress in a truss. In Fig. 2.13, a flaw is modeled by a break in the bond between two molecules (the interaction line is removed). In reality, a microcrack will have numerous bonds broken. By introducing a crack, the flow of stress is diverted, or redistributed to the adjacent bonds. This redistribution of stress will precipitate further breakage and a "domino effect" in crack propagation, until the material is completely fractured. Thus, a molecular

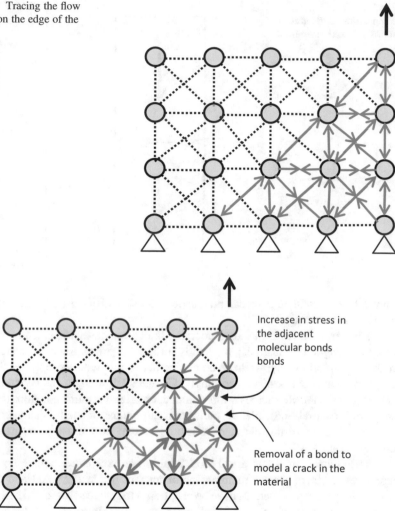

Fig. 2.12 Tracing the flow of stress on the edge of the lattice

Increase in stress in the adjacent molecular bonds bonds

Removal of a bond to model a crack in the material

Fig. 2.13 Modeling a crack in the material and the increase in stress

model (similar to Fig. 2.11) is useful in providing a physical understanding of the stress concentrates at the crack tip, and is why the crack growth tends to be unstoppable in brittle materials.

Another important concept is how stresses are transferred across flawed and cracked materials. In Fig. 2.14, a material is cracked across the entire section. Intuitively, a compression force may be transferred across the cracked section, but the same cannot be said for a tensile force. Flawed materials are able to cope better with compressive forces. Therefore, considering that construction materials are inherently flawed, structures perform better in pure compression. However, this

Fig. 2.14 Transferring
compressions and tensions
across a flawed material

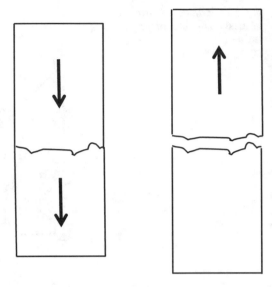

may not be possible, or practical, but the designer should at least attempt to minimize the tension stresses.

All materials have flaws and microcracks, and these flaws dramatically reduce the tensile strength of materials; especially brittle materials, such as concrete and masonry. Unfortunately, we have no choice but to live with these materials, which forces us to alter our designs accordingly. Greater shell strength and economy can be achieved if the tensile stresses are eliminated, or at least minimized. Admittedly, this may not be entirely possible, depending on our choice of shell. Whatever the case may be, the most durable and robust structures are those that are in pure compression.

The Pantheon in Rome and Hagia Sophia are excellent examples of durable structures. Both of these structures are of ancient origin (125 AD and 537 AD, respectively). These structures have been exposed to severe seismic and metrological events, yet they stand in remarkably good condition. Interestingly, both are pure compression structures (Goshima 2011), which is the secret of their longevity.

2.5 The Flow of Stress in Flat and Curved Walls

The study of the flow of stress in shells is drawn from the flow of stress in flat plates. In shells, the flow of stress is similar or identical—a shell is simply a curved plate. The comparison is increasingly valid with shells of a large diameter. Furthermore, the study is restricted to membrane stresses, since the stresses in shells are primarily membrane.

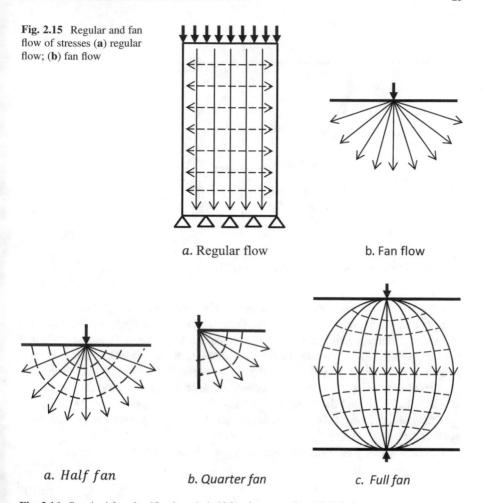

Fig. 2.15 Regular and fan flow of stresses (**a**) regular flow; (**b**) fan flow

a. Regular flow   b. Fan flow

a. Half fan   *b. Quarter fan*   *c. Full fan*

Fig. 2.16 Standard fan classifications (**a** half fan, **b** quarter fan, **c** full fan)

The flow of stress in a plate can be broken down into two primary types of flow: regular (or parallel flow) and fan flow. Regular flow is usually associated with uniform distributed loads, as depicted in Fig. 2.15a. The solid lines represent the compression stresses, and the dashed lines are tensile stresses. As seen in the figure, the flow of stress is parallel (compressions) and perpendicular (tensions) to the direction of the load. At 90° to the compression stresses are tensions, which exist in plated structures due to Poisson's effect (Allen and Zalewski 2010; Fung and Tong 2001).

The fan flow in Fig. 2.15b may be further broken down into three classifications (Fig. 2.16): (a) half fan, (b) quarter fan, and (c) full fan.

Stress flow, as mentioned previously, is analogous with the flow of water through a channel. Stress flow, unlike the flow of water, is not visible. We only see the effects of stress, if the stress is excessive. The flow of water, however, is a useful tool to

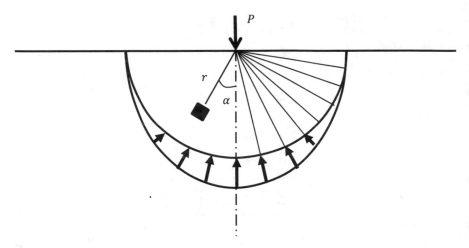

Fig. 2.17 Boussinesq-Flamant distribution of stress

observe how stress flows within solid objects. The direction of water flowing down a narrow channel is parallel, which is the same as stress flowing down the axis of a column. Similarly, when a channel of water opens into a field, the water spreads in a fan-like pattern; this flow characteristic is identical to the spread of stress from a point load into a wall or plate. Furthermore, the stress flow around openings may be examined by placing solid objects in a channel, and observing the increase in current and turbulence around these objects. Admittedly, the comparison is not perfect—fluid flow and stresses in solids do not act exactly the same (water is incompressible), but it is a useful tool in the study of stress flow.

Fan stresses may be approximated by using the Boussinesq-Flamant equation:

$$\sigma = \frac{2P\cos\alpha}{\pi t r} \tag{2.10}$$

where P is the load, α is the angle from the line of action, t is the thickness, and r is radius to the point where the stress is determined. The stress distribution is the highest along the line of action (i.e., $\alpha = 0$) and dissipates with increasing angle. The concentrated stress distribution along the line of action is an important observation.

If we solve for the resultant force of each quarter of the fan, the resultants are located at 32.5° from the line of action (see Fig. 2.18). The majority of the flow of force is therefore primarily within an angle of 65°, and this is the reason why designers often assume the flow of stress is restricted within a range of 60°.

Regular and fan flow do not necessarily act independently, but often act together, or rather transition from one type of flow to another. In Fig. 2.19, the flow of stress is depicted in a wall, from a point load applied at the top to the supports at the bottom. A point load will spread into a half fan, but will transition to regular, or parallel flow. The stresses from parallel flow will then divide into quarter fans near the supports. The solid lines are the principal or primary compressive stresses. Also depicted in the

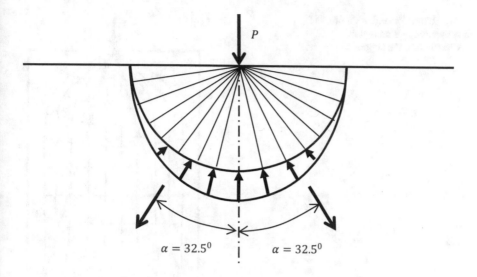

Fig. 2.18   Resultant forces of each quarter

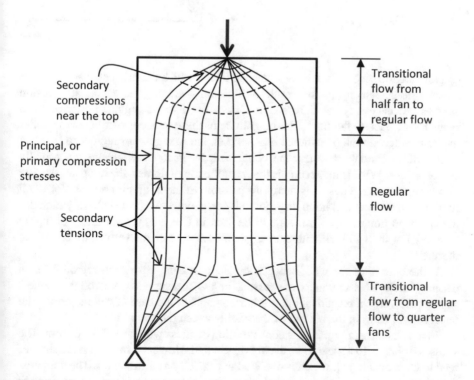

Fig. 2.19   The combination of fans and regular flow

Fig. 2.20 Concentration of
compression thrusts flow
directly into the supports

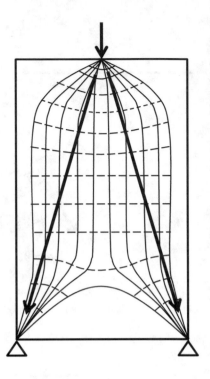

diagram are the secondary tensions (dashed lines), which are orientated at 90° to the compressions, which are the result of Poisson's effect. The secondary tension stresses are necessary for the transition from fan flow to regular flow and from regular flow back to fan flow. The magnitude of these tensions equilibrate and alter the flow pattern in the transitional regions (i.e., fan to regular or regular to fan).

It should be noted that near the point load, the secondary stresses are actually compressions. What is interesting is that near the top of domes, the hoop stresses are also compressions. Thus, it is therefore possible to have a pure compression shell (both in the vertical and hoop directions) if the geometry of the shell only includes the upper portion of the form. As will be seen in Chap. 4, the geometry of a pure compression shell will fall within an angle of 51.83°, from the vertical to the base of the shell.

At the bottom of the wall, between the supports, the unstressed region forms a catenary arch. This region is unstressed, unless the supports are rollers. Rollers support the vertical reactions, but not horizontal. For the case of rollers, horizontal tensions will form in the bottom of the wall between the supports.

What is lacking in Fig. 2.19 is concentration of stress within the flow pattern. The stresses in Fig. 2.19 appear to be evenly distributed. However, in reality, the stresses tend to be concentrated, which is depicted in Fig. 2.20 as thrust-lines. The compression thrusts will take the most direct route; thus, a concentration of force will split and flow directly from the point load into the supports (similar to the Boussinesq-Flamant equation).

Fig. 2.21 Idealized truss to
represent stress flow

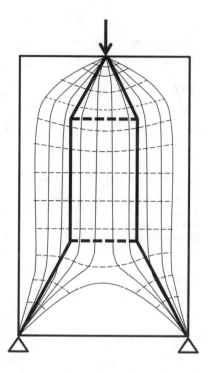

Since the stress tends to concentrate and flow toward the supports, flow patterns can be constructed from any number of load and support conditions. Sometimes it is easier to represent the stress flow as a truss. This is a very useful method to determine the magnitudes of the principal compression thrusts and the equilibrating tensions in the transition regions. An idealized truss, representing the stress flow, is given in Fig. 2.21.

As before, the solid lines represent the path of the principal compressions, which should be orientated so that the path is as direct as possible from the point load to the supports. The dashed lines are the tensions, which equilibrate the truss and permits a change in the force direction. The truss representation allows one to determine the magnitude of the compression thrusts and the equilibrating tensions in the transition regions.

As mentioned previously, load will tend to flow in the most direct route to the supports. If the load is not placed over a support, the load will split. This concept is depicted in Fig. 2.22 for a single point load (a) and for multiple point loads (b).

If multiple point loads are applied to the top of a wall to simulate a uniform distributed load, and the individual compressive thrusts are summed, the resultant compressive thrust, or thrust-line, takes the shape of a catenary arch (Fig. 2.23). (Catenary arching is substantiated by the funicular point load method of Chap. 7.) If the wall or plate is on rollers (Fig. 2.24), a tension tie is required to resist the tension thrust-line at the base.

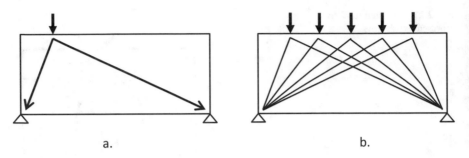

a.   b.

Fig. 2.22 Flow of compression thrusts for a single and multiple point loads

Fig. 2.23 Catenary thrust-
line forms for the case of a
uniform distributed load

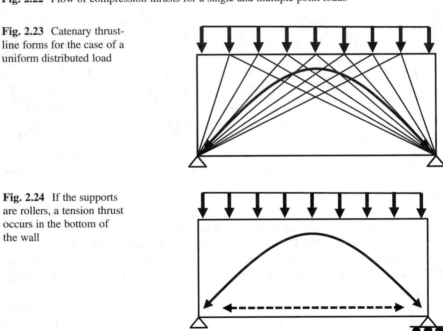

Fig. 2.24 If the supports
are rollers, a tension thrust
occurs in the bottom of
the wall

   Arching action is a well-known phenomenon that occurs in walls, beams, and
slabs (Ockleston 1958: 7; Ruddle and Rankin 2003; McDowell et al. 1997).
Masonry arches are common throughout the world, and some are more than a
thousand years old, yet many remain stable and serviceable. The key to their
structural strength is matching of the geometrical shape with the flow of the stress,
which takes the form of an arch (Fig. 2.25). Arching action is the natural flow of
stress (i.e., how the stress flows naturally, if uninhibited by the structural geometry).
   Engineers of the past commonly used hanging chains as a tool to design arches
and domes. The physical properties of a hanging chain are ideally suited to determine the shape of optimal structures; chains are only capable of resisting a tension
force, and free of bending moments and shears. This is ideal, since structural

Fig. 2.25 The flow of stress in a masonry arch

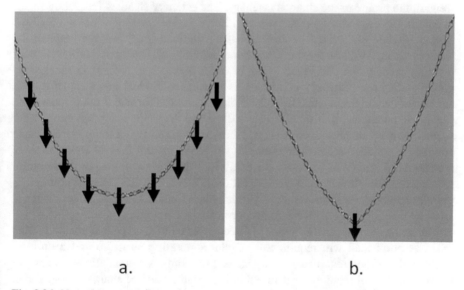

a.   b.

Fig. 2.26 Natural structural shapes for a continuous load (**a**) and a point load (**b**)

members that are subjected to bending are highly uneconomical (Sect. 2.3). Although a hanging chain is in pure tension, if the links are locked and the chain is turned upright, the force is reversed and in pure compression. A pure compression structure is advantageous; the stresses are lower (since bending is eliminated) and most construction materials are able to resist compressions far better than tensions, especially concrete and masonry (Sect. 2.4).

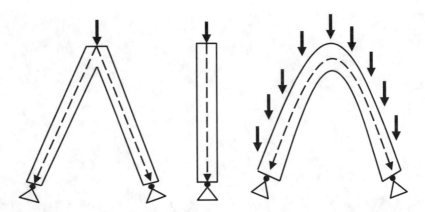

Fig. 2.27 Basic optimal shapes in pure compression

In Fig. 2.26, two types of loading are applied to a chain—a continuous load and a point load. If a chain is only subjected to a uniform distributed load (i.e., self-weight), the profile of the chain is catenary. The catenary shape is frequently found in arches, domes, buttresses, and bridges. Some notable examples are the St. Luis arch, St. Peters, and St Paul's cathedrals. If a point load is applied, the shape of the chain is triangular. Trusses are a common example of structural optimization of resisting point loads. However, what is often overlooked, which is the key to understanding the efficiency of trusses, is that the individual members of a truss are free of bending moments. In both of these cases (uniform and point loads), the chain is in pure tension. If the links are locked and the shape is flipped upright, the stresses will be in pure compression. Thus, with these two examples, the basic optimal shapes for structures are defined, where the natural flow of stress matches the structural shape. In addition, we can add a third optimal shape: the column, where loads flow directly into the support (Fig. 2.27). The efficiency of columns is visually apparent; columns are usually much more slender than beams; beams resist loads by bending and are therefore deeper in framed structures.

The shape of an optimal structure is directly related to the type and position of the loads. A point load will require a triangular supporting structure and a uniform distributed load will require a catenary shaped structure. A combination of point and uniform loads will require a structure that is a combination of shapes. The dome at St. Paul's Cathedral in London is an example of the combination of the two types of loads. The dome is constructed of masonry and must resist its self-weight, a uniformly distributed load. In addition, St. Paul's dome must resist a lantern (point load), which is placed on top of the dome. The combination of both types of loads necessitates a conical, or pointed shaped dome as the structural support. In Fig. 2.28, the conical internal dome is the main supporting structure. The outer and inner circular domes are decorative and nonstructural.

Fig. 2.28 Christopher
Wen's structural drawings
of St. Paul's Cathedral in
London

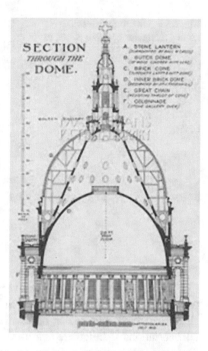

Fig. 2.29 The repelling of
stress (**a**) and the attraction
of stress (**b**)

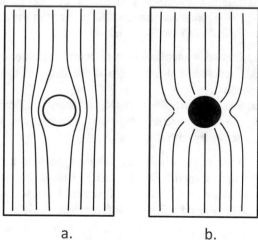

a.   b.

2.6   The Flow of Stress around Openings

A well-known principle of structural analysis is that stiffness attracts stress; conversely, a lack of stiffness repels stress.

In Fig. 2.29a, a hole is placed in a plate. The hole will have zero stiffness, and the stress flowing through the plate will be repelled away from the hole. However, in

Fig. 2.30 Maximum stress
around a circular opening

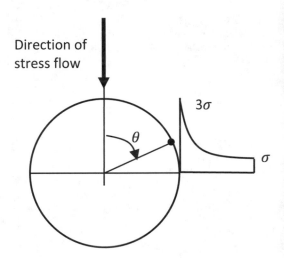

Fig. 2.29b, the hole is filled with a very stiff disk, and the stress in the plate will flow
to the material of much higher stiffness. A stiff disk is usually not introduced into
shell structures, but rather holes. The objective is therefore to minimize stress
concentrations if holes are introduced in shells (Salvin 1961; Timoshenko 1959).
The stress concentrations occur on the sides of the holes.

Different stress patterns will occur with different shapes of holes. General rules
for hole design are listed below:

1. Corners—angular corners cause higher stresses than rounded corners.
2. Shape—less stress occurs if the shape of the hole is stream-line to the flow of
   stress.
3. Size—the smaller the hole, less stress disturbance will occur.
4. Stress magnitudes—the maximum stress occurs at the edge of the hole, which
   decreases with distance.

Most of these rules are intuitive, without knowing the actual stress caused by
different shaped holes, and the best choice is usually selected without any calculations. However, for design, it is useful to know the magnitude of stress increases of
various shaped holes to reinforce these regions:

Circular Holes

The maximum stress occurs at $\theta = 90°$ and $270°$ (see Fig. 2.30)

$$\sigma_{max} = 3\sigma \tag{2.11}$$

Rectangular Hole with Rounded Corners

The maximum stress is dependent on the aspect (length to width) of the hole and
radius of the rounded corner.

Fig. 2.31 Maximum stress flow around a rectangular hole

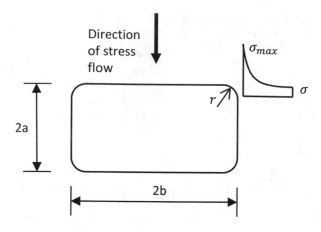

Table 2.3 Maximum stresses for various configurations of rectangular holes

Configuration	a/b	r	σ_{max}
Square with minimum radius	1	a/10	4.50 σ
Square with larger radius	1	a/2	2.97 σ
Rectangle with minimum radius	0.5	a/10	6.09 σ
Rectangle with minimum radius	0.2	a/10	8.34 σ
Circle (maximum radius)	1	a	3.00 σ

$$\sigma_{max} = \left[C_1 + C_2 \left(\frac{a}{b} \right) + C_3 \left(\frac{a}{b} \right)^2 + C_4 \left(\frac{a}{b} \right)^3 \right] \tag{2.12}$$

$$C_1 = 14.815 - 22.308 \sqrt{\frac{r}{2a}} + 16.298 \left(\frac{r}{2a} \right)$$

$$C_2 = -11.201 - 13.789 \sqrt{\frac{r}{2a}} + 19.200 \left(\frac{r}{2a} \right)$$

$$C_3 = 0.2020 + 54.620 \sqrt{\frac{r}{2a}} - 54.748 \left(\frac{r}{2a} \right)$$

$$C_2 = 3.232 - 32.530 \sqrt{\frac{r}{2a}} + 30.964 \left(\frac{r}{2a} \right)$$

range,

$$0.05 \leq \frac{r}{2a} \leq 0.5$$

$$0.2 \leq \frac{a}{b} \leq 1.0$$

Table 2.4 Maximum stress for various configurations of the ellipse

Configuration	a/b	r	σ_{max}
Flow against major axis	0.01	a/100	21 σ
Flow against major axis	0.1	a/10	7.32 σ
Circle	1	a	3 σ
Flow against minor axis	2	2a	2.41 σ
Flow against minor axis	5	5a	1.89 σ
Flow against minor axis	10	10a	1.63 σ

Fig. 2.32 Maximum concentrated stress in an elliptical hole

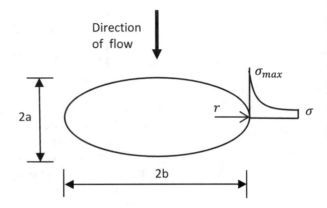

The maximum stress will occur at the rounded corners (see Fig. 2.31). To indicate the influence of shape, Table 2.3 gives the stresses for various configurations of rectangular holes.

Elliptical Hole

The equation for an elliptical hole is very versatile and is applicable to stress flows against and parallel to each of the axis. The ellipse is one of the most efficient holes to reduce stress concentrations, but also may be the worst, depending on the direction of stress flow and the ratio of the minor and major axis. If the minor axis is in the direction of the stress flow, the hole is more streamlined and therefore a lower concentration of stress is observed. The equation for the maximum stress is given as Eq. 2.13, and the results of various configurations are given in Table 2.4, and the location of the stress is shown in Fig. 2.32.

$$\sigma_{max} = \left(1 + 2\sqrt{\frac{a}{r}}\right)\sigma \tag{2.13}$$

$$r = \frac{a^2}{b} \tag{2.14}$$

range,

$$0 < \frac{a}{b} < 10$$

2.7 Exercises

2.7.1 The compression stress on a wall is 12 MPa. For the holes given below, determine the maximum stress and location around the opening. Comment concerning the most efficient shaped holes to minimize stress concentrations in shell walls.

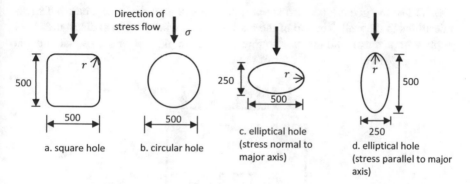

a. square hole   b. circular hole   c. elliptical hole (stress normal to major axis)   d. elliptical hole (stress parallel to major axis)

2.7.2 At some point in a membrane structure, the stress in the x and y directions are in pure shear. Using the differential element of stresses given below:

(a) Determine the maximum tension stress in the membrane.
(b) Draw a sketch of the differential element, and the orientation of principal stresses.
(c) On the element, draw the potential crack, if the material is concrete.
(d) If the material has a concrete strength of 40 MPa, comment on the probability of cracking. e. Using a simplified molecular lattice to represent a solid material, explain shear deformations.

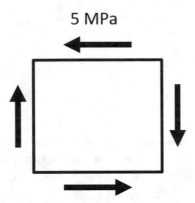

5 MPa

2.7.3 For the given loads and support conditions, graphically estimate the stress trajectories in the wall. The top of the wall is loaded with a uniform distributed load (applied in patches), and the two bottom corners are pin supports. A square hole is placed approximately in the center of the plate.

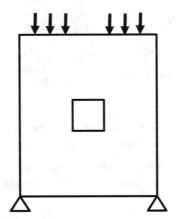

2.7.4 Provide a structure of curved or linear members for the loads given below. Two foundations are given to support the structure. Draw a pure compression structure to support the given loads, using the hanging chain analogy. Firstly, assume that the self-weight of the support structure is negligible compared to the forces (P). Secondly, assume that the self-weight is significant, and must be considered in the configuration of the support structure.

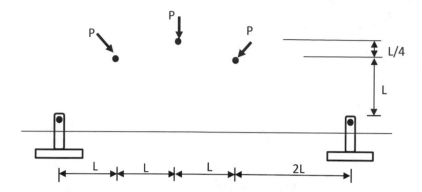

2.7.5 Explain the basic principles of structural optimization, in terms of stress flow and material efficiency.

2.7.6 Explain why shell structures are so efficient and time enduring.

References

Allen, E. & Zalewski, W., 2010. *Form and Forces, Expressive Structures.* New Jersey: John Wiley and Son.

Billington, D. P., 1982. *Thin Shell Concrete Structures.* 2 ed. New York: McGraw-Hill.

Chatterjee, B. K., 1971. *Theory and Design of Concrete Structures.* Calcutta: Oxford and IBH Publishing Co.

Domone, P. & Illston, J., 2010. *Concrete Materials: Their Nature and Behavior.* 4 ed. New York: Spoon Press.

Fung, Y. C. & Tong, P., 2001. *Classical and Computational Solid Mechanics.* New Jersey: World Scientific Publishing Co.

Goshima, R., 2011. Investigation of the Cross-section of the Pantheon Dome through Catenary Mechanism. *Journal of the Association for Shell and Spatial Structures,* 52(167), pp. 19-23.

Kong, F. K. & Evans, R. H., 1987. *Reinforced and Prestressed Concrete.* 3 ed. United Kingdom: Van Nostrand Reinhold.

McDowell, E., Hansen, N. W. & McHenry, D., 1997. Arching Action Theory of Masonry Walls. *The American Society of Civil Engineers,* 82(ST2), pp. 1-18.

Ockleston, A. J., 1958. Arching Action in Reinforced Concrete. *The Structural Engineer,* 36(6), pp. 197-201.

Ohlsen, E. et al., 1976. *Mechanics of Materials.* New York: John Wiley and Sons.

Popovics, S., 1998. *Strength and Related Properties of Concrete.* New York: John Wiley and Sons.

Roylance, D., 2001. *Introduction to Fracture Mechanics.* Cambridge, MA: Department of Material Science and Engineering, MIT.

Ruddle, M. E. & Rankin, G. I. B., 2003. *Arching Action-Flexural and Shear Strength Enhancements in Rectangular and Tee Beams.* s.l., Proceedings of the Institution of Civil Engineers, Structures and Building Journal.

Salvin, G. N., 1961. *Stress Concentrations Around Holes.* s.l.:Pergamon Press.

Timoshenko, S., 1958. *Strength of Materials.* New York: Van Nostrand Reinhold Co.

Timoshenko, S., 1959. *Theory of Plates and Shells.* 2 ed. New York: McGraw-Hill Book Co.

Zingoni, A., 1997. *Shell Structures in Civil and Mechanical Engineering.* Great Britain: Thomas Telford.

Chapter 3
Cylindrical Shells

Abstract This chapter develops the membrane (in-plane stresses) and deformation theories of cylindrical shells for various loading conditions (liquid and bulk solids), assuming the base is unrestrained. The theory for the boundary effects (support restraint) is also developed, for fixed and pinned supports. To solve for the effect of boundary conditions, compatibility equations are formulated to solve the base reactions, which, in turn, are used to distribute the stresses in the shell. The membrane and boundary solutions are then superimposed to solve for the stresses, bending moments, and shears at any point in the shell. The mathematical derivation is fully developed for this shell.

3.1 Introduction

The cylindrical shell is the most common shell and is used extensively to retain granular materials (i.e., grain, coal, cement, etc.) and liquids (i.e., water). In agricultural and industrial applications, the cylindrical shell is referred to as a silo. The water retaining shells are commonly referred to as circular reservoirs.

When designing a cylindrical shell, the two main stresses are hoop (i.e., circumferential) and vertical. The source of the vertical stress is the self-weight of the shell walls, the roof structure, and perhaps frictional forces on the walls due to the discharge/filling of materials. The hoop stress is caused by the lateral pressure on the walls from the retained materials or liquids. Many codes of practice and design guides only consider these two stresses when designing cylindrical shells. These stresses are called membrane stresses since they act within the plane of the shell. A balloon filled with air is an example of a membrane shell; all the stresses are in-plane. However, this assumption is only valid if the shell does not have a roof (or a roof that restrains the top of the shell) and if the base is free to slide laterally. An open reservoir, resting on a sliding base, is perhaps the only type of shell where the stresses are purely membrane. All other shells are attached at the base (fixed or pinned) and many have some sort of roof structure. This fixity at the base and roof causes additional stresses in the shell, referred to in shell theory as boundary effects.

Boundary effects cause bending moments and shears. The effect may be substantial, and the resulting stresses could be much higher than those generated by membrane stresses. Thus, it is essential that the shell designer is able to determine these stresses and design accordingly. However, the calculation of the boundary effects are significantly more complicated than the calculation of the membrane stresses, which explains why many designers choose to ignore boundary effects. Needless to say, boundary stresses do exist. Accounting for boundary effects is especially important in the design of water retaining structures, where an incorrect assessment could result in cracking, leakage, and deterioration of the structure.

The equations and notations are based on Billington (1982), which is widely read and adapted in many publications on shells and have become standard terms. This chapter also relies heavily on the work by Wilson (2005), Timoshenko (1959), Flugge (1960), and Chatterjee (1971). These books provide a wealth of theoretical and practical information on the design of cylindrical shells.

3.2   The Membrane Theory of Cylindrical Shells

Stresses in a cylindrical shell are defined on a differential element. The location of the differential element is given in Fig. 3.1, and the differential stresses are given in Fig. 3.2.

Fig. 3.1 Location of the differential element in the cylindrical shell

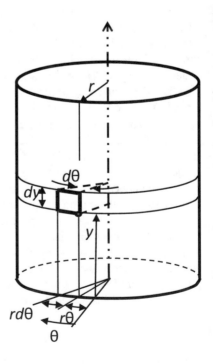

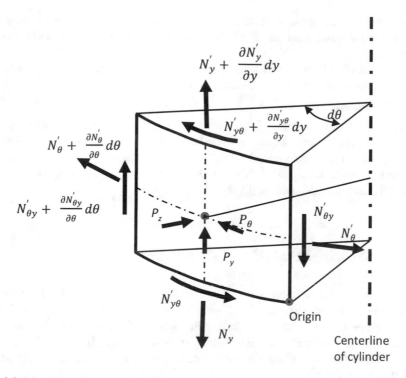

Fig. 3.2 Membrane stresses on the differential element

A differential element represents a point in the cylindrical walls, at an intersection of vertical and hoop strips. Although the square element depicted in Fig. 3.1 is proportionally large to the size of the shell, differential elements are infinitesimally small. These elements are used to describe the distribution of stresses, expressed in terms of differential equations. Differential equations are useful for continuous stress systems and to account for infinitesimal changes in stress. Fundamentally, differential elements are a mathematical convenience, which may not be apparent due to the enormity of the derivation.

Differential elements are a graphical and mathematical method of depicting the stress distribution in a shell. In shell theory, the values N_θ', , $N_{\theta y}'$ and $N_{y\theta}'$ (Fig. 3.2) are referred to as stress resultants, and the units are force per unit length (kN/m or kips/ft). The stress N_θ' is the circumferential stress (commonly called the hoop stress), which is the result of the retained materials or liquids contained in the shell. The value N_y' is vertical stress, which is the result of the self-weight of the shell walls. The other two stresses ($N_{\theta y}'$ and N_y') are the in-plane shear stresses, which occur if nonsymmetrical loads are present. It should be noted that all of the stress terms have a prime as a superscript, which indicates that the stresses are membrane and derived from membrane theory.

Membrane stresses are in-plane stresses, and the membrane theory assumes that only in-plane stresses exist; bending moments and shears are not accounted for in the analysis.

An inspection of Fig. 3.2 will show that the stresses change from one edge to the opposite edge. If we consider the circumferential stress N'_θ in the horizontal direction, the stress changes to $N'_\theta + \frac{\partial N'_\theta}{\partial \theta} d\theta$ on the opposite end. The added term $\left(\frac{\partial N'_\theta}{\partial \theta}\right)$ accounts for change in stress from one edge of the element to the other, in the circumferential direction. When multiplied by a small angle $d\theta$, the additional stress across the element is determined $(\partial N'_\theta)$, albeit an increase or decrease. Thus, the differential element graphically and mathematically represents how the stresses change in the circumferential and vertical directions.

3.2.1   Circumferential (Hoop) Equilibrium

The circumferential, or hoop stress (N'_θ), is determined by formulating equilibrium equations (i.e., by summing forces). Figure 3.3 is a view of the differential element, looking down from the top of the shell. Although the differential element has eight internal stresses and three external pressures on the element, only five are included in Fig. 3.3 for clarity (only those stresses that contribute to the equilibrium equation are included in the figure).

Summing forces along the x–x' line,

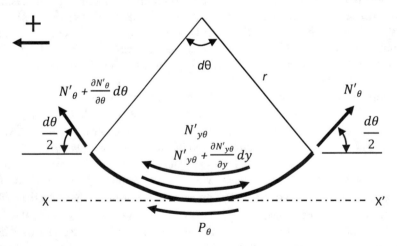

Fig. 3.3 Differential element stresses that contribute to the equilibrium equations in the circumferential direction

$$\left(N'_\theta + \frac{\partial N'_\theta}{\partial \theta} d\theta\right) dy \cos\left(\frac{d\theta}{2}\right) - N'_\theta dy \cos\left(\frac{d\theta}{2}\right) + \left(N'_{y\theta} + \frac{\partial N'_{y\theta}}{\partial y} dy\right) d\theta r$$
$$- N'_{y\theta} d\theta r + P_\theta (d\theta r) dy = 0 \qquad (3.1)$$

The stresses are multiplied by the lengths of the differential element to solve for the forces. Trigonometry is also used to solve for the components of force along the x–x' direction.

Since the differential element is infinitesimally small, and consequentially $d\theta$ is small, we can assume $\cos\left(\frac{d\theta}{2}\right) \approx 1$. Expanding Eq. 3.1,

$$N'_\theta dy + \frac{\partial N'_\theta}{\partial \theta} d\theta dy - N'_\theta dy + N'_{y\theta} d\theta r + \frac{\partial N'_{y\theta}}{\partial y} d\theta dyr - N'_{y\theta} d\theta r + P_\theta d\theta dyr = 0$$

$$(3.2)$$

Equation 3.2 may be simplified by canceling the first and third terms, the fourth and sixth terms, and by dividing by $d\theta dy$. The result is the first equilibrium equation.

$$\frac{\partial N'_\theta}{\partial \theta} + \frac{\partial N'_{y\theta}}{\partial y} r + P_\theta r = 0 \qquad (3.3)$$

3.2.2 Vertical Equilibrium

The next step is to sum forces in the vertical direction (y-direction). Again, Fig. 3.4 only shows the internal and external stresses that contribute to the vertical equilibrium equations.

Summing forces about the line x–x',

$$\left(N'_y + \frac{\partial N'_y}{\partial y} dy\right) d\theta r - N'_y d\theta r + \left(N'_{\theta y} + \frac{\partial N'_{\theta y}}{\partial \theta} d\theta\right) dy - N'_{\theta y} dy + P_y d\theta dyr = 0$$

$$(3.4)$$

Expanding Eq. 3.4,

$$N'_y d\theta r + \frac{\partial N'_y}{\partial y} dy d\theta r - N'_y d\theta r + N'_{\theta y} dy + \frac{\partial N'_{\theta y}}{\partial \theta} d\theta dy - N'_{\theta y} dy + P_y d\theta dyr = 0$$

$$(3.5)$$

Fig. 3.4 Differential
element stresses that
contribute to the equilibrium
equation in the vertical
direction

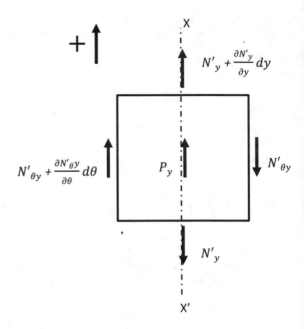

Fig. 3.4 Differential element stresses that contribute to the equilibrium equation in the vertical direction

Eliminating terms one and three, four and six, and dividing each side of the equation by $d\theta dy$, the equilibrium equation in vertical direction is simplified.

$$\frac{\partial N'_y}{\partial y}r + \frac{\partial N'_{\theta y}}{\partial \theta} + P_y r = 0 \tag{3.6}$$

3.2.3   Radial Equilibrium

The radial direction is the z-direction. Radial equilibrium is determined by summing of the forces normal to the surface of the differential element. Equilibrium in this direction only includes two internal stresses and one external pressure (see Fig. 3.5).
Summing forces along the line x–x',

$$\left(N'_\theta + \frac{\partial N'_\theta}{\partial \theta}d\theta\right)dy\sin\left(\frac{d\theta}{2}\right) + N'_\theta dy\sin\left(\frac{d\theta}{2}\right) + P_z d\theta dyr = 0 \tag{3.7}$$

Since $\frac{d\theta}{2}$ is small, the $\sin\left(\frac{d\theta}{2}\right) \approx \frac{d\theta}{2}$.
Making this substitution and expanding Eq. 3.7,

Fig. 3.5 Differential
element stresses that
contribute to the equilibrium
equation in the radial
direction

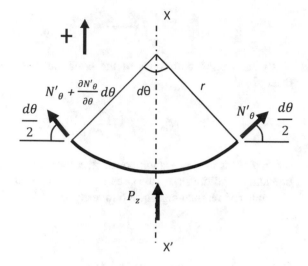

$$N'_\theta dy \frac{d\theta}{2} + \frac{\partial N'_\theta}{\partial \theta} dy \left(\frac{d\theta^2}{2} \right) + N'_\theta dy \left(\frac{d\theta}{2} \right) + P_z d\theta dy r = 0 \qquad (3.8)$$

The second term is what is called a higher order term. Terms are referred to as higher order terms when the magnitude is much smaller than the other terms, and therefore may be omitted to simplify the expression. The first and third terms are combined and the equation is divided by $d\theta dy$.

$$N'_\theta + P_z r = 0 \qquad (3.9)$$

3.2.4   General Membrane Stress Equations

The three equilibrium equations are collected and listed below:

$$\frac{\partial N'_\theta}{\partial \theta} + \frac{\partial N'_{y\theta}}{\partial y} r + P_\theta r = 0 \qquad (3.3)$$

$$\frac{\partial N'_y}{\partial y} r + \frac{\partial N'_{\theta y}}{\partial \theta} + P_y r = 0 \qquad (3.6)$$

$$N'_\theta + P_z r = 0 \qquad (3.9)$$

From these equations, the membrane stresses are solved.
Equation 3.3 is rearranged and expressed in terms of $\frac{\partial N'_{y\theta}}{\partial y}$.

$$\frac{\partial N'_{y\theta}}{\partial y} = -\frac{1}{r}\frac{\partial N'_{\theta}}{\partial \theta} - P_{\theta} \tag{3.10}$$

By integrating both sides of the equation, the membrane shear stress $N'_{y\theta}$ is determined.

$$N'_{y\theta} = \int \left(-\frac{1}{r}\frac{\partial N'_{\theta}}{\partial \theta} - P_{\theta} \right) dy + f_1(\theta) \tag{3.11}$$

where $f_1(\theta)$ is the integration constant, which represents the stresses caused by the boundary conditions (i.e., the stresses caused by fixity of the base or roof structure). Similarly, rearranging Eq. 3.6 in terms of N'_y, and dividing by the radius r,

$$\frac{\partial N'_y}{\partial y} = -\frac{1}{r}\frac{\partial N'_{\theta y}}{\partial \theta} - P_y \tag{3.12}$$

The membrane stress N'_y is solved by integration.

$$N'_y = \int \left(-\frac{1}{r}\frac{\partial N'_{\theta y}}{\partial \theta} - P_y \right) dy + f_2(\theta) \tag{3.13}$$

As before, $f_2(\theta)$ is the integration constant representing the boundary conditions. The last membrane stress is determined by rearranging Eq. 3.9.

$$N'_{\theta} = -P_z r \tag{3.14}$$

Equations 3.11, 3.13, and 3.14 are general stress equations, which are solved for a variety of surface loads (i.e., P_y, P_{θ}, and P_z).

Collecting the stress equations,

$$N'_{y\theta} = \int \left(-\frac{1}{r}\frac{\partial N'_{\theta}}{\partial \theta} - P_{\theta} \right) dy + f_1(\theta) \tag{3.11}$$

$$N'_y = \int \left(-\frac{1}{r}\frac{\partial N'_{\theta y}}{\partial \theta} - P_y \right) dy + f_2(\theta) \tag{3.13}$$

$$N'_{\theta} = -P_z r \tag{3.14}$$

It should be noted that he hoop stress (N'_{θ}) is independent of the vertical stress (N'_y) and the in-plane shears ($N'_{y\theta}$). The hoop stress is only dependent on the wall pressure, either from liquids or granular materials. The vertical and in-plane shears are interdependent.

3.2.5   *Membrane Stress Equations for Various Loadings on the Shell*

Since the primary uses of a cylindrical shell are silos and water reservoirs, the membrane stresses are solved for these types of loads. However, before we solve the equations for various loads, the general stress equations are simplified by making the following assumptions:

1. The boundary effects are ignored at this stage of the analysis. By doing so, the integration constants (f_1 and f_2) are dropped from the equations.
2. Although the stresses will change in the y-direction (vertical direction), the stresses will be constant in the θ-direction (hoop direction). This assumption implies symmetry of loading; a water reservoir will automatically fit this assumption, but with granular materials, the silo must be filled and discharged concentrically to ensure symmetrical loading of the shell walls. Thus, by assuming symmetrical loading, the partial differential terms in the hoop direction (i.e., $\frac{\partial N'_\theta}{\partial \theta}$ and $\frac{\partial N'_{\theta y}}{\partial \theta}$ terms) drop from the equations.
3. The only loads assumed on the shell are the vertical loads due to self-weight of the walls (P_y) and the lateral wall pressures from liquids or granular materials (P_z). The pressure in the hoop direction (P_θ) (called surface tractions) is also ignored and dropped from the stress equations.

Applying the above assumptions, Eq. 3.11 is eliminated and Eqs. 3.13 and 3.14 are simplified.

$$N'_y = -\int P_y dy \qquad (3.15)$$

$$N'_\theta = -P_z r \qquad (3.16)$$

Wall Stresses Due to Gravity Loads

The stresses due to gravity load will be zero at the top of the shell and maximum at the base. At any point y in the shell, the value of N'_y represents the force per unit length of the shell between elevations H and y. As mentioned previously, what we call a stress is actually a force per unit length.

Integrating Eq. 3.15 between y and H (see Fig. 3.6),

$$N'_y = -\int_y^H P_y dy = -P_y [y]_y^H \qquad (3.17)$$

The vertical stress in the shell walls is determined at any point y.

Fig. 3.6 The self-weight of
the shell walls

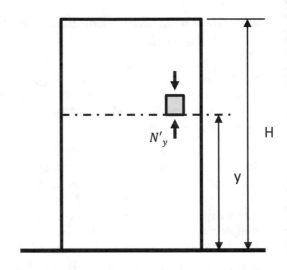

$$N'_y = -P_y(H - y) \tag{3.18}$$

If the shell wall is of constant thickness, N'_y may be expressed in the following form

$$N'_y = -\gamma h(H - y) \tag{3.19}$$

where γ is the weight of the shell walls per unit volume and h is the thickness of the shell. If the wall thickness varies,

$$N'_y = -\gamma \left[\frac{(h_t + h_b)}{2} + \frac{(h_t - h_b)y}{2H} \right] (H - y) \tag{3.20}$$

where h_t and h_b are the thicknesses at the top and bottom of the shell, respectively. If the values h_t and h_b are the same, Eq. 3.20 is identical to Eq. 3.19.

Circumferential (Hoop) Stresses Due to a Liquid Pressure

The distribution of a liquid pressure is triangular, as shown in Fig. 3.7. The hoop stress is constant and symmetrical around the shell at any given height y.

The normal pressure on the walls of the shell is defined by Eq. 3.21.

$$P_z = -\gamma(H - y) \tag{3.21}$$

where γ is the density of the liquid.

Substituting Eq. 3.21 into Eq. 3.16, the hoop stress is determined.

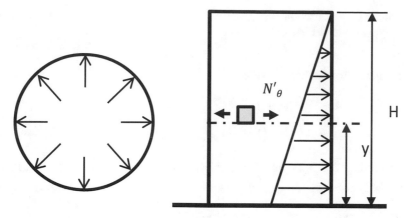

Fig. 3.7 Hoop stresses due to a liquid load

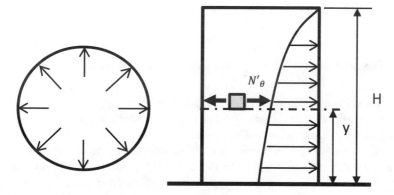

Fig. 3.8 Distribution of wall pressures due to granular materials

$$N'_\theta = \gamma(H - y)r \qquad (3.22)$$

Circumferential (Hoop) Stress Due to Granular Materials

The pressure distribution for granular materials is approximated by Janssen's equation (AS3774 1997) (Eq. 3.23). The distribution of pressures is illustrated in Fig. 3.8, and the equation for the distribution is given in Eq. 3.23.

$$P_z = -\frac{\gamma r_h \left[1 - e^{(y-H)/z_o}\right]}{\mu} \qquad (3.23)$$

where

γ = unit weight of the granular materials
$\mu = \tan(\phi_w)$
r_h= hydraulic radius

$\phi_w =$ angle of wall friction
$\phi_i =$ angle of internal friction
$z_o = \frac{r_h}{\mu k_j}$

k_j is the greater of the two equations.

$$k_j = (1 - \sin\phi_i)/(1 + \sin\phi_i)$$
$$kj = (1 - \sin^2\phi_i)/(1 + \sin^2\phi_i)$$

Substituting Eq. 3.23 into 3.16, the hoop equation is derived for granular materials.

$$N'_\theta = \frac{\gamma r_h \left[1 - e^{(y-H)/z_o}\right]}{\mu} r \qquad (3.24)$$

3.3   Displacement Theory for Membrane Stresses

The deformation of a differential element is illustrated in Fig. 3.9. The lower element is the differential element prior to the loading of the shell, and the upper element is the deformed element after loading. Since the loading of the shell is assumed symmetrical, the deformation from A to A' is identical to the deformation of D to D' (i.e., both A and D deform by w and v). Similarly, points B and C both deform by $w + \frac{\partial w}{\partial y} dy$ in the horizontal and $v + \frac{\partial v}{\partial y} dy$ in the vertical directions.

3.3.1   Strains, Displacements, and Stresses

The derivation begins by solving for the strains in the differential element. In Fig. 3.10, the vertical (3.10a) and the hoop (3.10b) strains are illustrated (left and center, respectively), as well as the rotations (3.10c). The dashed line represents the differential element after deformation.

The strains in the y-direction is equal to the change in length divided by the original length (Fig. 3.10a).

$$\varepsilon_y = \frac{v + \frac{\partial v}{\partial y} dy - v}{dy} = \frac{\partial v}{\partial y} \qquad (3.25)$$

In the hoop direction, the change in length is a function of the change in radius (Fig. 3.10b).

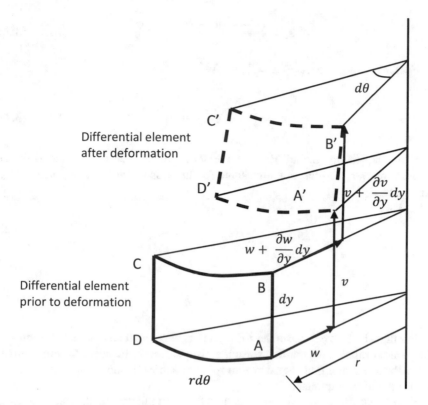

Fig. 3.9 Deformation of a differential element

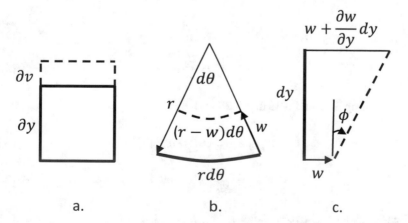

Fig. 3.10 Strains in the vertical and hoop directions and slope (left to right)

$$\varepsilon_\theta = \frac{(r-w)d\theta - rd\theta}{rd\theta} = -\frac{w}{r} \tag{3.26}$$

For small angles, $\tan\phi = \phi = \frac{\partial w}{\partial y}$. Thus,

$$\phi_y = \frac{w + \frac{\partial w}{\partial y}dy - w}{dy} = \frac{\partial w}{\partial y} \tag{3.27}$$

From plate theory (Szilard 1974), the relationship between stresses and strains are given. However, the subscripts are altered for the hoop (θ), vertical (y), and radial (z) directions.

$$\varepsilon_\theta = \frac{1}{E}\left[\sigma_\theta - \nu(\sigma_y + \sigma_z)\right] \tag{3.28}$$

$$\varepsilon_y = \frac{1}{E}\left[\sigma_y - \nu(\sigma_\theta + \sigma_z)\right] \tag{3.29}$$

$$\gamma_{\theta y} = \frac{1}{G}\tau_{\theta y} \tag{3.30}$$

For thin shells, the stress in the z-direction is very small compared to the stress in the y- and θ-directions. For this reason, σ_z is assumed to be equal to zero. Furthermore, the shear strains are equal to zero for symmetrical loading ($\gamma_{\theta y} = 0$). Equation 3.30 is therefore eliminated.

Since what we call stresses are actually a force per unit length, the true stresses are determined by dividing by the shell thickness.

$$\sigma_\theta = \frac{N'_\theta}{h} \tag{3.31}$$

$$\sigma_y = \frac{N'_y}{h} \tag{3.32}$$

Substituting 3.31 and 3.32 into 3.28 and 3.29,

$$\varepsilon_\theta = \frac{1}{Eh}\left[N'_\theta - \nu N'_y\right] \tag{3.33}$$

$$\varepsilon_y = \frac{1}{Eh}\left[N'_y - \nu N'_\theta\right] \tag{3.34}$$

Rearranging the above equations,

$$N'_\theta = \frac{Eh}{(1-\nu^2)}\left[\varepsilon_\theta + \nu\varepsilon_y\right] \tag{3.35}$$

$$N'_y = \frac{Eh}{(1 - \nu^2)} \left[\varepsilon_y + \nu\varepsilon_\theta\right] \tag{3.36}$$

If we substitute the strain Eqs. 3.25 and 3.26 into 3.35 and 3.36, the stresses are expressed in terms of the deformations.

$$N'_\theta = \frac{Eh}{(1 - \nu^2)} \left(-\frac{w}{r} + \nu\frac{\partial v}{\partial y}\right) \tag{3.37}$$

$$N'_y = \frac{Eh}{(1 - \nu^2)} \left(\frac{\partial v}{\partial y} - \nu\frac{w}{r}\right) \tag{3.38}$$

These are general stress equations, which may be solved for various loading conditions.

3.3.2   Displacements in the Shell Due to Various Loads

Displacements Due to Liquid Pressure

The hoop stress for a liquid pressure was previously determined, and listed below.

$$N'_\theta = \gamma(H - y)r \tag{3.22}$$

Equations 3.26 and 3.33 are used to determine the deformations.

$$\varepsilon_\theta = -\frac{w}{r} \tag{3.26}$$

$$\varepsilon_\theta = \frac{1}{Eh} \left[N'_\theta - \nu N'_y\right] \tag{3.33}$$

Setting the two above equations equal to each other,

$$-\frac{w}{r} = \frac{1}{Eh} \left[N'_\theta - \nu N'_y\right] \tag{3.39}$$

It is assumed that the hoop stress is independent of the vertical stress. For this reason, the value N'_y is eliminated from the above equation. The horizontal deformation is solved by rearranging 3.39 and substituting Eq. 3.22 for liquid load.

$$w = -\frac{r}{Eh} \left[\gamma(H - y)r\right]$$

or

$$w' = D_{10} = -\frac{\gamma r^2}{Eh}(H - y) \tag{3.40}$$

From Eq. 3.27,

$$\phi_y = \frac{\partial w}{\partial y} \tag{3.27}$$

The rotations are determined by taking the derivative of Eq. 3.40 with respect to y.

$$\phi'_y = D_{20} = \frac{\gamma r^2}{Eh} \tag{3.41}$$

The slope is therefore constant from $y = 0$ to $y = H$.

Displacements Due to Granular Materials

The hoop stress was determined by Eq. 3.24

$$N'_\theta = \frac{\gamma r_h \left[1 - e^{(y-H)/z_o} \right]}{\mu} r \tag{3.24}$$

Using the relationship 3.39, and assuming that the vertical stress is independent of the hoop stress, the horizontal deflection is determined by rearrangement and substituting Eq. 3.24.

$$-\frac{w}{r} = \frac{1}{Eh}\left[\frac{\gamma r_h \left[1 - e^{(y-H)/z_o}\right]}{\mu} r \right]$$

$$w' = D_{10} = -\frac{\gamma r^2 r_h}{Eh\mu}\left[1 - e^{(y-H)/z_o}\right] \tag{3.42}$$

The derivative of the above deflection equation is then taken to determine the slope.

$$\phi'_y = D_{20} = \frac{\gamma r^2 r_h}{Eh\mu z_o} e^{(y-H)/z_o} \tag{3.43}$$

The primes are added to the horizontal deformations and slopes to denote that these deformations are caused by membrane stresses.

3.4 Boundary Effects

The membrane theory assumes that the ends of the shell (i.e., base and roof) are unrestrained in translation and bending. This assumption is unrealistic, and not applicable to most water reservoirs and silos unless the base of the shell is on a sliding bearing. The majority of shells are attached to a foundation system. However, the shell may or may not have a roof, or the roof may be sufficiently flexible to ignore the effects of restrainment. Whatever the case may be, the roof system may also alter the distribution of stress in the shell.

The theory of boundary effects is far more complex than the membrane theory. For this reason, this theory is often ignored in many codes of practice and designer handbooks. Nonetheless, bending stresses and shears at the boundary may be significant. It is therefore possible that localized material failure at the boundary may occur, and a redistribution of stresses in the shell walls.

The bending moments (parallel to the edge), twisting moments (normal to the edge), and the out-of-plane shears on the differential element are illustrated in Fig. 3.11. A double arrow is used to symbolize the bending moments, and the direction of the bending moment is consistent with the right-hand-rule (the right thumb in the direction of the double arrows and fingers in the direction of the moment). In reality, the differential element should also include the membrane stresses (Fig. 3.2), but these stresses are omitted for clarity.

3.4.1 Circumferential Equilibrium

Circumferential, or hoop equilibrium, is determined in the same way as the membrane stresses—equilibrium equations are formulated by summing forces along the line x–x'. The only stresses that contribute to the circumferential equilibrium are the

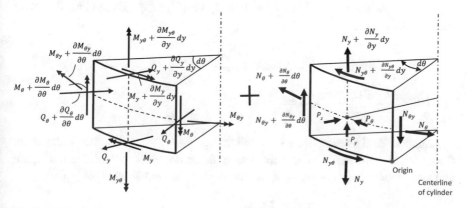

Fig. 3.11 Bending moments, twisting moments, and out-of-plane shears in a differential element

Fig. 3.12 Boundary
stresses on the differential
element that contribute to
the equilibrium equation in
the circumferential direction

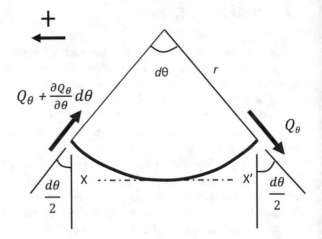

out-of-plane shears. Figure 3.12 is a view of the top surface of the differential
element.

The circumferential equilibrium equation will include the membrane stresses,
previously derived in Eq. 3.1, plus the components of the out-of-plane shears. It
should be noted that the membrane stresses (although the equations are the same) do
not include the prime, since these stresses are the result of the boundary effects. The
notation for stresses caused by boundary effects exclude the prime superscript.

$$\left(N_\theta + \frac{\partial N_\theta}{\partial \theta} d\theta\right) dy \cos\left(\frac{d\theta}{2}\right) - N_\theta dy \cos\left(\frac{d\theta}{2}\right) + \left(N_{y\theta} + \frac{\partial N_{y\theta}}{\partial y} dy\right) d\theta r$$

$$- N_{y\theta} d\theta r + P_\theta (d\theta r) dy - \left(Q_\theta + \frac{\partial Q_\theta}{\partial \theta} d\theta\right) dy \sin\left(\frac{d\theta}{2}\right) - Q_\theta dy \sin\left(\frac{d\theta}{2}\right) = 0$$

$$(3.44)$$

Since $\frac{d\theta}{2}$ is small, the $\cos\left(\frac{d\theta}{2}\right) \approx 1$ and $\sin\left(\frac{d\theta}{2}\right) \approx \frac{d\theta}{2}$. Expanding 3.44 and
incorporating these assumptions,

$$N_\theta dy + \frac{\partial N_\theta}{\partial \theta} d\theta dy - N_\theta dy + N_{y\theta} d\theta r + \frac{\partial N_{y\theta}}{\partial y} d\theta dy r - N_{y\theta} d\theta r + P_\theta d\theta dy r$$

$$- \frac{Q_\theta dy d\theta}{2} - \frac{\partial Q_\theta}{\partial \theta} \frac{d\theta^2 dy}{2} - \frac{Q_\theta dy d\theta}{2} = 0$$

$$(3.45)$$

Equation 3.45 is simplified by eliminating $\frac{\partial Q_\theta}{\partial \theta} \frac{d\theta^2 dy}{2}$ as a higher order term.
Canceling the first and third and the fourth and sixth terms, combining the eighth
and tenth terms, and dividing by $dyd\theta$, the equilibrium equation is simplified.

$$\frac{\partial N_\theta}{\partial \theta} + \frac{\partial N_{y\theta}}{\partial y} r + P_\theta r - Q_\theta = 0 \qquad (3.46)$$

3.4.2   Vertical Equilibrium

Since there are no boundary components in the y-direction, the equation for the vertical equilibrium is identical to Eq. 3.6. The only difference is that the membrane stresses are stresses that are influenced by restrainment at the boundaries, and therefore do not include the prime superscript.

$$\frac{\partial N_y}{\partial y}r + \frac{\partial N_{\theta y}}{\partial \theta} + P_y r = 0 \tag{3.47}$$

3.4.3   Radial Equilibrium

The equilibrium equation for equilibrium in the radial direction is solved by summing forces along the x–x' line of Fig. 3.13. Similar to circumferential equilibrium, the radial equilibrium equations are solved by summing the boundary effects stresses in the radial direction. Since the radial equilibrium equation will include the membrane stresses, the components of the out-of-plane shear stresses from Fig. 3.13 are added to Eq. 3.7.

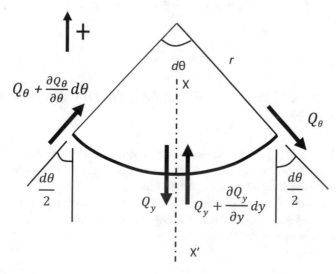

Fig. 3.13 Boundary stresses on the differential element that contribute to the radial equilibrium equations

$$\left(N_\theta + \frac{\partial N_\theta}{\partial \theta} d\theta\right) dy \sin\left(\frac{d\theta}{2}\right) + N_\theta dy \sin\left(\frac{d\theta}{2}\right) + P_z d\theta dyr$$

$$+ \left(Q_\theta + \frac{\partial Q_\theta}{\partial \theta} d\theta\right) dy \cos\left(\frac{d\theta}{2}\right) - Q_\theta dy \cos\left(\frac{d\theta}{2}\right) \qquad (3.48)$$

$$+ \left(Q_y + \frac{\partial Q_y}{\partial y} dy\right) r d\theta - Q_y r d\theta = 0$$

Expanding Eq. 3.48 and assuming $\cos\left(\frac{d\theta}{2}\right) \approx 1$ and $\sin\left(\frac{d\theta}{2}\right) \approx \frac{d\theta}{2}$,

$$N_\theta dy \frac{d\theta}{2} + \frac{\partial N_\theta}{\partial \theta} dy \left(\frac{d\theta^2}{2}\right) + N_\theta dy \left(\frac{d\theta}{2}\right) + P_z d\theta dyr + Q_\theta dy + \frac{\partial Q_\theta}{\partial \theta} d\theta dy$$

$$- Q_\theta dy + Q_y r d\theta + \frac{\partial Q_y}{\partial y} d\theta dyr - Q_y d\theta r = 0 \qquad (3.49)$$

Simplify by eliminating the higher order term, the fifth and seventh and the eighth and tenth terms. Also, combine the first and third terms.

$$N_\theta d\theta dy + P_z d\theta dyr + \frac{\partial Q_\theta}{\partial \theta} d\theta dy + \frac{\partial Q_y}{\partial y} d\theta dyr = 0 \qquad (3.50)$$

Dividing by $d\theta dy$,

$$N_\theta + P_z r + \frac{\partial Q_\theta}{\partial \theta} + \frac{\partial Q_y}{\partial y} r = 0 \qquad (3.51)$$

3.4.4 Moment Equilibrium Equation About the Horizontal Direction

Figure 3.14 illustrates only the stresses that contribute to the moment equilibrium equation about the horizontal line x–x'. Since no bending is produced by membrane stresses, membrane stresses are not part of the equations. The height of the differential element is dy. The moment arm is therefore $\frac{dy}{2}$. The cross depicts stresses going into the page, and dots depicts stresses going out of the page.

Summing moments about the line x–x',

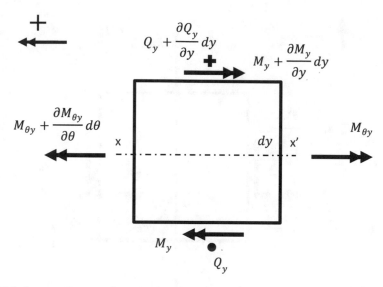

Fig. 3.14 Stresses that contribute to the moment equilibrium equation about the horizontal direction

$$M_y r d\theta - \left(M_y + \frac{\partial M_y}{\partial y} dy \right) r d\theta - M_{\theta y} dy + \left(M_{\theta y} + \frac{\partial M_{\theta y}}{\partial \theta} d\theta \right) dy$$

$$+ Q_y r d\theta \left(\frac{dy}{2} \right) + \left(Q_y + \frac{\partial Q_y}{\partial y} dy \right) r d\theta \left(\frac{dy}{2} \right)$$

$$= 0 \tag{3.52}$$

Expanding Eq. 3.52,

$$M_y r d\theta - M_y r d\theta - \frac{\partial M_y}{\partial y} d\theta dyr - M_{\theta y} dy + M_{\theta y} dy + \frac{\partial M_{\theta y}}{\partial \theta} d\theta dy$$

$$+ Q_y r d\theta \left(\frac{dy}{2} \right) + Q_y r d\theta \left(\frac{dy}{2} \right) + \frac{\partial Q_y}{\partial y} \frac{d\theta dy^2}{2} r$$

$$= 0 \tag{3.53}$$

Eliminating the first and second and the fourth and fifth terms and the higher order term $\left(\frac{\partial Q_y}{\partial y} \frac{dyd\theta^2}{2} r \right)$, the equation is simplified. Furthermore, combine the seventh and eighth terms and divide the equation by $dyd\theta$.

$$-\frac{\partial M_y}{\partial y} r + \frac{\partial M_{\theta y}}{\partial \theta} + Q_y r = 0 \tag{3.54}$$

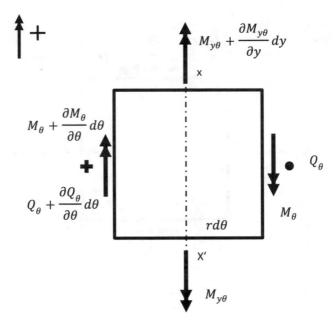

Fig. 3.15 Stresses that contribute to the moment equilibrium equation about the vertical direction

3.4.5   Moment Equilibrium Equation About the Vertical Direction

Figure 3.15 is the differential element with the bending moments and out-of-plane shears that contribute to the moments about the vertical line x–x′ (located in the middle of the element). For clarity, only the stresses that contribute are included in the diagram. The moment arm is equal to $\frac{rd\theta}{2}$.

Summing moments about the line x–x′,

$$\left(M_\theta + \frac{\partial M_\theta}{\partial \theta} d\theta\right) dy - M_\theta dy + \left(M_{y\theta} + \frac{\partial M_{y\theta}}{\partial y} dy\right) rd\theta - M_{y\theta} rd\theta$$

$$- Q_\theta dy\left(\frac{rd\theta}{2}\right) - \left(Q_\theta + \frac{\partial Q_\theta}{\partial \theta} d\theta\right) dy\left(\frac{rd\theta}{2}\right)$$

$$= 0 \tag{3.55}$$

Expanding Eq. 3.55,

$$M_\theta dy + \frac{\partial M_\theta}{\partial \theta} d\theta dy - M_\theta dy + M_{y\theta} r d\theta + \frac{\partial M_{y\theta}}{\partial y} d\theta dy r - M_{y\theta} r d\theta$$

$$- Q_\theta dy \left(\frac{rd\theta}{2}\right) - Q_\theta dy \left(\frac{rd\theta}{2}\right) - \frac{\partial Q_\theta}{\partial \theta} d\theta dy \left(\frac{rd\theta}{2}\right)$$

$$= 0 \qquad\qquad\qquad\qquad\qquad\qquad\qquad\qquad\qquad\qquad (3.56)$$

The last term of Eq. 3.56 is a higher order term (i.e., the $\frac{d\theta^2}{2}$ part makes the term much smaller in magnitude than the other terms), and therefore may be eliminated. Simplifying Eq. 3.56 by removing the first and third and the fourth and sixth terms, dividing by $d\theta dy$, and combining the seventh and eighth terms,

$$\frac{\partial M_\theta}{\partial \theta} + \frac{\partial M_{y\theta}}{dy} r - Q_\theta r = 0 \qquad\qquad\qquad\qquad\qquad (3.57)$$

3.4.6   Equilibrium Equation Summary for Symmetrical Loading

Collecting the equilibrium equations,

$$\frac{\partial N_\theta}{\partial \theta} + \frac{\partial N_{y\theta}}{\partial y} r + P_\theta r - Q_\theta = 0 \qquad\qquad\qquad (3.46)$$

$$\frac{\partial N_y}{\partial y} r + \frac{\partial N_{\theta y}}{\partial \theta} + P_y r = 0 \qquad\qquad\qquad\qquad (3.47)$$

$$N_\theta + P_z r + \frac{\partial Q_\theta}{\partial \theta} + \frac{\partial Q_y}{\partial y} r = 0 \qquad\qquad\qquad (3.51)$$

$$-\frac{\partial M_y}{\partial y} r + \frac{\partial M_{\theta y}}{\partial \theta} + Q_y r = 0 \qquad\qquad\qquad (3.54)$$

$$\frac{\partial M_\theta}{\partial \theta} + \frac{\partial M_{y\theta}}{dy} r - Q_\theta r = 0 \qquad\qquad\qquad\qquad (3.57)$$

The above five equations are the general equations for the boundary effects and includes the possibility of nonsymmetric loads. However, symmetrical loading is assumed (i.e., liquid loading and concentric loading/discharge of granular materials). By assuming symmetry, the shell will not be subjected to in-plane shears or twisting moments.

$$N_{y\theta} = N_{\theta y} = M_{\theta y} = M_{y\theta} = 0$$

Eliminating these stresses and moments,

$$\frac{\partial N_\theta}{\partial \theta} + P_\theta r - Q_\theta = 0$$

$$\frac{\partial N_y}{\partial y} r + P_y r = 0$$

$$N_\theta + P_z r + \frac{\partial Q_\theta}{\partial \theta} + \frac{\partial Q_y}{\partial y} r = 0$$

$$-\frac{\partial M_y}{\partial y} r + Q_y r = 0$$

$$\frac{\partial M_\theta}{\partial \theta} - Q_\theta r = 0$$

Furthermore, for symmetric loading, the stresses in the shell change in the y-direction. However, at any given y, the hoop stresses are constant at every angle θ. Therefore, all terms where the stress changes in the θ direction may also be eliminated from the expressions.

$$\frac{\partial M_\theta}{\partial \theta} = \frac{\partial N_\theta}{\partial \theta} = \frac{\partial Q_\theta}{\partial \theta} = 0$$

Applying the second set of assumptions,

$$P_\theta r - Q_\theta = 0$$

$$\frac{\partial N_y}{\partial y} r + P_y r = 0$$

$$N_\theta + P_z r + \frac{\partial Q_y}{\partial y} r = 0$$

$$-\frac{\partial M_y}{\partial y} r + Q_y r = 0$$

$$-Q_\theta r = 0$$

By inspection, the last equation falls away, which must be correct since a shear will only exist with a changing moment. Furthermore, since $Q_\theta = 0$ and with no surface tractions P_θ, the first equation will also fall away. Thus, the five equilibrium equations reduce to three for the case of symmetrical loading.

$$\frac{\partial N_y}{\partial y} r + P_y r = 0 \tag{3.58}$$

$$N_\theta + P_z r + \frac{\partial Q_y}{\partial y} r = 0 \tag{3.59}$$

$$-\frac{\partial M_y}{\partial y} r + Q_y r = 0 \tag{3.60}$$

3.4.7   Stress Equations for the Boundary Effects

Moments in the y-Direction

From Eqs. 3.58, 3.59, and 3.60, the vertical stresses (N_y), the vertical moments (M_y), the hoop stresses (N_θ), and the vertical shears (Q_y) are solved.

From plate theory (Szilard 1974), and replacing x with θ, the moment-curvature relationships are defined.

$$M_y = -D\left(\frac{\partial^2 w}{\partial y^2} + \nu \frac{\partial^2 w}{\partial \theta^2}\right) \tag{3.61}$$

$$M_\theta = -D\left(\frac{\partial^2 w}{\partial \theta^2} + \nu \frac{\partial^2 w}{\partial y^2}\right) \tag{3.62}$$

where

$$D = \frac{EI}{(1 - \nu^2)} \tag{3.63}$$

As mentioned above, for symmetrical loading, the stresses change in the y-direction, but not in the θ-direction. For this reason, it may be assumed that the curvature in the hoop direction is equal to zero $\left(\text{i.e., } \frac{\partial^2 w}{\partial \theta^2} = 0\right)$. Therefore, Eq. 3.61 becomes

$$M_y = -D\left(\frac{\partial^2 w}{\partial y^2}\right) \tag{3.64}$$

Moments in the Hoop Direction

Rearranging Eqs. 3.61 and 3.62 to express the curvature in terms of the moments,

$$\frac{\partial^2 w}{\partial y^2} = \frac{1}{EI}\left(-M_y + \nu M_\theta\right) \tag{3.65}$$

$$\frac{\partial^2 w}{\partial \theta^2} = \frac{1}{EI}\left(-M_\theta + \nu M_y\right) \tag{3.66}$$

Since it is assumed that the curvature in the θ-direction is zero,

$$\frac{\partial^2 w}{\partial \theta^2} = \frac{1}{EI}\left(-M_\theta + \nu M_y\right) = 0$$

or

$$\left(-M_\theta + \nu M_y\right) = 0 \tag{3.67}$$

Equation 3.67 rearranged,

$$M_\theta = \nu M_y \tag{3.68}$$

Moments in the θ-direction are not the result of shell deformations w but the result of Poisson's effect. The moment in the y-direction will cause a stress couple in the shell walls. These stresses will cause an opposite stress couple in the θ-direction, which is the source of Eq. 3.68.

Membrane Stresses in the Hoop Direction

Referring to the Eqs. 3.37 and 3.38 (without the primes since we are referring to stresses caused by boundary effects),

$$N_\theta = \frac{Eh}{(1-\nu^2)}\left(-\frac{w}{r} + \nu\frac{\partial v}{\partial y}\right) \tag{3.37}$$

$$N_y = \frac{Eh}{(1-\nu^2)}\left(\frac{\partial v}{\partial y} - \nu\frac{w}{r}\right) \tag{3.38}$$

As previously assumed, the vertical stresses in the y-direction are independent of the hoop stresses in the θ-direction (i.e., N_y does not influence the stresses N_θ). Based on this assumption, Eq. 3.38 is set equal to zero.

$$0 = \frac{Eh}{(1-\nu^2)}\left(\frac{\partial v}{\partial y} - \nu\frac{w}{r}\right) \tag{3.69}$$

Therefore,

$$\frac{\partial v}{\partial y} = \nu\frac{w}{r} \tag{3.70}$$

Substituting this relationship into Eq. 3.37 and simplifying,

$$N_\theta = \frac{Eh}{(1-\nu^2)}\left(-\frac{w}{r} + \nu^2\frac{w}{r}\right)$$

$$N_\theta = -\frac{Eh}{(1-\nu^2)}(1-\nu^2)\frac{w}{r}$$

$$N_\theta = -\frac{Ehw}{r} \tag{3.71}$$

Out-of-Plane Shears in the y-Direction

Using the general Eq. 3.60,

$$-\frac{\partial M_y}{\partial y}r + Q_y r = 0 \tag{3.60}$$

Rearranging 3.60, the out-of-plane shear is expressed in terms of the bending moments in the y-direction.

$$Q_y = \frac{\partial M_y}{\partial y} \tag{3.72}$$

Since $M_y = -D\left(\frac{\partial^2 w}{\partial y^2}\right)$ (Eq. 3.64), substitute this expression into 3.72,

$$Q_y = -D\frac{\partial^3 w}{\partial y^3} \tag{3.73}$$

3.5 Displacement Theory for the Boundary Effects

Using the second general equation for boundary effects (3.59) and the out-of-plane shear Eq. (3.72),

$$N_\theta + P_z r + \frac{\partial Q_y}{\partial y} r = 0 \tag{3.59}$$

$$Q_y = \frac{\partial M_y}{\partial y} \tag{3.72}$$

Substituting 3.72 into 3.59,

$$N_\theta + P_z r + \frac{\partial^2 M_y}{\partial y^2} r = 0 \tag{3.74}$$

Equation 3.74 gives the relationship between the hoop stress and bending moments. These stresses have been previously solved in terms of deformations.

$$M_y = -D\left(\frac{\partial^2 w}{\partial y^2}\right) \tag{3.64}$$

$$N_\theta = -\frac{Ehw}{r} \tag{3.71}$$

Substituting the bending moment Eq. (3.64) and the hoop stress Eq. (3.71) into 3.74.

$$-\frac{Ehw}{r} + P_z r - D\frac{\partial^4 w}{\partial y^4} r = 0 \tag{3.75}$$

Rearranging the above equation,

$$P_z = D\frac{\partial^4 w}{\partial y^4} + \frac{Ehw}{r^2} \tag{3.76}$$

If we say,

$$\beta^4 = \frac{Eh}{4r^2 D} = \frac{3(1-\nu^2)}{r^2 h^2} \tag{3.77}$$

We can express Eq. 3.76 in terms of β^4.

$$\frac{\partial^4 w}{\partial y^4} + 4\beta^4 w = \frac{P_z}{D} \tag{3.78}$$

which form is known and a well-established non-homogeneous differential equation for plates in bending (Jaeger 1969). The solution for this equation is given in Eq. 3.79.

$$w = e^{\beta y}(C_1 \cos \beta y + C_2 \sin \beta y) + e^{-\beta y}(C_3 \cos \beta y + C_4 \sin \beta y) + f(y) \qquad (3.79)$$

The first two terms are referred to as the complementary part of the solution and $f(y)$ is the particular part. The particular part is the membrane solution, which depends on the type of loading.

For liquid loading,

$$f(y) = w' = -\frac{\gamma r^2}{Eh}(H - y) \qquad (3.40)$$

For granular materials loading,

$$f(y) = w' = -\frac{\gamma r^2}{Eh\mu}\left[1 - e^{(y-H)/z_o}\right] \qquad (3.42)$$

The constants C_1, C_2, C_3, and C_4 are integration constants and are dependent on the boundary conditions. Two of the constants are related to the horizontal translations and the other two are the rotations. Furthermore, the horizontal translations are related to the out-of-plane shears and the rotations are related to the bending moments.

The complementary part of the solution is composed of two terms. The first term, $e^{\beta y}(C_1 \cos \beta y + C_2 \sin \beta y)$, increase as y increases. This is contrary to Saint-Venant's principle (Marshall and Nelson 1981), which states that a stress must diminish with increasing distance. The cylinder is assumed to be sufficiently long and the boundary effects (i.e., moments and shears) will diminish rapidly up the shell. The constants C_1 and C_2 may therefore be assumed equal to zero. The solution is therefore simplified.

$$w = e^{-\beta y}(C_3 \cos \beta y + C_4 \sin \beta y) \qquad (3.80)$$

Taking the first, second, and third derivatives,

$$\frac{\partial w}{\partial y} = \beta e^{-\beta y}[-C_3(\sin \beta y + \cos \beta y) + C_4(\cos \beta y - \sin \beta y)] \qquad (3.81)$$

$$\frac{\partial^2 w}{\partial y^2} = 2\beta^2 e^{-\beta y}[C_3 \sin \beta y - C_4 \cos \beta y] \qquad (3.82)$$

$$\frac{\partial^3 w}{\partial y^3} = 2\beta^3 e^{-\beta y}[C_3(\cos \beta y - \sin \beta y) + C_4(\sin \beta y + \cos \beta y)] \qquad (3.83)$$

At the base of the cylinder, the bending moment is equal to M_o. Using the bending moment equation in the y-direction,

$$M_y = -D\left(\frac{\partial^2 w}{\partial y^2}\right) \tag{3.64}$$

and substituting Eq. 3.82,

$$M_y = -D2\beta^2 e^{-\beta y}[C_3 \sin \beta y - C_4 \cos \beta y] \tag{3.84}$$

Solving for the moment at the base at $y = 0$,

$$M_o = 2D\beta^2 C_4 \tag{3.85}$$

Solving for C_4,

$$C_4 = \frac{M_o}{2D\beta^2} \tag{3.86}$$

From Eq. 3.73,

$$Q_y = -D\frac{\partial^3 w}{\partial y^3} \tag{3.73}$$

The shear is defined in terms of deformations by substituting Eq. 3.83.

$$Q_y = -D2\beta^3 e^{-\beta y}[C_3(\cos \beta y - \sin \beta y) + C_4(\sin \beta y + \cos \beta y)] \tag{3.87}$$

Setting $y = 0$ to determine the shear Q_o at the base of the shell,

$$Q_o = -2D\beta^3(C_3 + C_4) \tag{3.88}$$

Replacing C_4 with 3.86,

$$Q_o = -2D\beta^3\left(C_3 + \frac{M_o}{2D\beta^2}\right) \tag{3.89}$$

Rearranging 3.89 to solve for C_3,

$$C_3 = \frac{-1}{2\beta^3 D}(Q_o + \beta M_o) \tag{3.90}$$

The deformations at the base of the shell may now be solved. Substituting constants C_3 and C_4 into Eq. 3.80,

$$w = e^{-\beta y}\left(\frac{-1}{2\beta^3 D}(Q_o + \beta M_o)\ \cos\ \beta y + \frac{M_o}{2D\beta^2}\ \sin\ \beta y\right) \tag{3.91}$$

Similarly, solving for the rotations by substituting the constants into the slope Eq. (3.81),

$$\frac{\partial w}{\partial y} = \beta e^{-\beta y}\left[\frac{1}{2\beta^3 D}(Q_o + \beta M_o)\ (\sin\beta y + \ \cos\ \beta y) + \frac{M_o}{2D\beta^2}(\ \cos\beta y - \ \sin\ \beta y)\right] \tag{3.92}$$

Solving for the horizontal deformations at the base $(y = 0)$.

$$w_o = \frac{-1}{2\beta^3 D}(Q_o + \beta M_o)$$

or expressed as,

$$w_o = \frac{-Q_o}{2\beta^3 D} + \frac{-M_o}{2\beta^2 D} = Q_o D_{11} + M_o D_{12} \tag{3.93}$$

where

$$D_{11} = \frac{1}{2\beta^3 D} \tag{3.94}$$

$$D_{12} = \frac{1}{2\beta^2 D} \tag{3.95}$$

Similarly, solving for the slope at $y = 0$.

$$\phi_o = \frac{\partial w}{\partial y} = \beta\left[\frac{1}{2\beta^3 D}(Q_o + \beta M_o) + \frac{M_o}{2D\beta^2}\right]$$

$$\phi_o = \frac{1}{2\beta^2 D}(Q_o + 2\beta M_o)$$

or expressed as

$$\phi_o = \frac{Q_o}{2\beta^2 D} + \frac{M_o}{\beta D} = Q_o D_{21} + M_o D_{22} \tag{3.96}$$

where

$$D_{21} = \frac{1}{2\beta^2 D} \tag{3.97}$$

$$D_{22} = \frac{1}{\beta D} \tag{3.98}$$

3.6 Compatibility Equations

Fixed Base

The solution of cylindrical shells is based on the formation of a set of compatibility equations. Compatibility requires that the deformations are continuous in the shell, and the displacements and rotations fit the boundary conditions at the base (Kassimali 2005). In Fig. 3.16, compatibility of the shell is depicted for the case of

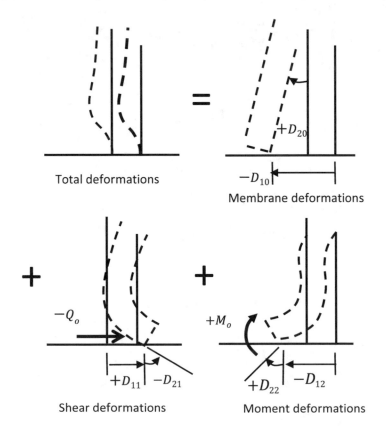

Fig. 3.16 Separation of deformations at the base of the cylinder for the fixed case

a fixed base. The top left figure represents the correct deformed shape, which is formed by the summation of the three deformations. Membrane stresses cause an error, but base reactions are applied (i.e., M_o and Q_o) to correct the deformations and ensure that compatibility is satisfied. The top right figure represents the deformations of the shell under pure membrane stresses. Under this condition, the shell will slide outward and rotate at the base, which is not correct if the base is fixed or pinned. To correct this error, a negative shear force (Q_o) and a positive bending moment (M_o) are applied to the base of the shell so that the lateral deformations at the base and slope sum to zero.

The compatibility of deformations may be expressed in mathematical form, based on the sign convention of Fig. 3.16. For the case of a fixed based,

$$- D_{10} - (-Q_o)D_{11} - M_o D_{12} = 0 \tag{3.99}$$

$$D_{20} + (-Q_o)D_{21} + M_o D_{22} = 0 \tag{3.100}$$

Equation 3.99 is the compatibility equation for horizontal deformations and Eq. 3.100 is the compatibility equation of the slope at the base of the shell. The two equations have three terms—the first terms are the membrane equations and the second and third terms are the deformations due to boundary effects. Both equations are set to zero, since for a fixed base the horizontal deformations and the slope at the base must be equal to zero.

The two compatibility equations have two unknowns (Q_o and M_o), and therefore a solution is possible. Setting the equations in matrix form,

$$\begin{bmatrix} D_{11} & D_{12} \\ D_{21} & D_{22} \end{bmatrix} \begin{Bmatrix} -Q_o \\ M_o \end{Bmatrix} = \begin{Bmatrix} -D_{10} \\ -D_{20} \end{Bmatrix} \tag{3.101}$$

The boundary shear and moment is solved by inverting the 2×2 matrix.

$$\begin{Bmatrix} -Q_o \\ M_o \end{Bmatrix} = \begin{bmatrix} D_{11} & D_{12} \\ D_{21} & D_{22} \end{bmatrix}^{-1} \begin{Bmatrix} -D_{10} \\ -D_{20} \end{Bmatrix} \tag{3.102}$$

$$\begin{Bmatrix} -Q_o \\ M_o \end{Bmatrix} = \frac{1}{(D_{11})D_{22} - (D_{12})(D_{21})} \begin{bmatrix} D_{22} & -D_{12} \\ -D_{21} & D_{11} \end{bmatrix} \begin{Bmatrix} -D_{10} \\ -D_{20} \end{Bmatrix} \tag{3.103}$$

Pinned Base

For the case of a pinned base, the concept is the same and Fig. 3.16 is valid, except that the bending moment is omitted. Only one compatibility equation is applicable. From Eq. 3.99,

$$D_{10} - Q_o D_{11} = 0 \tag{3.104}$$

The shear is solved directly from Eq. 3.104,

$$Q_o = \frac{D_{10}}{D_{11}} \tag{3.105}$$

$$M_o = 0$$

Sliding Base

For the case of a sliding bearing, $M_o = Q_o = 0$. The only equations that are applicable are the membrane equations.

Once the base shear and moment reactions are solved (M_o and Q_o), the deformations and stresses are determined at every point in the shell.

3.7   Steps in Solving the Deformations and Stresses in the Shell

The copious number of equations to solve cylindrical shells tend to muddle the solution. The designer is likely to get lost, trying to assemble the appropriate equations, and setting them in the correct sequence to solve even the most simple of problems. This section has therefore been added to provide a "cook book" approach to shell design; the necessary equations are set out in the correct order to calculate the stresses and deformations of the shell.

Because of the enormity of the solution, it is advisable to use a computer program (e.g., spread sheets or an alternative programing language).

1. **Solve for base reactions M_o and Q_o.**

Initial parameters		
	Diameter (L) Height (H) Thickness (h) Radius (r) Young's modulus (E) Poisson's ratio (ν)	
	$I = \frac{h^3}{12}$ (moment of inertia per unit width of shell)	3.106
	$D = \frac{EI}{(1-\nu^2)}$	3.63
	$\beta^4 = \frac{Eh}{4r^2D} = \frac{3(1-\nu^2)}{r^2h^2}$	3.77
	$\beta = \sqrt[4]{\beta^4}$	
	$\beta^2 = \left(\sqrt[4]{\beta^4}\right)^2$	
	$\beta^3 = \left(\sqrt[4]{\beta^4}\right)^3$	
Membrane deformations at the base (y = 0)		
Liquid loading	$D_{10} = -\frac{\gamma r^2}{Eh}(H-y)$	3.40
	$D_{20} = \frac{\gamma r^2}{Eh}$	3.41

(continued)

or		
Granular materials loading	$D_{10} = -\frac{\gamma r^2 r_h}{Eh\mu}\left[1 - e^{(y-H)/z_o}\right]$	3.42
	$D_{20} = \frac{\gamma r^2 r_h}{Eh\mu z_o}e^{(y-H)/z_o}$	3.43
Deformations at the base from boundary effects		
	$Q_o D_{11} = (-Q_o)\frac{1}{2\beta^3 D}$	3.94
	$M_o D_{12} = M_o\frac{1}{2\beta^2 D}$	3.95
	$Q_o D_{21} = (-Q_o)\frac{1}{2\beta^2 D}$	3.97
	$M_o D_{22} = M_o\frac{1}{\beta D}$	3.98
Compatibility equations (note sign change of Q_o in Sect. 3.6)		
Fixed base	$\begin{Bmatrix} -Q_o \\ M_o \end{Bmatrix} = \frac{1}{(D_{11})D_{22}-(D_{12})(D_{21})}\begin{bmatrix} D_{22} & -D_{12} \\ -D_{21} & D_{11} \end{bmatrix}\begin{Bmatrix} -D_{10} \\ -D_{20} \end{Bmatrix}$	3.103
or		
Pinned base	$Q_o = \frac{D_{10}}{D_{11}}$	3.105
	$M_o = 0$	
or		
Sliding base	$Q_o = 0$	
	$M_o = 0$	3.63

2. Solve for the constants C_3 and C_4.

$C_3 = -\frac{Q_o}{2\beta^3 D} - \frac{M_o}{2\beta^2 D}$	3.90
$C_4 = \frac{M_o}{2D\beta^2}$	3.86

3. Solve for the membrane deformations in the shell.

Liquid loading	$w' = -\frac{\gamma r^2}{Eh}(H - y)$	3.40
	$\phi'_y = \frac{\gamma r^2}{Eh}$	3.41
or		
Granular materialsloading	$w' = -\frac{\gamma r^2 r_h}{Eh\mu}\left[1 - e^{(y-H)/z_o}\right]$	3.42
	$\phi'_y = \frac{\gamma r^2 r_h}{Eh\mu z_o}e^{(y-H)/z_o}$	3.43

4. Solve for the deformations in the shell due to boundary effects.

$w = e^{-\beta y}(C_3 \cos \beta y + C_4 \sin \beta y)$	3.80
$\phi = \frac{\partial w}{\partial y} = \beta e^{-\beta y}[-C_3(\sin\beta y + \cos\beta y) + C_4(\cos\beta y - \sin\beta y)]$	3.81
$\frac{\partial^2 w}{\partial y^2} = 2\beta^2 e^{-\beta y}[C_3 \sin\beta y - C_4 \cos\beta y]$	3.82
$\frac{\partial^3 w}{\partial y^3} = 2\beta^3 e^{-\beta y}[C_3(\cos\beta y - \sin\beta y) + C_4(\sin\beta y + \cos\beta y)]$	3.83

5. Solve for the total deformations.

$w(total) = w' + w$	3.107
$\phi(total) = \phi_y' + \phi_y$	3.108

6. Solve for membrane stresses, moments, and shears in increments of y.

Membrane stresses		
Liquid loading	$N_y' = -\gamma h(H - y)$	3.19
	$N_\theta' = \gamma(H - y)r$	3.22
or		
Granular materials loading	$N_y' = -\gamma h(H - y)$	3.19
	$N_\theta' = \frac{\gamma r_h \left[1 - e^{(y-H)/z_o} \right]}{\mu} r$	3.24
Stresses from boundary effects		
	$M_y = -D\left(\frac{\partial^2 w}{\partial y^2}\right)$	3.64
	$M_\theta = \nu M_y$	3.68
	$N_\theta = -\frac{Ehw}{r}$	3.71
	$Q_y = -D\frac{\partial^3 w}{\partial y^3}$	3.73
Total stresses		
	$N_\theta(total) = N_\theta + N_\theta'$	3.109

Caution should be used that the correct material weight term (γ) is used in the above equations.

3.8   Worked Example

A 6-m diameter water tank is 6 m high and 150 mm thick (Fig. 3.17). The tank is constructed of reinforced concrete and has a concrete compressive strength of 30 MPa. The base is fixed and top of the shell is open. Determine the deformations and stresses in the shell.

1. Solve for the base reactions M_o and Q_o.

Initial parameters

$L = 6$ m
$H = 6$ m
$h = 0.15$ m
$r = 3$ m
$f_{cu} = 30$ N/mm^2
$E = 28(10^6)$ kN/m^2
$\nu = 0.15$

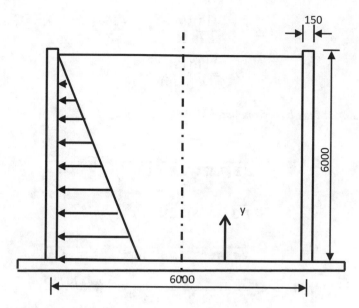

Fig. 3.17 Water tank with fixed supports

$\gamma_w = 10 \text{ kN/m}^3$ (unit weight of water)
$\gamma_m = 24 \text{ kN/m}^3$ (unit weight of concrete)

$$I = \frac{1\,(0.15)^3}{12} = 2.813\,(10^{-4})\,\text{m}^4 \tag{3.106}$$

$$D = \frac{E\,2.813\,(10^{-4})}{(1 - 0.15^2)} = E\,2.878\,(10^{-4}) = 8056.27 \tag{3.63}$$

$$\beta^4 = \frac{3\left[1 - (0.15)^2\right]}{3^2(0.15)^2} = 14.481 \tag{3.77}$$

$$\beta = \sqrt[4]{14.481} = 1.951$$

$$\beta^2 = 3.805$$

$$\beta^3 = 7.424$$

Membrane deformations at the base (y = 0) for liquid loading

$$D_{10} = \frac{-10\,(3)^2(6-0)}{E\,(0.15)} = \frac{-3600}{E} \tag{3.40}$$

$$D_{20} = \frac{10\,(3)^2}{E\,(0.15)} = \frac{600}{E} \tag{3.41}$$

Deformations at the base from boundary effects

$$Q_o D_{11} = Q_o \frac{1}{E\,2\,(7.424)\,2.878\,\left(10^{-4}\right)} = Q_o \frac{234.03}{E} \tag{3.94}$$

$$M_o D_{12} = M_o \frac{1}{E\,2\,(3.805)2.878\,\left(10^{-4}\right)} = M_o \frac{456.59}{E} \tag{3.95}$$

$$Q_o D_{21} = Q_o \frac{1}{E\,2\,(3.805)2.878\,\left(10^{-4}\right)} = Q_o \frac{456.59}{E} \tag{3.97}$$

$$M_o D_{22} = M_o \frac{1}{E\,(1.951)2.878\,\left(10^{-4}\right)} = M_o \frac{1780.95}{E} \tag{3.98}$$

Compatibility equations

$$\begin{bmatrix} 234.09 & 456.65 \\ 456.65 & 1781.64 \end{bmatrix} \frac{1}{E} \left\{ \begin{matrix} -Q_o \\ M_o \end{matrix} \right\} = \left\{ \begin{matrix} -3600 \\ -600 \end{matrix} \right\} \tag{3.101}$$

$$\left\{ \begin{matrix} Q_o \\ M_0 \end{matrix} \right\} = \begin{bmatrix} 0.00854 & -0.00219 \\ -0.00219 & 0.00112 \end{bmatrix} \frac{1}{E} \left\{ \begin{matrix} -3600 \\ -600 \end{matrix} \right\} = \left\{ \begin{matrix} -29.44 \\ 7.21 \end{matrix} \right\} \tag{3.103}$$

Reactions at the base,

$$Q_o = -29.44 \text{ kN/m}$$

$$M_o = 7.21 \text{ kN.m/m}$$

The negative shear indicates that the direction of Q_o is according to Fig. 3.16.

2. **Solve for the constants C_3 and C_4.**

$$C_3 = \frac{-[-29.44 + 1.951(7.21)]}{2\,(7.424)\,8056.27} = 1.29\left(10^{-4}\right) \tag{3.90}$$

$$C_4 = \frac{7.21}{2\,(8056.27)\,3.805} = 1.18(10^{-4}) \tag{3.86}$$

3. **Solve for the membrane deformations in the shell.**

   The deformations are solved at $y = 0$ for liquid loading.

   $$w' = D_{10} = -1.29(10^{-4}) \tag{3.40}$$

   $$\phi'_y = D_{20} = 2.14(10^{-5}) \tag{3.41}$$

4. **Solve for the deformations in the shell due to boundary effects.**

$$w = e^{-1.951(0)}\left[1.285(10^{-4})\cos\,1.951(0) + 1.176(10^{-4})\sin 1.951(0)\right]$$
$$= 1.29\,(10^{-4}) \tag{3.80}$$

$$\phi = 1.951e^{-1.951(0)}[-1.285(10^{-4})\{\sin 1.951(0) + \cos\,1.951(0)\}$$
$$+ 1.176(10^{-4})\{\cos 1.951(0) - \sin\,1.951(0)\}]$$
$$= -2.14(10^{-5}) \tag{3.81}$$

$$\frac{\partial^2 w}{\partial y^2} = 2(1.951)^2 e^{-1.951(0)}\left[1.285(10^{-4})\sin 1.951(0) - 1.176(10^{-4})\cos 1.951(0)\right]$$
$$= -8.95(10^{-4}) \tag{3.82}$$

$$\frac{\partial^3 \partial w}{\partial y^3} = 2(1.951)^3 e^{-1.951(0)}\{1.285(10^{-4})[\cos 1.951(0) - \sin\,1.951(0)]$$
$$+ 1.176(10^{-4})(\sin 1.951(0) + \cos 1.951(0))\}$$
$$= 3.65(10^{-3}) \tag{3.83}$$

5. **Solve for the total deformations.**

$$w(total) = w' + w = -1.29(10^{-4}) + 1.29(10^{-4}) = 0 \tag{3.107}$$

$$\phi(total) = \phi'_y + \phi_y = 2.14(10^{-5}) - 2.14(10^{-5}) = 0 \tag{3.108}$$

At the base, the horizontal deformations and the slope must sum to zero for a fixed base.

6. Solve for the membrane stresses, moments, and shears in increments *y*.

Membrane stresses for liquid loading at y = 0

$$N'_y = -24(0.15)(6 - 0) = -21.60 \text{ kN/m} \tag{3.20}$$

$$N'_\theta = 10(6 - 0)3 = 180.00 \text{ kN/m} \tag{3.22}$$

Stresses from boundary effects at y = 0

$$M_y = -8058.40(-8.98)(10^{-4}) = 7.21 \text{ kN.m/m} \tag{3.64}$$

$$M_\theta = 0.15(7.24) = 1.08 \text{ kN.m/m} \tag{3.68}$$

$$N_\theta = -\frac{28(10^6)(0.15)1.29(10^{-4})}{3} = -180.00 \text{ kN/m} \tag{3.71}$$

$$Q_y = -(8058.40)\, 3.66(10^{-3}) = -29.44 \text{ kN/m} \tag{3.73}$$

Total stresses at y = 0

$$N_{(total)} = 180.00 - 180.00 = 0 \text{ kN/m} \tag{3.109}$$

Distribution of deformations and stresses in the shell

The deformations and stresses are given in Tables 3.1 and 3.2 at 1 m increments, and graphical representations are given in Figs. 3.18, 3.19, 3.20 and 3.21. It should be noted that the boundary effects, which cause bending and shears at the support,

Table 3.1 Deformations in the shell

y (m)	w' (m)	w (m)	w(total) (m)	ϕ'	ϕ	$\varnothing$(total)	$\frac{\partial w^2}{\partial^2 y}$	$\frac{\partial w^3}{\partial^3 y}$
0	−1.29E-04	1.29E-04	0.00E+00	2.14E-05	−2.14E-05	0.00E+00	−8.95E-04	3.65E-03
1	−1.07E-04	8.75E-06	−9.84E-05	2.14E-05	−6.23E-05	−4.08E-05	1.76E-04	−2.14E-04
2	−8.57E-05	−3.52E-06	−8.92E-05	2.14E-05	7.00E-06	2.84E-05	−5.12E-07	−5.13E-05
3	−6.43E-05	1.95E-07	−6.41E-05	2.14E-05	5.20E-07	2.19E-05	−3.51E-06	9.74E-06
4	−4.29E-05	5.06E-08	−4.28E-05	2.14E-05	−1.96E-07	2.12E-05	3.81E-07	9.50E-09
5	−2.14E-05	−9.27E-09	−2.14E-05	2.14E-05	1.02E-08	2.14E-05	3.08E-08	−1.98E-07
6	0.00E+00	−4.58E-11	−4.58E−11	2.14E-05	2.89E-09	2.14E-05	-1.09E-08	2.07E-08

Table 3.2 Stresses in the shell

y (m)	N'_y (kN/m)	N'_θ (kN/m)	N_θ (kN/m)	N_θ(total) (kN/m)	M_y (kN.m/m)	M_θ (kN.m/m)	Q_y (kN/m)
0	−21.60	180.00	−180.00	0.00	7.21	1.08	−29.44
1	−18.00	150.00	−12.24	137.76	−1.42	−0.21	1.73
2	−14.40	120.00	4.93	124.93	0.00	0.00	0.41
3	−10.80	90.00	−0.27	89.73	0.03	0.00	−0.08
4	−7.20	60.00	−0.07	59.93	0.00	0.00	0.00
5	−3.60	30.00	0.01	30.01	0.00	0.00	0.00
6	0.00	0.00	0.00	0.00	0.00	0.00	0.00

Fig. 3.18 Horizontal deflections in the water tank

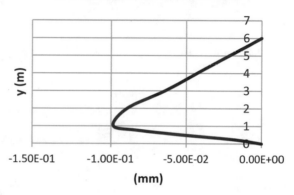

w **Horizontal Deflections**

Fig. 3.19 Hoop stresses in the water tank

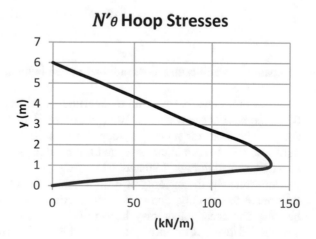

N'_θ **Hoop Stresses**

are restricted to regions near the base and rapidly dissipate. The rate of dissipation is dependent on the stiffness of the shell walls. In concrete shells, the boundary effects influence is about a third of the shell. If the walls are constructed of steel, the

Fig. 3.20 Bending
moments in the water tank

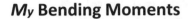

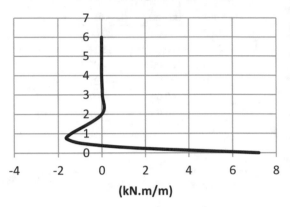

Fig. 3.21 Shear stresses in
the water tank

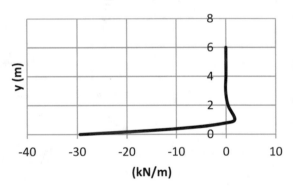

dissipation is significantly more rapid and often influences only about a tenth of the
height of the shell.

A few quick checks are useful to determine if the calculation appears reasonable.
Since the base of the walls are fixed, the lateral deformation and slope must equal
zero. Furthermore, the sum of the hoop stress for the membrane and boundary effects
must also sum to zero at the base, and the bending and shears must look similar in
shape illustrated in Figs. 3.20 and 3.21.

Unfortunately, the majority of design guides and codes of practice are based on
membrane theory only, due to the complexity of the boundary theory. However,
bending and shear forces may be significant at the base of the shell and cause
structural damage if not accounted for. This is of particular importance in water
retaining shells. The boundary effects will also alter the distribution of hoop stresses,
which increases in the middle third of the shell.

Since the moments and shears change rapidly near the base of the shell, these forces should be calculated at smaller increments to determine a more accurate representation of the graph patterns.

3.9 Exercises

3.9.1 Using a spreadsheet or alternative programing language, program the membrane theory. Using the membrane theory, solve for the membrane stresses (N'_y and N'_θ) and the deformations (w' and ϕ'_y) for the worked example in Sect. 3.8 and tabulate in increments of 500 mm.

3.9.2 Using the spreadsheet or computer program in question 3.9.1, include the boundary solution equations. Resolve the worked example in Sect. 3.8 for a fixed support. Determine the distribution of stresses (N_y and N_θ), the deformations (w and ϕ_y), and the bending moments and shear distribution (M_y, M_θ, and Q_y). Tabulate in increments of 500 mm. Furthermore, make the following checks:

(a) Do the horizontal deformations and slopes at the base equal to zero?
(b) Do the hoop stresses cancel to zero at the base?
(c) Do the graphs of the moments, shears, in-plane stresses, and deformations have the correct form?
(d) Are the moments and shears confined to the lower third of the shell, and are the peak deformations and hoops stresses near the base of the shell?

3.9.3 Referring to the worked example in Sect. 3.8, what are the in-plane shears ($N'_{\theta y}$ and $N'_{y\theta}$) and the out-of-plane shears (Q_y) in the shell?

3.9.4 Considering the distribution of stresses and forces determined in 3.9.2, demonstrate how the shell will potentially crack in hoop and bending stresses, if constructed of reinforced concrete. Use a sketch to illustrate the location and orientation of cracks.

3.9.5 Resolve for the worked example in Sect. 3.8 for the case of a pinned support. Determine the distribution of stresses (N_y and N_θ), the deformations (w and ϕ_y), and the bending moments and shear distribution (M_y, M_θ and Q_y).

3.9.6 A 30 m high silo is 250 mm thick and has a 4 m radius (see Fig. 3.22). The silo is filled with maize and the base is fixed. Assume the roof is a lightweight material and may be neglected in the analysis. Assume that the bulk density of the maize is 7.5 kN/m^3 and the angle of internal friction and wall friction is approximately 30°, and E $= 28$E6 kN/m^2. Also, assume that the average height of the maize is at the top of the silo wall. Determine the distribution of stresses (N_y and N_θ), moments (M_y and M_θ), shears (Q_y), and deformations (w and ϕ_y) in the walls of the silo.

3.9.7 Repeat the example problem in Sect. 3.8, but with a 250 mm monolithic roof slab (Fig. 3.23). Determine the resultant stresses, bending moments, shears, and deformations in the shell walls.

Fig. 3.22 Maize silo details

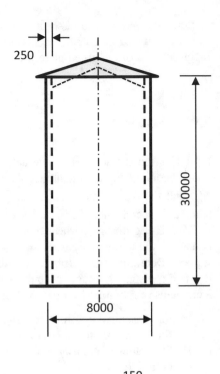

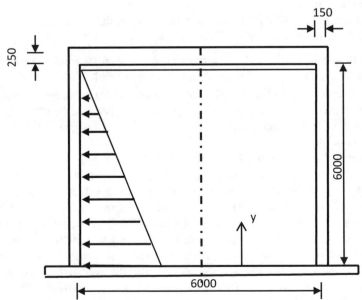

Fig. 3.23 Water tank with roof

Most slab coverings will not be connected to the walls of the tank, but rest on a sliding bearing because of thermal effects. This will only effect the vertical stress in the wall, which is a constant stress added to the self-weight of the walls. In this case, however, the slab is connected monolithically to the walls. To solve this problem, reconsider the compatibility equations at the top of the wall.

References

AS3774, 1997. *Supplement 1, Loads on Bulk Solid Containers-Commentary*. NSW, Australia: Standards Australia.

Billington, D. P., 1982. *Thin Shell Concrete Structures*. 2 ed. New York: McGraw-Hill.

Chatterjee, B. K., 1971. *Theory and Design of Concrete Structures*. Calcutta: Oxford and IBH Publishing Co.

Flugge, W., 1960. *Stresses in Shells*. Berlin: Springer-Verag.

Jaeger, L. G., 1969. *Elementary Theory of Elastic Plates*. Oxford: Thomson.

Kassimali, A., 2005. *Structural Analysis*. Canada: Thomson.

Marshall, W. T. & Nelson, H. M., 1981. *Structures*. 2 ed. London: Pittman.

Szilard, R., 1974. *Theory and Analysis of Plates*. New Jersey: Prentice-Hall.

Timoshenko, S., 1959. *Theory of Plates and Shells*. 2 ed. New York: McGraw-Hill Book Co.

Wilson, A., 2005. *Practical Design of Concrete Shells*. Texas: Monolithic Dome Institute.

Chapter 4
Circular Domes

Mitchell Gohnert

Abstract This chapter develops the membrane and deformation theories of circular domes for various loading conditions such as uniform vertical, horizontal, and pressure loading, assuming the base is unrestrained. The theory for the boundary effects is also developed for fixed and pinned supports. To solve for the effects of boundary conditions, compatibility equations are formulated to solve the base reactions, which, in turn, are used to distribute the stresses in the dome. The membrane and boundary solutions are then superimposed to solve for the stresses, bending moments, and shears at any point in the shell. Theories for loads applied to skylight openings and ring tensions at the base are also developed. The mathematical derivation is fully developed for both membrane and boundary theories.

4.1 Introduction

The word dome is a generic word to describe the upper half of a rounded vault, which may be circular, parabolic, catenary, or elliptical. In this chapter, only the circular dome will be derived.

The root of the word dome (or *Domus*) is Latin, which means "house." In antiquity, the word *Domus Dei* was more common, translated as "House of God." This was a fitting description, since the majority of domes were found in religious buildings. However, the application to cathedrals and mosques is not exclusive. In more recent times, domes have been used extensively in government buildings, theaters, sport arenas, and places of public gatherings. All of these buildings have in common a need to cover a large space, without the hindrance of column supports. Free open spaces, minimal materials, and the superior strength of domes make this option a viable and favorable solution.

The sturdiness and robustness of domes are mystifying, and they seem to possess eternal properties. This perception is not unfounded. Many of the ancient structures which stand today are domes. The Pantheon in Rome and Hagia Sofia in Turkey are

M. Gohnert
University of the Witwatersrand, Johannsesburg, South Africa

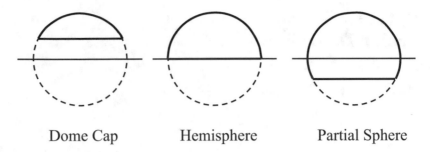

Dome Cap   Hemisphere   Partial Sphere

Fig. 4.1 Various profiles of a circular dome

perfect examples: both structures are approaching two millennia, yet they are structurally sound and in remarkably good condition.

Domes are doubly curved shells and have positive Gaussian curvature. The double curvature adds strength over shells that have single curvature (i.e., vaults) (Wilson 2005). Two-way slab systems act similarly and provide an understanding to this important concept. It is well known that two-way slabs carry load in two directions and are significantly stronger and deflect less than one-way slabs (Wood 1961); the same can be said of domes. However, the strength of the dome lies in how loads are carried by the shell walls. Domes carry loads primarily by membrane action. In the meridian direction (i.e., vertically), the stresses are compressive. In the hoop direction (i.e., horizontally), the stresses are compressive near the top, but switch to tensile stresses in the lower half of a hemisphere. Similar to the cylindrical shells, the base is subjected to boundary effects: moments and shears in the region of the foundation or ring beam. The derivation of boundary effects is complex, but the resulting stresses may be significant, especially in shells constructed of rigid thick materials. For this reason, the derivation of boundary effects is included in this chapter.

In Fig. 4.1, three possible profiles of the circular dome are illustrated. The theory presented in this chapter is only applicable to hemispheres and low profile spherical segments (i.e., dome caps). The high profile spherical segment (i.e., partial sphere) attracts significantly more stress than the other profiles. Other than water tanks, high profile spheres are rarely used as dome structures (Roy 2004).

As will be seen, a flat dome is in pure compression (in the hoop and meridian directions), which is a favorable characteristic. However, flat domes have a high horizontal thrust and therefore require a substantial ring beam to resist the tensions caused by the thrust. Steeper domes, such as a hemisphere, have a reduced horizontal thrust, but hoop tensions occur in the lower half of the shell.

Similar to the cylindrical shell, the equations and notations are largely based on the work by Billington (1982). The derivation also relies on the works of Wilson (2005), Chatterjee (1971), Timoshenko (1959), and Flugge (1960), which provide extensive theoretical and practical information on the analysis of circular domes.

4.2   The Membrane Theory of the Circular Dome

The development of the membrane theory for circular domes is similar to cylindrical shell: Equilibrium equations are formulated by summing forces in each direction. Differential elements are used as a tool to enable the formulation of these equations. As seen in Fig. 4.2, differential elements are formed by two intersecting strips: one segmental strip in the meridian direction and the other in the hoop direction. The purpose of the element is to depict the in-plane membrane forces. The edge lengths of the element are given in Fig. 4.3, and a side view, defining the radial geometry, is illustrated in Fig. 4.4.

4.2.1   Equilibrium in the Hoop Direction (θ-Direction)

In Fig. 4.5, only the stresses that contribute to the equilibrium equation are illustrated.

Prior to formulating the equilibrium equation, the area of the differential element must be solved. Referring to Fig. 4.3, the area of the differential element is determined by taking the average length of the top and bottom edges and multiplying by the length of the side of the element.

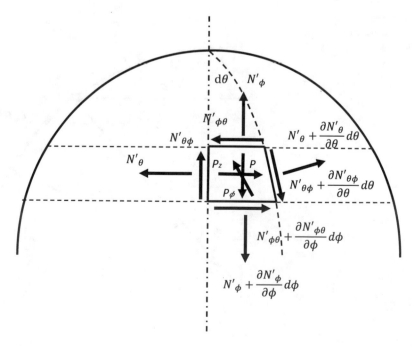

Fig. 4.2 Membrane stresses on a circular dome differential element

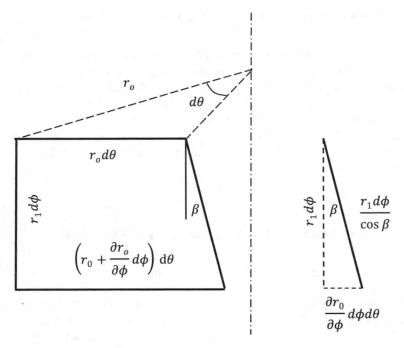

Fig. 4.3 Geometry of the differential element

Fig. 4.4 Side view of the differential element defining the radial geometry

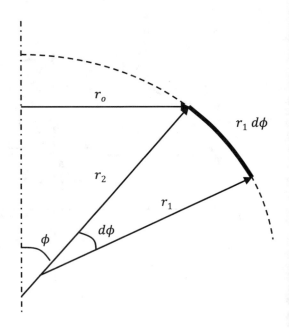

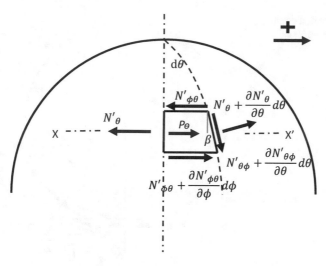

Fig. 4.5 Stresses that contribute to the equilibrium equation in the hoop direction

Area of the differential element $= \left[\dfrac{r_o d\theta + \left(r_o + \frac{\partial r_o}{\partial \phi} d\phi\right) d\theta}{2} \right] r_1 d\phi$.

Expanding,

$$= \left[\frac{r_o d\theta}{2} + \frac{r_o d\theta}{2} + \frac{\partial r_o}{\partial \phi} \frac{d\phi d\theta}{2} \right] r_1 d\phi = r_o r_1 d\theta d\phi + \frac{\partial r_o}{\partial \phi} \frac{d\phi^2 d\theta r_1}{2}$$

The above expression is simplified by eliminating the higher order term (the second term). Higher order terms are terms where the magnitude of the term is significantly less than the other terms. Thus, these terms may be eliminated, since they contribute little to the overall equation.

$$\text{Area of the differential element} = r_o r_1 d\theta d\phi \qquad (4.1)$$

Referring to Fig. 4.6, the equilibrium equation is formulated by summing forces along the x–x′ line,

$$-N'_\theta r_1 d\phi + \left(N'_{\theta\phi} + \frac{\partial N'_{\theta\phi}}{\partial \phi} d\theta\right) \frac{r_1 d\phi}{\cos\beta} \sin\beta \, \cos d\theta + \left(N'_\theta + \frac{\partial N'_\theta}{\partial \theta} d\theta\right) \frac{r_1 d\phi}{\cos\beta} \cos\beta \, \cos d\theta -$$
$$N'_{\phi\theta} r_o d\theta \, \cos\frac{d\theta}{2} + \left(N'_{\phi\theta} + \frac{\partial N'_{\phi\theta}}{\partial \phi} d\phi\right) \left(r_o + \frac{\partial r_o}{\partial \phi} d\phi\right) d\theta \, \cos\frac{d\theta}{2} + P_\theta r_o r_1 d\theta d\phi \, \cos\frac{d\theta}{2} = 0 \qquad (4.2)$$

The equilibrium equation is formulated by multiplying the stress by the edge length of the differential element to obtain the total force along the edge. Likewise, the surface pressure P_θ is multiplied by the surface area of the differential element

Fig. 4.6 Direction of edge stresses in the differential element

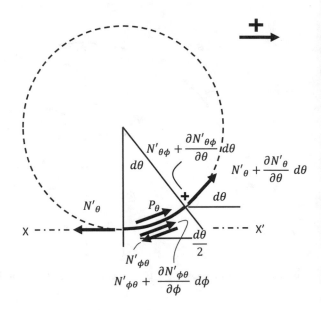

(Eq. 4.1). The *sin* and *cos* terms are included in Eq. 4.2 to solve the components of stress along the x–x′ line (see Figs. 4.5 and 4.6).

Referring to the triangle of Fig. 4.3,

$$\frac{\sin \beta}{\cos \beta} = \tan \beta = \frac{\frac{\partial r_o}{\partial \phi} d\phi d\theta}{r_1 d\phi} \tag{4.3}$$

Since the angle $d\theta$ is very small, the $\cos d\theta \approx \cos \frac{d\theta}{2} \approx 1$. This assumption is applied to Eq. 4.2 and the $\tan\beta$ is replaced by the expression in 4.3.

Expanding Eq. 4.2 and including the simplifications,

$$-N'_\theta r_1 d\phi + N'_{\theta\phi} r_1 d\phi \left(\frac{\frac{\partial r_o}{\partial \phi} d\phi d\theta}{r_1 d\phi} \right) + \frac{\partial N'_{\theta\phi}}{\partial \theta} d\theta r_1 d\theta \left(\frac{\frac{\partial r_o}{\partial \phi} d\phi d\theta}{r_1 d\theta} \right) + N'_\theta r_1 d\phi + \frac{\partial N'_\theta}{\partial \theta} d\theta r_1 d\phi -$$

$$N'_{\phi\theta} r_o d\theta + N'_{\phi\theta} r_o d\theta + N'_{\phi\theta} \frac{\partial r_o}{\partial \phi} d\phi d\theta + \frac{\partial N'_{\phi\theta}}{\partial \phi} d\phi r_o d\theta + \frac{\partial N'_{\phi\theta}}{\partial \phi} d\phi \frac{\partial r_o}{\partial \phi} d\phi d\theta +$$

$$P_\theta r_o r_1 d\theta d\phi = 0$$

$$\tag{4.4}$$

The above equation is further simplified by eliminating the first and fourth and the sixth and the seventh terms, and by removing the higher order terms (third and tenth terms). The eighth and ninth terms are combined using the product rule.

$$f'g + fg' = (fg)' \tag{4.5}$$

$$N'_{\phi\theta} \frac{\partial r_o}{\partial \phi} d\phi d\theta + \frac{\partial N'_{\phi\theta}}{\partial \phi} d\phi r_o d\theta = \frac{\partial \left(N'_{\phi\theta} r_o \right)}{\partial \phi} d\phi d\theta \tag{4.6}$$

Making these changes,

$$N'_{\theta\phi} \frac{\partial r_o}{\partial \phi} d\phi d\theta + \frac{\partial N'_{\theta}}{\partial \theta} r_1 d\phi d\theta + \frac{\partial \left(N'_{\phi\theta} r_o \right)}{\partial \phi} d\phi d\theta + P_\theta r_o r_1 d\theta d\phi = 0 \tag{4.7}$$

Divide by $d\theta d\phi$.

$$N'_{\theta\phi} \frac{\partial r_o}{\partial \phi} + \frac{\partial N'_{\theta}}{\partial \theta} r_1 + \frac{\partial \left(N'_{\phi\theta} r_o \right)}{\partial \phi} + P_\theta r_o r_1 = 0 \tag{4.8}$$

4.2.2   Equilibrium in the Meridian Direction (ϕ-Direction)

To solve for the equilibrium equation in the meridian direction, only the stresses that contribute to this equation are given in Fig. 4.7. A side view of the differential element is given in Fig. 4.8. Summing forces,

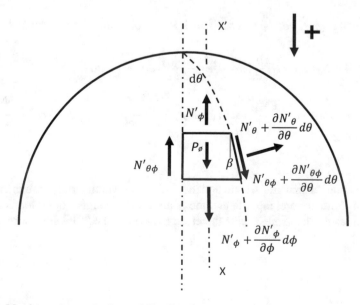

Fig. 4.7 Membrane stresses in the meridian direction

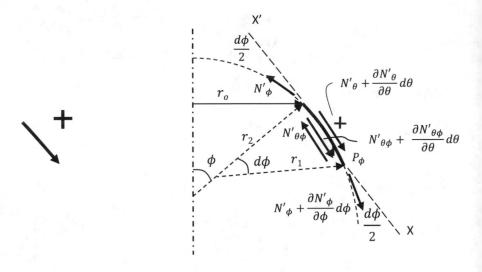

Fig. 4.8 Side view of the orientation of stresses in the meridian direction

$$-N'_\phi r_o d\theta \, \cos \frac{d\phi}{2} + \left(N'_\phi + \frac{\partial N'_\phi}{\partial \phi} d\phi\right)\left(r_o + \frac{\partial r_o}{\partial \phi} d\phi\right) d\theta \, \cos \frac{d\phi}{2} - N'_{\theta\phi} r_1 d\phi +$$

$$\left(N'_{\theta\phi} + \frac{\partial N'_{\theta\phi}}{\partial \theta} d\theta\right)\left(\frac{r_1 d\phi}{\cos \beta}\right)\cos \beta - \left(N'_\theta + \frac{\partial N'_\theta}{\partial \theta} d\theta\right)\left(\frac{r_1 d\phi}{\cos \beta}\right)\sin \beta + P_\phi r_1 r_o d\theta d\phi = 0$$

$$(4.9)$$

For small angles, $\cos d\phi \approx \cos \frac{d\phi}{2} \approx 1$. Replace $\sin\beta / \cos \beta$ with 4.3 and expand.

$$-N'_\phi r_o d\theta + N'_\phi r_o d\theta + N'_\phi \frac{\partial r_o}{\partial \phi} d\phi d\theta + r_o \frac{\partial N'_\phi}{\partial \phi} d\phi d\theta + \frac{\partial N'_\phi}{\partial \phi} d\phi \frac{\partial r_o}{\partial \phi} d\phi d\theta -$$

$$N'_{\theta\phi} r_1 d\phi + N'_{\theta\phi} r_1 d\phi + \frac{\partial N'_{\theta\phi}}{\partial \theta} d\theta d\phi \, r_1 - N'_\theta r_1 d\phi \frac{\frac{\partial r_o}{\partial \phi} d\phi d\theta}{r_1 d\phi} - \frac{\partial N'_\theta}{\partial \theta} d\theta d\phi \, r_1 \frac{\frac{\partial r_o}{\partial \phi} d\phi d\theta}{r_1 d\phi} +$$

$$P_\phi r_1 r_o d\theta d\phi = 0$$

$$(4.10)$$

Eliminating the first and second and the sixth and seventh terms, the expression is simplified. Furthermore, the fifth and the tenth terms are higher order terms and are also eliminated. The expression is therefore reduced to the following form:

$$N'_\phi \frac{\partial r_o}{\partial \phi} d\phi d\theta + r_o \frac{\partial N'_\phi}{\partial \phi} d\phi d\theta + \frac{\partial N'_{\theta\phi}}{\partial \theta} d\theta r_1 d\phi - N'_\theta \frac{\partial r_o}{\partial \phi} d\phi d\theta$$

$$+ P_\phi r_1 r_o d\theta d\phi$$
$$= 0 \qquad\qquad (4.11)$$

Combining the first and second terms using the chain rule (4.5) and dividing by $d\theta d\phi$, the equilibrium equation in the meridian direction is solved.

$$\frac{\partial \left(N'_\phi r_o\right)}{\partial \phi} + \frac{\partial N'_{\theta\phi}}{\partial \theta} r_1 - N'_\theta \frac{\partial r_o}{\partial \phi} + P_\phi r_1 r_o = 0 \qquad\qquad (4.12)$$

4.2.3   Equilibrium in the Radial Direction (z-Direction)

The stresses that contribute to the equilibrium equation in the radial direction are illustrated in Fig. 4.9, and the side view of the differential element is given in Fig. 4.10. Summing forces along the $x - x'$ line,

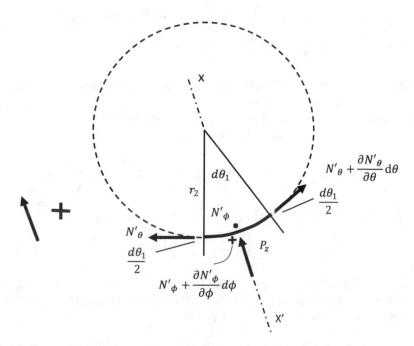

Fig. 4.9 Stresses that contribute to the equilibrium equation in the radial direction

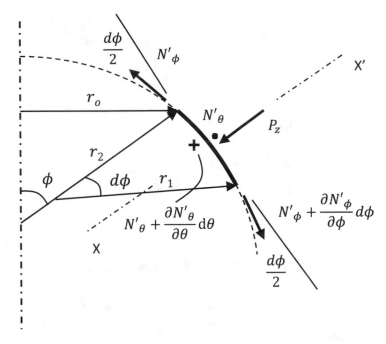

Fig. 4.10 Side view perspective of the radial stresses

$$N'_\phi r_o d\theta \sin\frac{d\phi}{2} + \left(N'_\phi + \frac{\partial N'_\phi}{\partial \phi}d\phi\right)\left(r_o + \frac{\partial r_o}{\partial \phi}d\phi\right)d\theta \sin\frac{d\phi}{2} + N'_\theta r_1 d\phi \sin\frac{d\theta_1}{2} +$$

$$\left(N'_\theta + \frac{\partial N'_\theta}{\partial \theta}d\theta\right)\frac{r_1 d\phi}{\cos\beta}\cos\beta \sin\frac{d\theta_1}{2} + P_z r_1 r_o d\theta d\phi = 0$$

(4.13)

where the lengths of the differential element are according to Fig. 4.3 and the summation is in the same direction as r_2.

For small angles, we assume that $\sin\frac{d\theta}{2} \approx \frac{d\theta}{2}$ and $\sin\frac{d\phi}{2} \approx \frac{d\phi}{2}$.

Expanding Eq. 4.13 and simplifying,

$$N'_\phi r_o d\theta\left(\frac{d\phi}{2}\right) + N'_\phi r_o d\theta\left(\frac{d\phi}{2}\right) + N'_\phi\frac{\partial r_o}{\partial \phi}d\phi d\theta\left(\frac{d\phi}{2}\right) + \frac{\partial N'_\phi}{\partial \phi}d\phi r_o d\theta\left(\frac{d\phi}{2}\right) +$$

$$\frac{\partial N'_\phi}{\partial \phi}d\phi\frac{\partial r_o}{\partial \phi}d\phi d\theta\left(\frac{d\phi}{2}\right) + N'_\theta r_1 d\phi\left(\frac{d\theta_1}{2}\right) + N'_\theta r_1 d\phi\left(\frac{d\theta_1}{2}\right) + \frac{\partial N'_\theta}{\partial \theta}d\theta r_1 d\phi\left(\frac{d\theta_1}{2}\right) +$$

$$P_z r_1 r_o d\theta d\phi = 0$$

(4.14)

The third, fourth, fifth, and eight terms are higher order terms and therefore may be eliminated from the equation. In addition, the first and second and the sixth and seventh terms are combined.

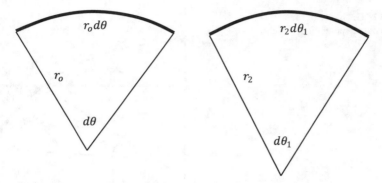

Fig. 4.11 Relationship between the radius and arc length of the differential element

$$N'_\phi r_o d\theta d\phi + N'_\theta r_1 d\phi d\theta_1 + P_z r_1 r_o d\theta d\phi = 0 \tag{4.15}$$

From Fig. 4.11, the arc lengths of the top edge of the differential element are the same and may be equated (refer to Fig. 4.10).

$$r_2 d\theta_1 = r_o d\theta \tag{4.16}$$

Rearranged,

$$d\theta_1 = \frac{r_o d\theta}{r_2} \tag{4.17}$$

Substituting 4.17 into 4.15 and dividing by $d\theta d\phi r_o r_1$.

$$\frac{N'_\phi}{r_1} + \frac{N'_\theta}{r_2} + P_z = 0 \tag{4.18}$$

4.2.4   Summary of Equilibrium Equations and Simplifications

Collecting the three equilibrium equations,

$$N'_{\theta\phi}\frac{\partial r_o}{\partial \phi} + \frac{\partial N'_\theta}{\partial \theta}r_1 + \frac{\partial \left(N'_{\phi\theta}r_o\right)}{\partial \phi} + P_\theta r_o r_1 = 0 \tag{4.8}$$

Fig. 4.12 Geometry of the
differential element

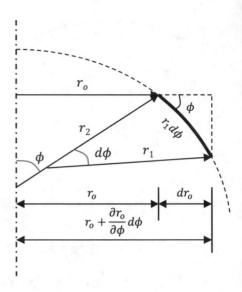

$$\frac{\partial (N'r_o)}{\partial \phi} + \frac{\partial N'_{\theta\phi}}{\partial \theta} r_1 - N'_\theta \frac{\partial r_o}{\partial \phi} + P_\phi r_1 r_o = 0 \qquad (4.12)$$

$$\frac{N'_\phi}{r_1} + \frac{N'_\theta}{r_2} + P_z = 0 \qquad (4.18)$$

If we assume that the dome is symmetrically loaded, all of the terms which have $\partial \theta$ will vanish; the stress does not change in the hoop direction. Symmetrical loading will also imply that the in-plane shear stresses are equal to zero (i.e., $N'_{\phi\theta} = N'_{\theta\phi} = 0$). Furthermore, we also assume that the surface pressure applied in the hoop direction (surface tractions) is equal to zero (i.e., $P_\theta = 0$). These assumptions will eliminate Eq. 4.8, and the second term of 4.12. The equilibrium equations reduce to the following form:

$$\frac{N'_\phi}{r_1} + \frac{N'_\theta}{r_2} + P_z = 0 \qquad (4.18)$$

$$\frac{\partial \left(N'_\phi r_o \right)}{\partial \phi} - N'_\theta \frac{\partial r_o}{\partial \phi} + P_\phi r_1 r_o = 0 \qquad (4.19)$$

From Fig. 4.12,

$$r_1 d\phi \cos \phi = dr_o$$

or

$$\frac{dr_o}{d\phi} = r_1 \cos \phi \tag{4.20}$$

Substituting 4.20 into 4.19,

$$\frac{\partial \left(N'_\phi r_o \right)}{\partial \phi} - N'_\theta r_1 \cos \phi + P_\phi r_1 r_o = 0 \tag{4.21}$$

If we rearrange 4.18,

$$N'_\theta = -r_2 \left(\frac{N'_\phi}{r_1} + P_z \right) \tag{4.22}$$

Referring to Fig. 4.10,

$$r_o = r_2 \sin \phi \tag{4.23}$$

Rearranging 4.23,

$$r_2 = \frac{r_o}{\sin \phi} \tag{4.24}$$

Substituting 4.24 into 4.22,

$$N'_\theta = -\frac{r_o}{\sin \phi} \left(\frac{N'_\phi}{r_1} + P_z \right) \tag{4.25}$$

Substitute 4.25 into 4.21.

$$\frac{\partial \left(N'_\phi r_o \right)}{\partial \phi} + \frac{r_o}{\sin \phi} \left(\frac{N'_\phi}{r_1} + P_z \right) r_1 \cos \phi + P_\phi r_1 r_o = 0 \tag{4.26}$$

Multiply each side of the equation by $\sin\phi$.

$$\frac{\partial \left(N'_\phi r_o \right)}{\partial \phi} \sin \phi + r_o \left(\frac{N'_\phi}{r_1} + P_z \right) r_1 \cos \phi + P_\phi r_1 r_o \sin \phi = 0 \tag{4.27}$$

Expanding 4.27,

$$\frac{\partial\left(N'_{\phi}r_o\right)}{\partial\phi}\sin\phi + N'_{\phi}r_o\cos\phi + P_z r_o r_1\cos\phi + P_{\phi}r_o r_1\sin\phi = 0 \qquad (4.28)$$

Integrating the above expression from 0 to ϕ, combine the last two terms and rearrange.

$$\int_0^{\phi}\frac{\partial\left(N'_{\phi}r_o\right)}{\partial\phi}\sin\phi d\phi + \int_0^{\phi}N'_{\phi}r_o\cos\phi\,d\phi$$
$$= -\frac{1}{2\pi}\int_0^{\phi}\left(P_z\cos\phi + P_{\phi}\sin\phi\right)2\pi r_o r_1 d\phi \qquad (4.29)$$

Solve the first term by integration by parts. The general formula is given below:

$$\int u\,dv = uv - \int v\,du \qquad (4.30)$$

If we assume,

$$u = \sin\phi$$
$$du = \cos\phi$$
$$dv = \frac{\partial\left(N'_{\phi}r_o\right)}{\partial\phi}d\phi$$
$$v = N'_{\phi}r_o$$

Substituting the above into Eq. 4.29,

$$N'_{\phi}r_o\sin\phi - \int_0^{\phi}N'_{\phi}r_o\cos\phi\,d\phi + \int_0^{\phi}N'_{\phi}r_o\cos\phi\,d\phi$$
$$= -\frac{1}{2\pi}\int_0^{\phi}\left(P_z\cos\phi + P_{\phi}\sin\phi\right)2\pi r_o r_1 d\phi \qquad (4.31)$$

Eliminating the second and third terms,

$$N'_{\phi}r_o\sin\phi = -\frac{1}{2\pi}\int_0^{\phi}\left(P_z\cos\phi + P_{\phi}\sin\phi\right)2\pi r_o r_1 d\phi \qquad (4.32)$$

Divide each side by $r_o\sin\phi$,

Fig. 4.13 Components of external pressures

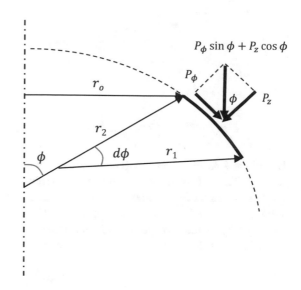

$$N'_\phi = -\frac{1}{2\pi r_o \sin\phi} \int_o^\phi (P_z \cos\phi + P_\phi \sin\phi)(2\pi r_o) r_1 d\phi \qquad (4.33)$$

Evaluating 4.33, the term $(2\pi r_o)$ sums the stress in the hoop direction and $\int_0^\phi r_1 d\phi$ sums the stress in the meridian direction. Furthermore, the term $P_z \cos\phi + P_\phi \sin\phi$ is the vertical resultant of the two external pressures P_z and P_ϕ (see Fig. 4.13). If we assume R is the total load of the shell above the angle ϕ,

$$R = \int_o^\phi (P_z \cos\phi + P_\phi \sin\phi)(2\pi r_o) r_1 d\phi \qquad (4.34)$$

Substituting 4.34 into 4.33, the general membrane stress equation in the meridian direction is defined.

$$N'_\phi = -\frac{R}{2\pi r_o \sin\phi} \qquad (4.35)$$

Substituting 4.35 into 4.25,

$$N'_\theta = -\frac{r_o}{\sin\phi}\left(\frac{-R}{r_1 2\pi r_o \sin\phi} + P_z\right) \qquad (4.36)$$

Rearranged, the general membrane hoop stress equation is similarly determined.

$$N'_\theta = \frac{R}{2\pi r_1 \sin^2 \phi} - \frac{P_z r_o}{\sin \phi} \tag{4.37}$$

4.2.5 Uniform Vertical Loading

Equations 4.35 and 4.37 are general equations that may be applied to any type of loading. The first loading is a uniformly distributed load, which is evenly applied over the surface of the dome. A uniform vertical load is the loading from the self-weight of the shell or any evenly applied load that is affected by gravity. Thus, this loading may include live loads and dead loads.

Referring to Fig. 4.14,

$$P_z = q \cos \phi \tag{4.38}$$

$$P_\phi = q \sin \phi \tag{4.39}$$

Furthermore, since the shape is circular,

$$r_1 = r_2 = a \tag{4.40}$$

$$r_o = a \sin \phi \tag{4.41}$$

Substituting equations 4.38 and 4.39 into 4.34, replace r_o with 4.41 and r_1 with a,

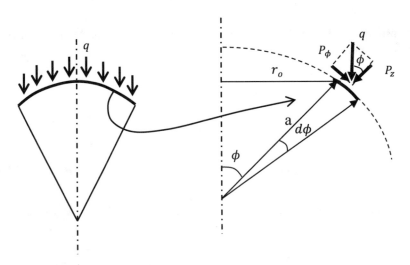

Fig. 4.14 Vertical load on a differential element

$$R = \int_o^{\phi} (q \cos^2 \phi + q \sin^2 \phi) 2\pi a^2 \sin \phi \, d\phi \qquad (4.42)$$

Since

$$\cos^2 \phi + \sin^2 \phi = 1 \qquad (4.43)$$

Equation 4.42 reduces to

$$R = \int_o^{\phi} 2q\pi a^2 \sin \phi \, d\phi \qquad (4.44)$$

Separating the integration constants,

$$R = 2q\pi a^2 \int_o^{\phi} \sin \phi \, d\phi \qquad (4.45)$$

Integrate the above expression from 0 to ϕ.

$$R = 2q\pi a^2 (1 - \cos \phi) \qquad (4.46)$$

Substituting 4.46 into 4.35 and replacing r_o in the denominator with $a \sin \phi$,

$$N'_{\phi} = -\frac{2q\pi a^2 (1 - \cos \phi)}{2\pi a \, \sin^2 \phi} \qquad (4.47)$$

Simplifying the expression,

$$N'_{\phi} = -\frac{qa(1 - \cos \phi)}{\sin^2 \phi} \qquad (4.48)$$

Using the trigonometric identity,

$$(1 + \cos \phi) = \frac{\sin^2 \phi}{(1 - \cos \phi)} \qquad (4.49)$$

the final form of the equation is determined.

$$N'_{\phi} = -\frac{qa}{(1 + \cos \phi)} \qquad (4.50)$$

Substituting 4.46 into 4.37, replace r_o with $a \sin \phi$, r_1 with a, and P_z with $q \cos \phi$.

$$N'_\theta = \frac{2q\pi a^2 (1 - \cos\phi)}{2\pi a \, \sin^2\phi} - \frac{q\cos\phi \, a\sin\phi}{\sin\phi} \tag{4.51}$$

Simplifying,

$$N'_\theta = \frac{qa(1 - \cos\phi)}{\sin^2\phi} - aq\cos\phi \tag{4.52}$$

Using the same trigonometric identity, the equation for the hoop stress is simplified.

$$N'_\theta = aq \left(\frac{1}{1 + \cos\phi} - \cos\phi \right) \tag{4.53}$$

The above equations are valid for a circular dome, with the meridian angle ranging from 0 to 90°. Solving for the stresses of a hemisphere.
If $\phi = 0$,

$$N'_\phi = -\frac{aq}{2} \text{ and } N'_\theta = -\frac{aq}{2}$$

If $\phi = 90°$,

$$N'_\phi = -aq \text{ and } N'_\theta = +aq$$

Furthermore, if we set $\left(\frac{1}{1 + \cos\phi} - \cos\phi \right) = 0$ and solve for ϕ,

$$\phi = 51.83°$$

This angle is the point at which the dome changes from pure compressions to a shell with hoop tensions (Fig. 4.15).

A pure compression structure has many advantages, especially with materials that are prone to tension cracking; concrete and masonry are two such materials. For this reason, it is common practice to use shallow domes as a reservoir roof. If in pure compression, the structure is far more durable, crack free, and more impermeable to the ingress of moisture. If the shell is constructed of masonry, a pure compression structure is essential.

The vertical, or meridian stress, may be broken down into components of stress in the vertical and horizontal directions at the base of the shell (see Fig. 4.16). From pressure vessel theory (Higdon 1976), the horizontal pressure multiplied by the radius of the shell is equal to the ring tension in the shell.

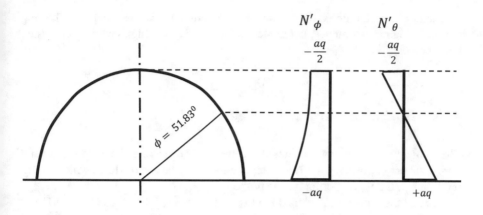

Fig. 4.15 Vertical and hoop stresses in a hemisphere (compression is negative)

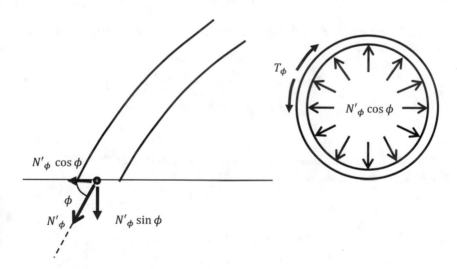

Fig. 4.16 Components of the meridian stress at the base of the shell

$$T_\phi = -N'_\phi \cos \phi \, r_o \qquad (4.54)$$

Since $r_o = a \sin \phi$,

$$T_\phi = -N'_\phi \cos \phi \, a \sin \phi \qquad (4.55)$$

The ring tension is resisted by a ring beam at the base of the shell. As will be seen later, the boundary effect contributes little to the N'_ϕ value. Thus, the ring tension equation is applicable to domes designed with membrane and boundary theories.

The profile of the dome determines the magnitude of the ring tension. The ring tension is high in flat domes, but reduce in magnitude with high profile domes (i.e., hemispheres).

4.2.6 Horizontal Vertical Loading

The second type of loading considered is a uniformly distributed load, applied over a horizontal projection (Fig. 4.17). An application could be a snow load, which would only stick to the upper portion of the dome.

Because the uniform distributed load is applied to a reduced area of the differential element, the value of P on the differential element is reduced accordingly.

$$P \cos \phi$$

The external loads in the ϕ and z-directions are therefore defined in terms of a horizontal projection.

$$P_z = P \cos \phi \, (\cos \phi) = P \cos^2 \phi \qquad (4.56)$$

$$P_\phi = P \cos \phi \sin \phi \qquad (4.57)$$

$$P_\theta = 0 \qquad (4.58)$$

The total load on the shell, at an angle of ϕ, is equal to load P multiplied by the area of the horizontal projection.

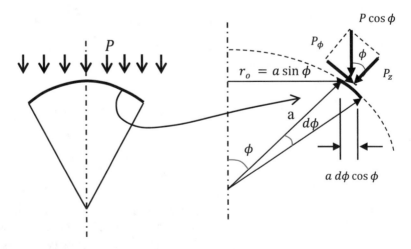

Fig. 4.17 Vertical loading over a horizontal projection

$$R = P\pi r_o^2 \tag{4.59}$$

Since $r_o = a \sin \phi$,

$$R = P\pi a^2 \sin^2 \phi \tag{4.60}$$

Substituting 4.60 into 4.35 and replacing r_o with $a \sin \phi$,

$$N_\phi' = -\frac{P\pi a^2 \sin^2 \phi}{2\pi (a \sin \phi) \sin \phi} \tag{4.61}$$

Simplifying the above expression,

$$N_\phi' = -\frac{Pa}{2} \tag{4.62}$$

Substituting 4.56 and 4.60 into the general Eq. 4.37 and replacing $r_o = a \sin \phi$ and $r_1 = a$,

$$N_\theta' = \frac{P\pi a^2 \sin^2 \phi}{2\pi a \sin^2 \phi} - \frac{(P \cos^2 \phi) a \sin \phi}{\sin \phi} \tag{4.63}$$

Simplifying,

$$N_\theta' = \frac{Pa}{2} - Pa \cos^2 \phi \tag{4.64}$$

or

$$N_\theta' = Pa \left(\frac{1}{2} - \cos^2 \phi \right) \tag{4.65}$$

Multiply each side by -2.

$$-2N_\theta' = Pa \left(2 \cos^2 \phi - 1 \right) \tag{4.66}$$

Substitute the following identity.

$$\cos 2\phi = 2 \cos^2 \phi - 1 \tag{4.67}$$

Dividing by -2, the equation for the hoop stress is determined.

$$N_\theta' = \frac{-Pa}{2} \cos 2\phi \tag{4.68}$$

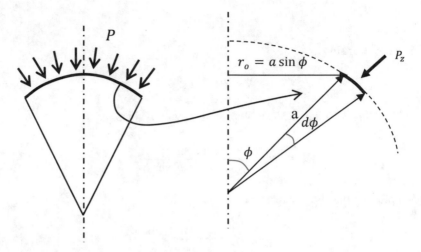

Fig. 4.18 Uniform pressure applied to the surface of the dome

4.2.7   Uniform Pressure

This type of loading is applied normal to the surface of the shell (see Fig. 4.18), which is usually caused by an internal pressure. The loading, however, is similar to the uniform vertical load case for a flat dome.

For uniform pressure, only the load in the z-direction is considered.

$$P_z = P$$

$$P_\phi = 0$$

$$P_\theta = 0$$

Using the Eq. 4.34 for the total external load and substituting $P_\phi = 0$, $P_z = P$, $r_o = a \sin \phi$, and $r_1 = a$,

$$R = \int_o^\phi (P \cos \phi)(2\pi a \sin \phi) a \, d\phi \tag{4.69}$$

Separating the integration constants,

$$R = Pa^2 \pi \int_o^\phi \cos \phi \sin \phi \, d\phi \tag{4.70}$$

Integrating the expression from 0 to ϕ,

$$R = 2Pa^2\pi \left(\frac{\sin^2 \phi}{2} \right) \tag{4.71}$$

or

$$R = Pa^2\pi \, \sin^2 \phi \tag{4.72}$$

Substituting the value of R into Eq. 4.35 and replacing $r_o = a \sin \phi$,

$$N'_\phi = -\frac{Pa^2\pi \, \sin^2 \phi}{2\pi(a \sin \phi) \sin \phi} \tag{4.73}$$

Simplifying the expression,

$$N'_\phi = \frac{-Pa}{2} \tag{4.74}$$

Replacing R, P_z, r_o, and r_1 of Eq. 4.37,

$$N'_\theta = \frac{Pa^2\pi \, \sin^2 \phi}{2\pi a \, \sin^2 \phi} - \frac{P(a \sin \phi)}{\sin \phi} \tag{4.75}$$

Simplifying the above expression,

$$N'_\theta = \frac{Pa}{2} - Pa$$

or

$$N'_\theta = -\frac{Pa}{2} \tag{4.76}$$

4.2.8   *Circular Openings at the Apex of the Shell*

A skylight is a common feature at the apex of the dome. A circular hole is placed at the top of the shell and placed concentrically. The Pantheon in Rome is perhaps the most famous of these types of structures. The skylight opening may be of any diameter, yet the shell will remain structurally stable. This characteristic simplifies construction, especially when building with masonry. If the walls of the shell are constructed in successive rings, the structure is stable during construction and does not require formwork. Most openings, however, are covered with a skylight or lantern. A lantern is an architectural feature that is placed on top of a shell roof to

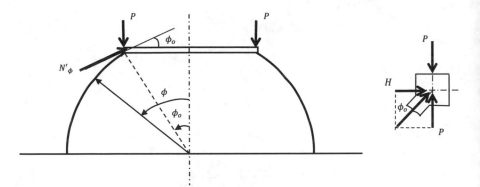

Fig. 4.19 Dome with a circular opening at the apex of the shell

allow natural light into the space below. These edifices often incorporate a steeple. A skylight is a contemporary version of a lantern, albeit less ornamental. Lanterns and skylights cause a continuous load around the opening (see Fig. 4.19) and may necessitate a ring beam.

As depicted in Fig. 4.19, the load P is a uniform line load that circles the opening. This load will generate a horizontal thrust H due to the angle of the supporting walls.

Since

$$\tan \phi_o = \frac{P}{H}$$

$$H = \frac{P}{\tan \phi_o} \tag{4.77}$$

$$H \sin \phi_o = P \cos \phi_o$$

The compression in the ring may therefore be defined as

$$C_o = Hr_o = Ha \sin \phi_o = a P \cos \phi_o \tag{4.78}$$

where ϕ_o is the angle from the vertical to the edge of the opening.

For the skylight calculation, we assume

$$P_z = P_\phi = P_\theta = 0$$

Since the shell will be subjected to gravity and live loads, this assumption is incorrect. However, we can analyze the skylight loads separately and superimpose the stresses.

The total load is equal to

$$R = P2\pi r_o = P2\pi a \sin \phi_o \tag{4.79}$$

Substitute Eqs. 4.41 and 4.79 into 4.35 and 4.37.

$$N'_\phi = -\frac{P2\pi a \sin\phi_o}{2\pi r_o \sin^2\phi} \tag{4.80}$$

$$N'_\phi = -\frac{P \sin\phi_o}{\sin^2\phi} \tag{4.81}$$

$$N'_\theta = \frac{P2\pi a \sin\phi_o}{2\pi a \sin^2\phi} \tag{4.82}$$

$$N'_\theta = \frac{P \sin\phi_o}{\sin^2\phi} \tag{4.83}$$

The effects of an oculus, or hole at the apex, must be separately considered. The weight of the removed cap (above the hole) must be subtracted from the hoop and meridian stress Eqs. (4.35) and (4.37).

4.3 Displacement Theory

Typically, deflections are relatively small in shells. If the shell is thought of as a series of meridian strips and horizontal rings, the interaction between them provides insight in why the deflections are small (Torroja 1958). As the meridian strips deflect downward, the hoop rings counteracts this movement by hoop compressions. Likewise, where the meridian strips rise, tensions form in the hoop rings to counteract outward movement. This interaction lessens the deflections to such an extent that designs are rarely controlled by deflections. This, of course, is a generalization and may not hold true for shells with a high radius to thickness ratio or composed of flexible material.

Referring to Fig. 4.20, the solid line is the differential element in the original undeformed state. The dashed line is the same differential element after deformation. As illustrated, the element has deformed radially inward by the amount w and downward along the meridian by the amount v.

4.3.1 Circumferential, or Hoop Strain

The circumferential strain may be defined as the change in the circumferential length.

$$\epsilon_\theta = \frac{2\pi\Delta r_o}{2\pi r_o} \tag{4.84}$$

Referring to Fig. 4.20,

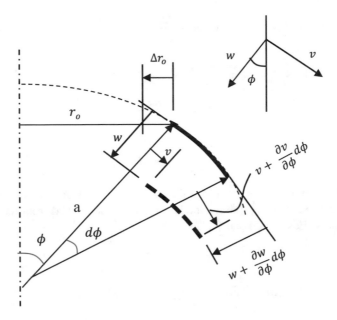

Fig. 4.20 Displacements of the differential element

$$\Delta r_o = v \cos \phi - w \sin \phi \tag{4.85}$$

Substituting 4.85 into 4.84,

$$\varepsilon_\theta = \frac{2\pi (v \cos \phi - w \sin \phi)}{2\pi r_o} \tag{4.86}$$

Expanding 4.86,

$$\varepsilon_\theta = \frac{2\pi v \cos \phi}{2\pi r_o} - \frac{2\pi w \sin \phi}{2\pi r_o} \tag{4.87}$$

$$\varepsilon_\theta = \frac{v \cos \phi}{r_o} - \frac{w \sin \phi}{r_o} \tag{4.88}$$

Since $r_o = r_2 \sin \phi$,

$$\varepsilon_\theta = \frac{v \cos \phi}{r_o} - \frac{w}{r_2} \tag{4.89}$$

4.3.2 Meridional, or Vertical Strains

Influence of v

If we initially consider the strain due to the influence of v, the change in the length of the element, in the vertical direction, is the difference between the deformations (refer to Fig. 4.21).

$$\Delta r_1 d\phi = \left(v + \frac{\partial v}{\partial \phi} d\phi\right) - v = \frac{\partial v}{\partial \phi} d\phi \tag{4.90}$$

The strain is defined as the change in length divided by the original length.

$$\varepsilon_\phi = \frac{\Delta r_1 d\phi}{r_1 d\phi} \tag{4.91}$$

Substituting 4.90 into 4.91,

$$\varepsilon_\phi = \frac{\frac{\partial v}{\partial \phi} d\phi}{r_1 d\phi} = \frac{dv}{r_1 d\phi} \tag{4.92}$$

Influence of w

The expansion or contraction of the shell will influence the strains in the meridional direction. As can be seen in Fig. 4.22, the contraction of the differential element will cause a compressive strain in the element. The change in the length of the differential element is the difference in the arc length, represented as $\Delta r_1 d\phi$.

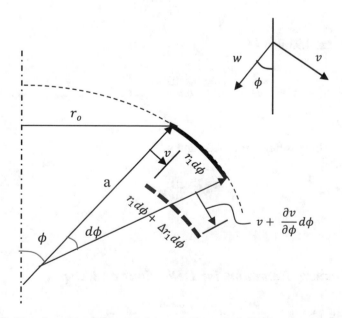

Fig. 4.21 Deformation of the differential element in the vertical direction due to the influence of v

Fig. 4.22 Deformation of
the differential element in
the vertical direction due to
the influence of w

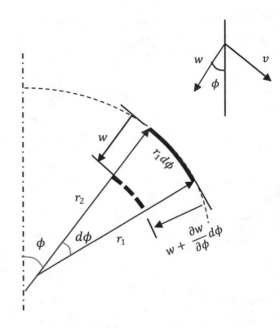

$$\Delta r_1 d\phi = (r_1 - w)d\phi - r_1 d\phi = -wd\phi \tag{4.93}$$

Since the strain is the change in length over the original length,

$$\varepsilon_\phi = \frac{\Delta r_1 d\phi}{r_1 d\phi} \tag{4.94}$$

Substituting 4.93 into 4.94,

$$\varepsilon_\phi = \frac{-wd\phi}{r_1 d\phi} = -\frac{w}{r_1} \tag{4.95}$$

Total Strain

The total strain is the sum of 4.92 and 4.95.

$$\varepsilon_\phi = \frac{dv}{r_1 d\phi} - \frac{w}{r_1} \tag{4.96}$$

4.3.3   General Equations for Deflections and Slopes

Rearranging 4.89,

$$\frac{w}{r_2} = \frac{v\cos\phi}{r_o} - \varepsilon_\theta$$

Solving for w,

$$w = \frac{r_2 v\cos\phi}{r_o} - r_2\varepsilon_\theta \tag{4.97}$$

Rearranging 4.96,

$$\frac{w}{r_1} = \frac{dv}{r_1 d\phi} - \varepsilon_\phi \tag{4.98}$$

Solving for w,

$$w = \frac{r_1 dv}{r_1 d\phi} - r_1\varepsilon_\phi = \frac{dv}{d\phi} - r_1\varepsilon_\phi \tag{4.99}$$

Equating 4.97 and 4.99,

$$\frac{r_2 v\cos\phi}{r_o} - r_2\varepsilon_\theta = \frac{dv}{d\phi} - r_1\varepsilon_\phi \tag{4.100}$$

Substituting $r_o = r_2 \sin\phi$,

$$\frac{r_2 v\cos\phi}{r_2 \sin\phi} - r_2\varepsilon_\theta = \frac{dv}{d\phi} - r_1\varepsilon_\phi \tag{4.101}$$

$$v\cot\phi - r_2\varepsilon_\theta = \frac{dv}{d\phi} - r_1\varepsilon_\phi \tag{4.102}$$

Rearranged,

$$\frac{dv}{d\phi} - v\cot\phi = r_1\varepsilon_\phi - r_2\varepsilon_\theta \tag{4.103}$$

Use Eqs. 3.33 and 3.34 from cylindrical shell theory (Billington 1982), but change the subscripts to be consistent with circular dome geometry.

$$\varepsilon_\theta = \frac{1}{Eh}\left(N'_\theta - \nu N'_\phi\right) \tag{4.104}$$

$$\varepsilon_\phi = \frac{1}{Eh}\left(N'_\phi - \nu N'_\theta\right) \tag{4.105}$$

Substituting the above equations into 4.103,

$$\frac{dv}{d\phi} - v\cot\phi = r_1\left(\frac{1}{Eh}\left[N'_\phi - \nu N'_\theta\right]\right) - r_2\left(\frac{1}{Eh}\left[N'_\theta - \nu N'_\phi\right]\right) \tag{4.106}$$

Rearranging the above equation,

$$\frac{dv}{d\phi} - v\cot\phi = \frac{1}{Eh}\left[N'_\phi(r_1 + \nu r_2) - N'_\theta(r_2 + \nu r_1)\right] \tag{4.107}$$

The above equation may be expressed as

$$\frac{dv}{d\phi} - v\cot\phi = f(\phi) \tag{4.108}$$

where

$$f(\phi) = \frac{1}{Eh}\left[N'_\phi(r_1 + \nu r_2) - N'_\theta(r_2 + \nu r_1)\right] \tag{4.109}$$

The above linear differential equation is similar to the mathematical form:

$$\frac{dy}{dx} + P(x)y = f(x) \tag{4.110}$$

A solution to this differential equation is available, expressed as Eq. 4.111 in terms of x and y.

$$y = e^{-\int P(x)dx} \int e^{\int P(x)dx} f(x) + ce^{-\int P(x)dx} \tag{4.111}$$

If we make the following substitutions:

$$\frac{dy}{dx} = \frac{dv}{d\phi}$$

$$P(x) = -\cot\phi$$

$$y = v$$

$$f(x) = f(\phi)$$

The solution is therefore

$$v = e^{-\int(-\cot\phi)d\phi} \int e^{\int(-\cot\phi)d\phi} f(\phi) + ce^{-\int(-\cot\phi)d\phi} \tag{4.112}$$

Since

$$\int \cot \phi \, d\phi = \ln|\sin \phi|$$
(4.113)

and

$$e^{\ln|\sin \phi|} = \sin \phi$$
(4.114)

$$e^{-\ln|\sin \phi|} = \frac{1}{\sin \phi}$$
(4.115)

the solution becomes,

$$v = \sin \phi \int \frac{f(\phi)}{\sin \phi} + c \sin \phi$$
(4.116)

or

$$v = \sin \phi \left(\int \frac{f(\phi)}{\sin \phi} + c \right)$$
(4.117)

Replacing $f(\phi)$ with equation 4.109,

$$v = \sin \phi \left(\int \frac{\frac{1}{Eh}\left[N'_\phi(r_1 + \nu r_2) - N'_\theta(r_2 + \nu r_1) \right]}{\sin \phi} + c \right)$$
(4.118)

Separating the constants,

$$v = \frac{\sin \phi \left[N'_\phi(r_1 + \nu r_2) - N'_\theta(r_2 + \nu r_1) \right]}{Eh} \int \left(\frac{1}{\sin \phi} + c \right)$$
(4.119)

where c is an integration constant determined from boundary conditions.

From Eq. 4.97,

$$w = \frac{r_2 v \cos \phi}{r_o} - r_2 \varepsilon_\theta$$
(4.97)

Replacing $r_o = r_2 \sin \phi$,

$$w = \frac{r_2 v \cos \phi}{r_2 \sin \phi} - r_2 \varepsilon_\theta$$

$$w = v \cot \phi - r_2 \varepsilon_\theta \tag{4.120}$$

Replacing ε_θ with Eq. 4.104,

$$w = v \cot \phi - \frac{r_2}{Eh}\left(N'_\theta - \nu N'_\phi\right) \tag{4.121}$$

Meridian Rotations

Figure 4.23 is an exaggeration of the deformations of the differential element. In reality, the deformations are small, which permits the following assumption:

$$\Delta'_v r_1 = v \tag{4.122}$$

Rearranged,

$$\Delta'_v = \frac{v}{r_1} \tag{4.123}$$

Similarly, with the w deformations,

$$\Delta'_w = \frac{\left(w + \frac{\partial w}{\partial \phi} d\phi\right) - w}{r_1 d\phi} \tag{4.124}$$

$$\Delta'_w = \frac{\partial w}{r_1 \partial \phi} \tag{4.125}$$

The meridian rotation is therefore the sum of the two rotations.

$$\Delta'_\phi = \frac{v}{r_1} + \frac{\partial w}{r_1 \partial \phi} \tag{4.126}$$

Horizontal Deformations

The differential element undergoing deformations will either expand or contract under load. This expansion/contraction will cause the differential element to undergo a strain, defined by the change in length divided by the original length (see Fig. 4.24).

$$\varepsilon_\theta = \frac{\left(r_o + \Delta'_H\right)d\theta - r_o d\theta}{r_o d\theta} \tag{4.127}$$

Multiply each side by r_o.

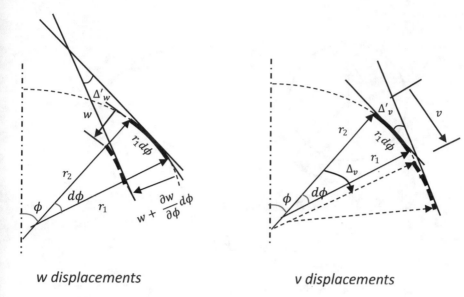

w displacements   v displacements

Fig. 4.23 Rotations in the meridian direction

$$r_o \varepsilon_\theta = r_o \frac{(r_o + \Delta'_H) d\theta - r_o d\theta}{r_o d\theta} \tag{4.128}$$

Expanding and simplifying the above expression,

$$r_o \varepsilon_\theta = r_o \frac{r_o d\theta + \Delta'_H d\theta - r_o d\theta}{r_o d\theta} = \Delta'_H \tag{4.129}$$

ε_θ was previously defined in Eq. 4.104.

$$\varepsilon_\theta = \frac{1}{Eh} \left(N'_\theta - \nu N'_\phi \right) \tag{4.104}$$

Substituting $r_o = r_2 \sin \phi$ and Eq. 4.104 into 4.129, the general equation for horizontal deformations is determined.

$$\Delta'_H = \frac{r_2 \sin \phi}{Eh} \left(N'_\theta - \nu N'_\phi \right) \tag{4.130}$$

At the base of the shell, the vertical deformation is assumed to be zero. Thus, $v = 0$. This will eliminate the first term of Eq. 4.126.

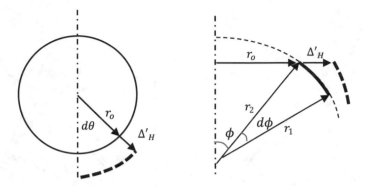

Fig. 4.24 Horizontal deformation of a differential element

$$\Delta'_\phi = \frac{\partial w}{r_1 \partial \phi}$$

Substitute 4.121 into the above equation.

$$\Delta'_\phi = \frac{\partial}{r_1 \partial \phi} \left[v \cot \phi - \frac{r_2}{Eh} \left(N'_\theta - \nu N'_\phi \right) \right] \tag{4.131}$$

Take the derivative of 4.131 with respect to ϕ.

$$\Delta'_\phi = \frac{dv}{d\phi} \frac{\cot \phi}{r_1} - \frac{v \csc^2 \phi}{r_1} - \frac{d\left[\frac{r_2}{Eh} \left(N'_\theta - \nu N'_\phi \right) \right]}{r_1 d\phi} \tag{4.132}$$

Since $v = 0$ at the base,

$$\Delta'_\phi = \frac{dv}{d\phi} \frac{\cot \phi}{r_1} - \frac{d\left[\frac{r_2}{Eh} \left(N'_\theta - \nu N'_\phi \right) \right]}{r_1 d\phi} \tag{4.133}$$

Using Eq. 4.107,

$$\frac{dv}{d\phi} - v \cot \phi = \frac{1}{Eh} \left[N'_\phi (r_1 + \nu r_2) - N'_\theta (r_2 + \nu r_1) \right] \tag{4.107}$$

and simplify since $v = 0$.

$$\frac{dv}{d\phi} = \frac{1}{Eh} \left[N'_\phi (r_1 + \nu r_2) - N'_\theta (r_2 + \nu r_1) \right] \tag{4.134}$$

Substituting the above equation into 4.133,

$$\Delta'_\phi = \left\{ \frac{1}{Eh} \left[N'_\phi(r_1 + \nu r_2) - N'_\theta(r_2 + \nu r_1) \right] \right\} \frac{\cot \phi}{r_1} - \frac{d \left[\frac{r_2}{Eh} \left(N'_\theta - \nu N'_\phi \right) \right]}{r_1 d\phi} \quad (4.135)$$

The horizontal deflection was previously defined as 4.130.

$$\Delta'_H = \frac{r_2 \sin \phi}{Eh} \left(N'_\theta - \nu N'_\phi \right) \quad (4.130)$$

Rearranged,

$$\frac{\Delta'_H}{\sin \phi} = \frac{r_2}{Eh} \left(N'_\theta - \nu N'_\phi \right)$$

Placing the above equation into 4.135, the general equation for membrane rotations are determined.

$$\Delta'_\phi = \left\{ \frac{\cot \phi}{Ehr_1} \left[N'_\phi(r_1 + \nu r_2) - N'_\theta(r_2 + \nu r_1) \right] \right\} - \frac{d}{r_1 d\phi} \left(\frac{\Delta_H}{\sin \phi} \right) \quad (4.136)$$

Equations 4.130 and 4.136 are general equations to determine the horizontal and slope deformations. These two equations are now applied to the various loading conditions.

4.3.4   Deformations for the Case of Uniformly Distributed Load (Gravity Load)

For the case of the uniformly distributed load, the stresses were previously determined as:

$$N'_\phi = - \frac{aq}{(1 + \cos \phi)} \quad (4.50)$$

$$N'_\theta = aq \left(\frac{1}{1 + \cos \phi} - \cos \phi \right) \quad (4.53)$$

Substituting the above equations into 4.130 and replacing $r_1 = r_2 = a$,

$$\Delta'_H = \frac{a \sin \phi}{Eh} \left[aq \left(\frac{1}{1 + \cos \phi} - \cos \phi \right) + \nu \frac{aq}{(1 + \cos \phi)} \right] \quad (4.137)$$

Simplify.

$$\Delta'_H = D_{10} = \frac{a^2 q}{Eh} \left(\frac{1+v}{1+\cos\,\phi} - \cos\phi \right) \sin\phi \qquad (4.138)$$

The general equation for the slope is Eq. 4.136. Substitute the above equation into 4.136 and replace $r_1 = r_2 = a$.

$$\Delta'_\phi = \left\{ \frac{\cot\phi}{Eha} \left[N'_\phi(a+va) - N'_\theta(a+va) \right] \right\} - \frac{d}{a\,d\phi}$$
$$\times \left(\frac{\frac{a^2 q}{Eh} \left(\frac{1+v}{1+\cos\,\phi} - \cos\phi \right) \sin\phi}{\sin\phi} \right) \qquad (4.139)$$

$$\Delta'_\phi = \left\{ \frac{\cot\phi}{Eh} \left[N'_\phi(1+v) - N'_\theta(1+v) \right] \right\} - \frac{aq}{Eh} \frac{d}{d\phi}$$
$$\times \left(\frac{1+v}{1+\cos\phi} - \cos\phi \right) \qquad (4.140)$$

Substitute 4.50 and 4.53 to replace N'_ϕ and N'_θ.

$$\Delta'_\phi = \left\{ \frac{\cot\phi}{Eh} \left[-\frac{aq}{(1+\cos\phi)}(1+v) - aq\left(\frac{1}{1+\cos\phi} - \cos\phi \right)(1+v) \right] \right\}$$
$$- \frac{aq}{Eh} \frac{d}{d\phi} \left(\frac{1+v}{1+\cos\phi} - \cos\phi \right) \qquad (4.141)$$

Manipulating the first term and take the derivative of the second term,

$$\Delta'_\phi = \frac{(1+v)aq\cot\phi}{Eh} \left\{ \left[-\frac{1}{(1+\cos\phi)} - \left(\frac{1}{1+\cos\phi} - \cos\phi \right) \right] \right\}$$
$$- \frac{aq}{Eh} \left[\frac{(1+v)(\sin\phi)}{(1+\cos\phi)^2} + \sin\phi \right] \qquad (4.142)$$

$$\Delta'_\phi = \frac{(1+v)aq\cot\phi}{Eh} \left(-\frac{2}{(1+\cos\phi)} + \cos\phi \right) - \frac{aq}{Eh}$$
$$\times \left[\frac{(1+v)\sin\phi}{(1+\cos\phi)^2} + \sin\phi \right] \qquad (4.143)$$

$$\Delta'_\phi = \frac{aq}{Eh}$$

$$\times \left[-\frac{2\,(1+v)\cot\phi}{(1+\cos\phi)} + (1+v)\cot\phi\cos\phi - \frac{(1+v)\sin\phi}{(1+\cos\phi)^2} - \sin\phi \right]$$

$$\hspace{10cm} (4.144)$$

$$\Delta'_\phi = D_{20} = \frac{aq}{Eh}\,(2+v)\sin\phi \hspace{3cm} (4.145)$$

The sign was changed so that a downward uniform load corresponds to a positive rotation.

4.3.5 Deformations for the Case of a Horizontal Uniformly Distributed Load

Using the previously defined stresses for the case of a horizontally applied uniformly distributed load,

$$N'_\phi = -\frac{Pa}{2} \hspace{4cm} (4.62)$$

$$N'_\theta = -\frac{Pa}{2}\cos 2\phi \hspace{3cm} (4.68)$$

Substituting these equations into 4.130 and set $r_1 = r_2 = a$.

$$\Delta'_H = \frac{a\sin\phi}{Eh}\left(\frac{-Pa}{2}\cos 2\phi + v\frac{Pa}{2} \right) \hspace{2cm} (4.146)$$

$$\Delta'_H = D_{10} = \frac{a^2 P \sin\phi}{2Eh}\,(-\cos 2\phi + v\,) \hspace{2cm} (4.147)$$

Substitute 4.62, 4.68, and 4.147 into 4.136, and set $r_1 = r_2 = a$.

$$\Delta'_\phi = \left\{ \frac{\cot\phi}{Eha}\left[-\frac{Pa}{2}(a+va) + \frac{Pa}{2}\cos 2\phi(a+va) \right] \right\} - \frac{d}{ad\phi}$$

$$\times \left(\frac{\frac{a^2 P \sin\phi}{2Eh}\,(-\cos 2\phi + v\,)}{\sin\phi} \right) \hspace{3cm} (4.148)$$

$$\Delta'_\phi = \left\{ \frac{ap\,(1+v)\,\cot\phi}{2Eh}[-1+\cos 2\phi] \right\} - \frac{aP}{2Eh}\frac{d}{d\phi}(-\cos 2\phi + v\,) \hspace{1cm} (4.149)$$

$$\Delta'_\phi = \left\{ \frac{ap\ (1+\nu)\ \cot\phi}{2Eh} [-1 + \cos 2\phi] \right\} - \frac{aP}{2Eh}(2\sin 2\phi) \tag{4.150}$$

$$\Delta'_\phi = \frac{ap}{2Eh} [-(1+\nu)\cot\phi + (1+\nu)\cot\phi\cos 2\phi - 2\sin 2\phi] \tag{4.151}$$

$$\Delta'_\phi = -\frac{ap}{2Eh} [(1+\nu)\cot\phi(1 - \cos 2\phi) + 2\sin 2\phi] \tag{4.152}$$

Using the trigonometric identities,

$$1 - \cos 2\phi = 2\ \sin^2\phi \tag{4.153}$$

$$\sin 2\phi = 2\sin\phi\cos\phi \tag{4.154}$$

and substituting into 4.152,

$$\Delta'_\phi = -\frac{ap}{2Eh} 2 \left[(1+\nu)\frac{\cos\phi}{\sin\phi}\ \sin^2\phi + 2\sin\phi\cos\phi \right] \tag{4.155}$$

$$\Delta'_\phi = -\frac{ap}{2Eh}(2\sin\phi\cos\phi)(3+\nu) \tag{4.156}$$

$$\Delta'_\phi = D_{20} = -\frac{ap}{2Eh}(3+\nu)\sin 2\phi \tag{4.157}$$

4.3.6 Deformations for the Case of a Uniform Pressure

The stresses for the case of a uniform pressure:

$$N'_\phi = -\frac{Pa}{2} \tag{4.74}$$

$$N'_\theta = -\frac{Pa}{2} \tag{4.76}$$

Substituting these stresses into Eqs. 4.130 and setting $r_1 = r_2 = a$,

$$\Delta'_H = \frac{a\sin\phi}{Eh}\left(-\frac{Pa}{2} + \nu\,\frac{Pa}{2} \right) \tag{4.158}$$

$$\Delta'_H = D_{10} = \frac{a^2 P\ \sin\phi}{2Eh}(-1+\nu) \tag{4.159}$$

Furthermore, substituting 4.74, 4.76, and 4.159 into 4.136,

$$\Delta'_\phi = \left\{ \frac{\cot\phi}{Eha} \left[\frac{-Pa}{2}(a+va) - \frac{-Pa}{2}(a+va) \right] \right\} - \frac{d}{ad\phi}$$

$$\times \left(\frac{\frac{a^2 P \, \sin\phi}{2Eh} (-1+v)}{\sin\phi} \right) \tag{4.160}$$

$$\Delta'_\phi = \left[\frac{\cot\phi}{Eha}(0) \right] - \frac{d}{ad\phi} \left(\frac{a^2 P(-1+v)}{2Eh} \right) \tag{4.161}$$

Since the derivative of a constant is zero,

$$\Delta'_\phi = D_{20} = 0 \tag{4.162}$$

4.4   Boundary Effects

If the shell is unrestrained at the boundary, the stresses will only be membrane. However, the base of the shell will most likely be connected to a foundation, or ring beam, which restrains movement in translation and bending. The stresses which are caused by this restrainment is what is referred to as boundary effects. The left-hand side of Fig. 4.25 illustrates the location and orientation of the moments and shears. The boundary effects will also produce membrane stresses, which are illustrated in the right-hand side of Fig. 4.25. The forces and stresses are separated for clarity, but the total stress is the combination of the two figures. It should be noticed that the membrane stresses, which are caused by boundary effects, are represented without a prime in the superscript.

The extent, or influence, of the boundary effects is directly related to the type of construction materials used and the thickness of the shell walls. A thin flexible material, such as steel sheets, may have negligible effect on the stresses. However, a thicker stiffer material, such as concrete, may influence as much as a third of the shell and cause significant stresses.

The centroid of the foundation, or ring beam, should be in line with axis of the shell walls (Fig. 4.26). If these lines do not intersect, additional moments will occur in the shell. The theory presented is based on the influence of base fixity and does not include the effects of foundation eccentricities. If the foundation is designed with an eccentricity, this will defeat the purpose and benefit of dome structures, since the offset will generate additional moments and reduce the capacity of the shell. An eccentricity is therefore not advisable.

If a ring beam is placed at 51°, the hoop tension in the beam is not zero (as suggested by Fig. 4.15), since the horizontal component of the meridian stresses causes a ring tension.

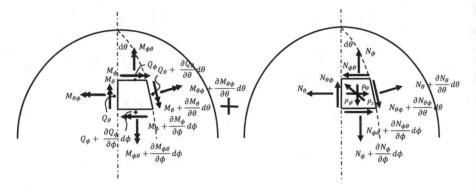

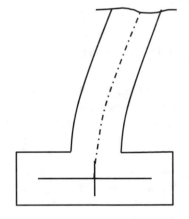

Fig. 4.25 Bending, shears, and membrane stresses in a differential element due to boundary effects

Fig. 4.26 Shell walls
intersecting the centroid of
the ring beam or foundation

The boundary theory is determined by solving for the equilibrium equations in
each of the directions.

4.4.1 Equilibrium in the Hoop Direction (θ-Direction)

In addition to the membrane stresses, only the shears contribute to the equilibrium
equation in the hoop direction (see Fig. 4.27). The equilibrium equation therefore
includes the membrane equilibrium Eq. 4.7, plus the out-of-plane shear. Each of the
stresses are multiplied by the length of the differential element (see Fig. 4.3).

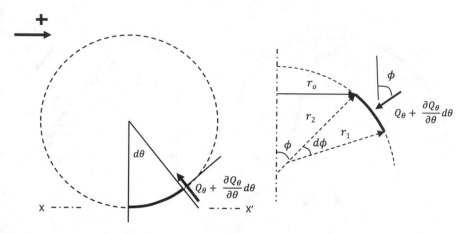

Fig. 4.27 Stresses that contribute to the equilibrium equation in the hoop direction

$$-\left(Q_\theta + \frac{\partial Q_\theta}{\partial \theta} d\theta\right)\left(\frac{r_1 d\phi}{\cos \beta}\right) \cos \beta \sin d\theta \sin \phi + N_{\theta\phi} \frac{\partial r_o}{\partial \phi} d\theta d\phi + \frac{\partial N_\theta}{\partial \theta} r_1 d\theta d\phi +$$
$$\frac{\partial (N_{\phi\theta} r_o)}{\partial \phi} d\theta d\phi + P_\theta r_o r_1 d\theta d\phi = 0 \qquad (4.163)$$

Expand the first term and simplify (assuming $\sin d\theta \approx d\theta$).

$$-Q_\theta r_1 d\phi \, d\theta \sin \phi - \frac{\partial Q_\theta}{\partial \theta} d\phi \, r_1 d\theta^2 \sin \phi + N_{\theta\phi} \frac{\partial r_o}{\partial \phi} d\theta d\phi + \frac{\partial N_\theta}{\partial \theta} r_1 d\theta d\phi +$$
$$\frac{\partial (N_{\phi\theta} r_o)}{\partial \phi} d\theta d\phi + P_\theta r_o r_1 d\theta d\phi = 0 \qquad (4.164)$$

The second term is a higher order term and therefore may be eliminated from the equation. Furthermore, the equation is simplified by dividing by $d\theta d\phi$.

$$-Q_\theta r_1 \sin \phi + N_{\theta\phi} \frac{\partial r_o}{\partial \phi} + \frac{\partial N_\theta}{\partial \theta} r_1 + \frac{\partial (N_{\phi\theta} r_o)}{\partial \phi} + P_\theta r_o r_1 = 0 \qquad (4.165)$$

4.4.2 Equilibrium in the Meridian Direction (φ-Direction)

In addition to the membrane stresses, the shear stresses that contribute to the equilibrium equations are depicted in Fig. 4.28. As illustrated, only the bottom out-of-plane shear contributes to the equation. The component of this shear along the x–x′ line is added to the membrane stress equation (Eq. 4.11).

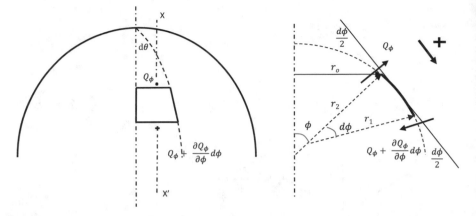

Fig. 4.28 Stresses that contribute to the equilibrium equation in the meridian direction

$$-Q_\phi(r_o d\theta)\sin\frac{d\phi}{2} - \left(Q_\phi + \frac{\partial Q_\phi}{\partial\phi}d\phi\right)\left(r_o + \frac{\partial r_o}{\partial\phi}d\phi\right)d\theta\sin\frac{d\phi}{2} + N_\phi\frac{\partial r_o}{\partial\phi}d\phi d\theta +$$

$$r_o\frac{\partial N_\phi}{\partial\phi}d\phi d\theta + \frac{\partial N_{\theta\phi}}{\partial\theta}d\theta r_1 d\phi - N_\theta\frac{\partial r_o}{\partial\phi}d\phi d\theta + P_\phi r_1 r_o d\theta d\phi = 0$$

$$(4.166)$$

Expanding the above equation and assuming $\sin\frac{d\phi}{2} \approx \frac{d\phi}{2}$,

$$-Q_\phi(r_o d\theta)\frac{d\phi}{2} - Q_\phi r_o\frac{d\phi}{2}d\theta - Q_\phi\frac{\partial r_o}{\partial\phi}\frac{d\phi^2}{2}d\theta - \frac{\partial Q_\phi}{\partial\phi}\frac{d\phi^2}{2}r_o d\theta -$$

$$\frac{\partial Q_\phi}{\partial\phi}\frac{d\phi^3}{2}\frac{\partial r_o}{\partial\phi}d\theta + N_\phi\frac{\partial r_o}{\partial\phi}d\phi d\theta + r_o\frac{\partial N_\phi}{\partial\phi}d\phi d\theta + \frac{\partial N_{\theta\phi}}{\partial\theta}d\theta r_1 d\phi - N_\theta\frac{\partial r_o}{\partial\phi}d\phi d\theta +$$

$$P_\phi r_1 r_o d\theta d\phi = 0$$

$$(4.167)$$

The third, fourth, and fifth terms are higher order terms, and therefore may be eliminated. The first and second terms are combined and the equation is divided by $d\theta d\phi$.

$$-Q_\phi r_o + N_\phi\frac{\partial r_o}{\partial\phi} + r_o\frac{\partial N_\phi}{\partial\phi} + \frac{\partial N_{\theta\phi}}{\partial\theta}r_1 - N_\theta\frac{\partial r_o}{\partial\phi} + P_\phi r_1 r_o = 0 \qquad (4.168)$$

The expression may be further simplified by applying the product rule, combining the second and third terms.

$$\frac{\partial (N_\phi r_o)}{\partial \phi} = N'_\phi \frac{\partial r_o}{\partial \phi} + r_o \frac{\partial N'_\phi}{\partial \phi} \tag{4.169}$$

Therefore,

$$-Q_\phi r_o + \frac{\partial (N_\phi r_o)}{\partial \phi} + \frac{\partial N_{\theta\phi}}{\partial \theta} r_1 - N_\theta \frac{\partial r_o}{\partial \phi} + P_\phi r_1 r_o = 0 \tag{4.170}$$

4.4.3   Stresses That Contribute to the Equilibrium Equation in the z-Direction

When summing the internal stresses, all four out-of-plane shears contribute to the equilibrium Eq. 4.15 (see Fig. 4.29). These stresses are added to the membrane stress equilibrium equation.

$$-Q_\phi r_o d\theta \cos \frac{d\phi}{2} + \left(Q_\phi + \frac{\partial Q_\phi}{\partial \phi} d\phi \right) \left(r_o + \frac{\partial r_o}{\partial \phi} d\phi \right) d\theta \cos \frac{d\theta}{2} - Q_\theta r_1 d\phi \cos \frac{d\theta}{2} +$$

$$\left(Q_\theta + \frac{\partial Q_\theta}{\partial \theta} d\theta \right) \frac{r_1 d\phi}{\cos \beta} \cos \frac{d\theta}{2} \cos \beta + N_\phi r_o \, d\phi d\theta + N_\theta r_1 d\phi d\theta_1 + P_z r_1 r_o d\phi d\theta = 0$$

$$\tag{4.171}$$

The expression is simplified by substituting $d\theta_1 = \frac{r_o \, d\theta}{r_2}$, $r_2 = \frac{r_o}{\sin \phi}$ and $\cos \frac{d\theta}{2} \approx \cos \frac{d\phi}{2} \approx 1$. Expand the above equation.

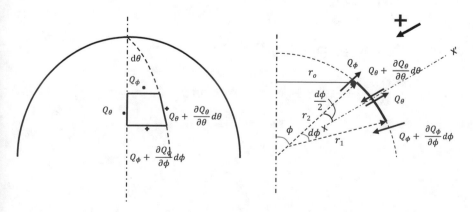

Fig. 4.29 Stresses that contribute to equilibrium in the z-direction

$$-Q_\phi r_o d\theta + Q_\phi r_o d\theta + Q_\phi \frac{\partial r_o}{\partial \phi} d\phi d\theta + \frac{\partial Q_\phi}{\partial \phi} r_o d\phi d\theta + \frac{\partial Q_\phi}{\partial \phi} d\phi \frac{\partial r_o}{\partial \phi} d\phi d\theta - Q_\theta r_1 d\phi +$$

$$Q_\theta r_1 d\phi + \frac{\partial Q_\theta}{\partial \theta} r_1 d\phi d\theta + N_\phi r_o d\phi d\theta + N_\theta r_1 \sin \phi d\phi d\theta + P_z r_1 r_o d\phi d\theta = 0$$

$$(4.172)$$

Eliminate the first and second terms and the sixth and seventh. The fifth term is a higher order term, which is eliminated. Equation 4.172 is therefore reduced to the following form:

$$Q_\phi \frac{\partial r_o}{\partial \phi} d\phi d\theta + \frac{\partial Q_\phi}{\partial \phi} r_o d\phi d\theta + \frac{\partial Q_\theta}{\partial \theta} r_1 d\phi d\theta + N_\phi r_o d\phi d\theta$$

$$+ N_\theta r_1 \sin \phi d\phi d\theta + P_z r_1 r_o d\phi d\theta$$

$$= 0 \qquad (4.173)$$

Using the product rule,

$$(fg)' = f'g + fg'$$

The first two terms of 4.173 may be combined.

$$Q_\phi \frac{\partial r_o}{\partial \phi} d\phi d\theta + \frac{\partial Q_\phi}{\partial \phi} r_o d\phi d\theta = \frac{\partial (Q_\phi r_o)}{\partial \phi} d\phi d\theta$$

Therefore,

$$\frac{\partial (Q_\phi r_o)}{\partial \phi} d\phi d\theta + \frac{\partial Q_\theta}{\partial \theta} r_1 d\phi d\theta + N_\phi r_o d\phi d\theta + N_\theta r_1 \sin \phi d\phi d\theta$$

$$+ P_z r_1 r_o d\phi d\theta$$

$$= 0 \qquad (4.174)$$

Divide the above equation by $d\phi d\theta$.

$$\frac{\partial (Q_\phi r_o)}{\partial \phi} + \frac{\partial Q_\theta}{\partial \theta} r_1 + N_\phi r_o + N_\theta r_1 \sin \phi + P_z r_1 r_o = 0 \qquad (4.175)$$

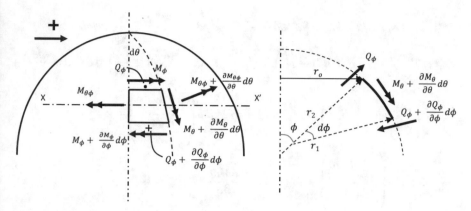

Fig. 4.30 Moments and shears that contribute to the moment equilibrium equation in the hoop direction

4.4.4 Equilibrium of Moments About the Hoop Direction

In reference to Fig. 4.30, only the moments and shears that contribute to the equilibrium equation are included in the diagram. In this case, the membrane forces are not included since they do not contribute to the moments about a center line of the differential element in the hoop direction. Summing moments,

$$
M_\phi r_o d\theta - \left(M_\phi + \frac{\partial M_\phi}{\partial \phi} d\phi\right)\left(r_o + \frac{\partial r_o}{\partial \phi} d\phi\right) d\theta + Q_\phi r_o d\theta \left(\frac{r_1 d\phi}{2}\right) \cos\frac{d\phi}{2} +
$$
$$
\left(Q_\phi + \frac{\partial Q_\phi}{\partial \phi} d\phi\right)\left(r_o + \frac{\partial r_o}{\partial \phi} d\phi\right) d\theta \left(\frac{r_1 d\phi}{2}\right) \cos\frac{d\phi}{2} - M_{\theta\phi} r_1 d\phi
$$
$$
+ \left(M_\theta + \frac{\partial M_\theta}{\partial \theta} d\theta\right)\left(\frac{r_1 d\phi}{\cos\beta}\right) \sin\beta +
$$
$$
\left(M_{\theta\phi} + \frac{\partial M_{\theta\phi}}{\partial \theta} d\theta\right)\left(\frac{r_1 d\phi}{\cos\beta}\right) \cos\beta = 0
$$

$$(4.176)$$

where the lengths of the differential element are taken from Fig. 4.3.

Expand the above equation, assume $\cos\frac{d\phi}{2} \approx 1$ and $\tan\beta = \frac{\frac{\partial r_o}{\partial \phi} d\phi d\theta}{r_1 d\phi}$.

$$M_\phi r_o d\theta - M_\phi r_o d\theta - M_\phi \frac{\partial r_o}{\partial \phi} d\phi d\theta - \frac{\partial M_\phi}{\partial \phi} r_o d\phi d\theta - \frac{\partial M_\phi}{\partial \phi} d\phi \frac{\partial r_o}{\partial \phi} d\phi d\theta +$$

$$Q_\phi r_o d\theta \left(\frac{r_1 d\phi}{2}\right) + Q_\phi r_o d\theta \left(\frac{r_1 d\phi}{2}\right) + Q_\phi \frac{\partial r_o}{\partial \phi} d\phi d\theta \left(\frac{r_1 d\phi}{2}\right) + \frac{\partial Q_\phi}{\partial \phi} r_o d\phi d\theta \left(\frac{r_1 d\phi}{2}\right) +$$

$$\frac{\partial Q_\phi}{\partial \phi} d\phi \frac{\partial r_o}{\partial \phi} d\phi d\theta \left(\frac{r_1 d\phi}{2}\right) - M_{\theta\phi} r_1 d\phi + M_\theta r_1 d\phi \left(\frac{\frac{\partial r_o}{\partial \phi} d\phi d\theta}{r_1 d\phi}\right) +$$

$$\frac{\partial M_\theta}{\partial \theta} d\theta r_1 d\phi \left(\frac{\frac{\partial r_o}{\partial \phi} d\phi d\theta}{r_1 d\phi}\right) + M_{\theta\phi} r_1 d\phi + \frac{\partial M_{\theta\phi}}{\partial \theta} d\theta r_1 d\phi = 0$$

$$(4.177)$$

Eliminating the higher order terms (i.e., fifth, eight, ninth, tenth, and thirteenth terms).

$$M_\phi r_o d\theta - M_\phi r_o d\theta - M_\phi \frac{\partial r_o}{\partial \phi} d\phi d\theta - \frac{\partial M_\phi}{\partial \phi} d\phi r_o d\theta + Q_\phi r_o d\theta \left(\frac{r_1 d\phi}{2}\right) +$$

$$Q_\phi r_o d\theta \left(\frac{r_1 d\phi}{2}\right) - M_{\theta\phi} r_1 d\phi + M_\theta r_1 d\phi \left(\frac{\frac{\partial r_o}{\partial \phi} d\phi d\theta}{r_1 d\phi}\right) +$$

$$M_{\theta\phi} r_1 d\phi + \frac{\partial M_{\theta\phi}}{\partial \theta} r_1 d\theta d\phi = 0$$

$$(4.178)$$

In the above equation, terms one and two and the seventh and ninth cancel each other out (equal but opposite signs). Furthermore, the eight term is simplified.

$$-M_\phi \frac{\partial r_o}{\partial \phi} d\phi d\theta - \frac{\partial M_\phi}{\partial \phi} r_o d\phi d\theta + Q_\phi r_o d\theta \left(\frac{r_1 d\phi}{2}\right) + Q_\phi r_o d\theta \left(\frac{r_1 d\phi}{2}\right)$$

$$+ M_\theta \frac{\partial r_o}{\partial \phi} d\phi d\theta + \frac{\partial M_{\theta\phi}}{\partial \theta} d\theta r_1 d\phi$$

$$= 0 \qquad (4.179)$$

Furthermore, use the product rule to combine the first two terms. The third and the fourth terms may also be added.

$$-\frac{\partial (M_\phi r_o)}{\partial \phi} d\phi d\theta + Q_\phi r_o r_1 d\theta d\phi + M_\theta \frac{\partial r_o}{\partial \phi} d\phi d\theta + \frac{\partial M_{\theta\phi}}{\partial \theta} r_1 d\theta d\phi = 0 \quad (4.180)$$

Divide by $d\phi d\theta$.

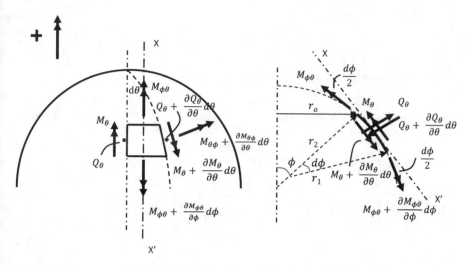

Fig. 4.31 Moments and shears that contribute to the equilibrium equation in the meridian direction

$$-\frac{\partial(M_\phi r_o)}{\partial\phi} + Q_\phi r_o r_1 + M_\theta\frac{\partial r_o}{\partial\phi} + \frac{\partial M_{\theta\phi}}{\partial\theta}r_1 = 0 \qquad (4.181)$$

4.4.5 Equilibrium of Moments About the Meridian Direction

Figure 4.31 depicts the moments and shears that contribute to the equilibrium equations about a vertical line that passes through the differential element. Only the moment and shears that contribute to the equilibrium equations are illustrated. No membrane forces contribute to this equation. Summing moments about the vertical center line,

$$M_{\phi\theta}r_o d\theta\left(\cos\frac{d\phi}{2}\right) - \left(M_{\phi\theta} + \frac{\partial M_{\phi\theta}}{\partial\phi}d\phi\right)\left(r_o + \frac{\partial r_o}{\partial\phi}d\phi\right)d\theta\left(\cos\frac{d\phi}{2}\right) + M_\theta r_1 d\phi -$$
$$\left(M_\theta + \frac{\partial M_\theta}{\partial\theta}d\theta\right)\frac{r_1 d\phi}{\cos\beta}\cos\beta + \left(M_{\theta\phi} + \frac{\partial M_{\theta\phi}}{\partial\theta}d\theta\right)\frac{r_1 d\phi}{\cos\beta}\sin\beta + Q_\theta r_1 d\phi\frac{r_o d\theta}{2} +$$
$$\left(Q_\theta + \frac{\partial Q_\theta}{\partial\theta}d\theta\right)\frac{r_1 d\phi}{\cos\beta}\cos\beta\frac{r_o d\theta}{2} = 0$$

$$(4.182)$$

where the lengths of the sides of the differential element are given in Fig. 4.3.
    Using Eq. 4.3,

$$\frac{\sin \beta}{\cos \beta} = \tan \beta = \frac{\frac{\partial r_o}{\partial \phi} d\phi d\theta}{r_1 d\phi} \tag{4.3}$$

Assuming $\cos \frac{d\phi}{2} \approx 1$ and expanding Eq. 4.182,

$$M_{\phi\theta} r_o d\theta - M_{\phi\theta} r_o d\theta - M_{\phi\theta} \frac{\partial r_o}{\partial \phi} d\phi d\theta - \frac{\partial M_{\phi\theta}}{\partial \phi} r_o d\phi d\theta - \frac{\partial M_{\phi\theta}}{\partial \phi} d\phi \frac{\partial r_o}{\partial \phi} d\phi d\theta + M_\theta r_1 d\phi -$$

$$M_\theta r_1 d\phi - \frac{\partial M_\theta}{\partial \theta} d\theta r_1 d\phi + M_{\theta\phi} r_1 d\phi \left(\frac{\frac{\partial r_o}{\partial \phi} d\phi d\theta}{r_1 d\phi} \right) + \frac{\partial M_{\theta\phi}}{\partial \theta} r_1 d\theta d\phi \left(\frac{\frac{\partial r_o}{\partial \phi} d\phi d\theta}{r_1 d\phi} \right) +$$

$$Q_\theta r_1 d\phi \left(\frac{r_o d\theta}{2} \right) + Q_\theta r_1 d\phi \left(\frac{r_o d\theta}{2} \right) + \frac{\partial Q_\theta}{\partial \theta} r_1 d\theta d\phi \left(\frac{r_o d\theta}{2} \right) = 0$$

$$\tag{4.183}$$

The fifth, tenth, and thirteenth terms are higher order terms and therefore may be eliminated from Eq. 4.183. The first, second, sixth, and seventh terms are also eliminated (equal but opposite terms).

$$-M_{\phi\theta} \frac{\partial r_o}{\partial \phi} d\phi d\theta - \frac{\partial M_{\phi\theta}}{\partial \phi} r_o d\phi d\theta - \frac{\partial M_\theta}{\partial \theta} r_1 d\theta d\phi + M_{\theta\phi} \frac{\partial r_o}{\partial \phi} d\phi d\theta$$

$$+ Q_\theta r_1 d\phi \left(\frac{r_o d\theta}{2} \right) - Q_\theta r_1 d\phi \left(\frac{r_o d\theta}{2} \right)$$

$$= 0 \tag{4.184}$$

Combining the last two terms and using the product rule to simplify,

$$-\frac{\partial \left(M_{\phi\theta} r_o \right)}{\partial \phi} d\phi d\theta = -M_{\phi\theta} \frac{\partial r_o}{\partial \phi} d\phi d\theta - \frac{\partial M_{\phi\theta}}{\partial \phi} r_o d\phi d\theta \tag{4.185}$$

Therefore,

$$-\frac{\partial \left(M_{\phi\theta} r_o \right)}{\partial \phi} d\phi d\theta - \frac{\partial M_\theta}{\partial \theta} r_1 d\theta d\phi + M_{\theta\phi} \frac{\partial r_o}{\partial \phi} d\phi d\theta + Q_\theta r_1 r_o d\phi d\theta = 0 \tag{4.186}$$

Divide by $d\theta d\phi$.

$$-\frac{\partial \left(M_{\phi\theta} r_o \right)}{\partial \phi} - \frac{\partial M_\theta}{\partial \theta} r_1 + M_{\theta\phi} \frac{\partial r_o}{\partial \phi} + Q_\theta r_1 r_o = 0 \tag{4.187}$$

Fig. 4.32 Radial geometric relationships of a differential element

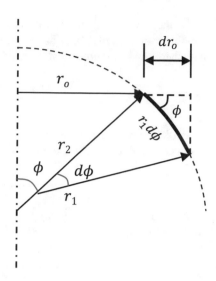

4.4.6 Simplifying the General Equilibrium Equations

Collecting the five equilibrium equations,

$$-Q_\theta r_1 \sin \phi + N_{\theta\phi} \frac{\partial r_o}{\partial \phi} + \frac{\partial N_\theta}{\partial \theta} r_1 + \frac{\partial (N_{\phi\theta} r_o)}{\partial \phi} + P_\theta r_o r_1 = 0 \qquad (4.165)$$

$$-Q_\phi r_o + \frac{\partial (N_\phi r_o)}{\partial \phi} + \frac{\partial N_{\theta\phi}}{\partial \theta} r_1 - N_\theta \frac{\partial r_o}{\partial \phi} + P_\phi r_1 r_o = 0 \qquad (4.170)$$

$$\frac{\partial (Q_\phi r_o)}{\partial \phi} + \frac{\partial Q_\theta}{\partial \theta} r_1 + N_\phi r_o + N_\theta r_1 \sin \phi + P_z r_1 r_o = 0 \qquad (4.175)$$

$$-\frac{\partial (M_\phi r_o)}{\partial \phi} + Q_\phi r_o r_1 + M_\theta \frac{\partial r_o}{\partial \phi} + \frac{\partial M_{\theta\phi}}{\partial \theta} r_1 = 0 \qquad (4.181)$$

$$-\frac{\partial (M_{\phi\theta} r_o)}{\partial \phi} - \frac{\partial M_\theta}{\partial \theta} r_1 + M_{\theta\phi} \frac{\partial r_o}{\partial \phi} + Q_\theta r_1 r_o = 0 \qquad (4.187)$$

Then above equations are simplified by making the following assumptions: From Fig. 4.32, the following relationship may be formed:

$$r_1 d\phi \cos \phi = dr_o$$

or,

$$\frac{dr_o}{d\phi} = r_1 \cos \phi \qquad (4.188)$$

Furthermore, for symmetrical loading, the membrane and shear stresses and the bending moments may vary in the ϕ-direction but are constant in the θ-direction. It may therefore be assumed that the change in stress in the θ-direction is equal to zero.

$$\frac{\partial N_\theta}{\partial \theta} = \frac{\partial Q_\theta}{\partial \theta} = \frac{\partial M_\theta}{\partial \theta} = 0 \tag{4.189}$$

Furthermore, the in-plane membrane shears and the twisting moments are also assumed equal to zero, due to symmetrical loading.

$$N_{\phi\theta} = N_{\theta\phi} = M_{\phi\theta} = M_{\theta\phi} = 0 \tag{4.190}$$

Applying the above assumptions, the equilibrium equations reduce to the following form:

$$-Q_\theta r_1 \sin\phi + P_\theta r_o r_1 = 0 \tag{4.191}$$

$$\frac{\partial (N_\phi r_o)}{\partial \phi} - Q_\phi r_o - N_\theta r_1 \cos\phi + P_\phi r_1 r_o = 0 \tag{4.192}$$

$$\frac{\partial (Q_\phi r_o)}{\partial \phi} - N_\phi r_o + N_\theta r_1 \sin\phi + P_z r_1 r_o = 0 \tag{4.193}$$

$$-\frac{\partial (M_\phi r_o)}{\partial \phi} + Q_\phi r_o r_1 + M_\theta r_1 \cos\phi = 0 \tag{4.194}$$

$$Q_\theta r_1 r_o = 0 \tag{4.195}$$

Since Q_θ must be equal to zero, Eq. 4.195 is eliminated. Similarly, if $Q_\theta = 0$, then P_θ also must be equal to zero, and Eq. 4.191 is eliminated. The equilibrium equations are therefore reduced to three.

$$\frac{\partial (N_\phi r_o)}{\partial \phi} - Q_\phi r_o - N_\theta r_1 \cos\phi + P_\phi r_1 r_o = 0 \tag{4.192}$$

$$\frac{\partial (Q_\phi r_o)}{\partial \phi} - N_\phi r_o + N_\theta r_1 \sin\phi + P_z r_1 r_o = 0 \tag{4.193}$$

$$-\frac{\partial (M_\phi r_o)}{\partial \phi} + Q_\phi r_o r_1 + M_\theta r_1 \cos\phi = 0 \tag{4.194}$$

4.5   Displacement Theory of the Boundary Effects

The relationship between strain and deformation was previously solved.

$$\varepsilon_\theta = \frac{v \cos \phi}{r_o} - \frac{w}{r_2} \tag{4.89}$$

$$\varepsilon_\phi = \frac{dv}{r_1 d\phi} - \frac{w}{r_1} \tag{4.96}$$

From cylindrical shell theory, and changing the y subscript to ϕ,

$$N_\theta = \frac{Eh}{(1 - v^2)} \left[\varepsilon_\theta + v \varepsilon_\phi \right] \tag{3.35}$$

$$N_\phi = \frac{Eh}{(1 - v^2)} \left[\varepsilon_\phi + v \varepsilon_\theta \right] \tag{3.36}$$

Substituting Eqs. 4.89 and 4.96 into 3.35 and 3.36,

$$N_\theta = \frac{Eh}{(1 - v^2)} \left[\left(\frac{v \cos \phi}{r_o} - \frac{w}{r_2} \right) + v \left(\frac{dv}{r_1 d\phi} - \frac{w}{r_1} \right) \right] \tag{4.195}$$

$$N_\phi = \frac{Eh}{(1 - v^2)} \left[\left(\frac{dv}{r_1 d\phi} - \frac{w}{r_1} \right) + v \left(\frac{v \cos \phi}{r_o} - \frac{w}{r_2} \right) \right] \tag{4.196}$$

Referring to Fig. 4.33, the rotation of the differential element was previously defined by Eq. 4.126, in terms of deformations in the v and w directions.

Fig. 4.33 Slope at the top and bottom of the differential element

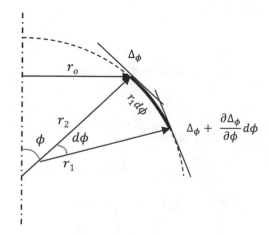

$$\Delta_\phi = \frac{v}{r_1} + \frac{\partial w}{r_1 \partial \phi} \tag{4.126}$$

Rotation at the bottom of the differential element is equal to the top rotation plus the change in rotation.

$$\overline{\Delta}_\phi = \Delta_\phi + \frac{\partial (\Delta_\phi)}{\partial \phi} d\phi \tag{4.197}$$

The curvature is the change in rotation with respect to the length of the differential element.

$$\chi_\phi = \frac{\overline{\Delta}_\phi - \Delta_\phi}{r_1 d\phi} \tag{4.198}$$

Substituting 4.197 into 4.198,

$$\chi_\phi = \frac{\left[\Delta_\phi + \frac{\partial (\Delta_\phi)}{\partial \phi} d\phi \right] - \Delta_\phi}{r_1 d\phi} \tag{4.199}$$

$$\chi_\phi = \frac{\partial (\Delta_\phi)}{r_1 \partial \phi} \tag{4.200}$$

Substituting Eq. 4.126,

$$\chi_\phi = \frac{\partial}{r_1 \partial \phi} \left(\frac{v}{r_1} + \frac{\partial w}{r_1 \partial \phi} \right) \tag{4.201}$$

Prior to deformation, the length of the top edge of the differential element is equal to:

$$r_o d\theta = r_2 \sin \phi \, d\theta \tag{4.202}$$

After deformation (refer to Fig. 4.34),

$$r_o d\theta = r_2' \sin (\phi + \Delta_\phi) \, d\theta \tag{4.203}$$

Equate 4.202 and 4.203,

$$r_2 \sin \phi \, d\theta = r_2' \sin (\phi + \Delta_\phi) \, d\theta \tag{4.204}$$

Using the trigonometric identity,

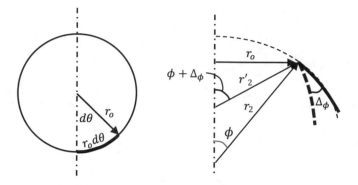

Fig. 4.34 Rotation of the differential element

$$\sin\left(\phi + \Delta_\phi\right) = \sin\phi\cos\Delta_\phi + \cos\phi\sin\Delta_\phi \tag{4.205}$$

Therefore,

$$r_2\sin\phi\,d\theta = r_2'\left(\sin\phi\cos\Delta_\phi + \cos\phi\sin\Delta_\phi\right)d\theta \tag{4.206}$$

Since Δ_ϕ is very small, we can make the following assumptions:

$$\cos\Delta_\phi \approx 1$$

$$\sin\Delta_\phi \approx \Delta_\phi$$

Equation 4.206 is therefore simplified.

$$r_2\sin\phi = r_2'\left(\sin\phi + \Delta_\phi\cos\phi\right) \tag{4.207}$$

Solving for r'_2,

$$r_2' = \frac{r_2\sin\phi}{\left(\sin\phi + \Delta_\phi\cos\phi\right)} \tag{4.208}$$

Since the curvature is defined as the inverse of the radius,

$$\frac{1}{r_2'} = \frac{\left(\sin\phi + \Delta_\phi\cos\phi\right)}{r_2\sin\phi} = \frac{\left(1 + \Delta_\phi\cot\phi\right)}{r_2} \tag{4.209}$$

Therefore, the curvature due to bending is the difference in curvatures after deformation.

$$\chi_\theta = \frac{1}{r_2'} - \frac{1}{r_2} \tag{4.210}$$

Substituting Eq. 4.209,

$$\chi_\theta = \frac{1}{r_2} + \frac{\Delta_\phi \cot \phi}{r_2} - \frac{1}{r_2} = \frac{\Delta_\phi \cot \phi}{r_2} \tag{4.211}$$

Since,

$$r_2 = \frac{r_o}{\sin \phi}$$

$$\chi_\theta = \frac{\Delta_\phi \cos \phi}{r_o} \tag{4.212}$$

Substituting Eq. 4.126 into 4.212,

$$\chi_\theta = \left(\frac{v}{r_1} + \frac{\partial w}{r_1 \partial \phi} \right) \frac{\cos \phi}{r_o} \tag{4.213}$$

From plate theory (Szilard 1974) (also see 3.61 and 3.62),

$$M_\theta = -D(\chi_\theta + v \chi_\phi) \tag{4.214}$$

$$M_\phi = -D(\chi_\phi + v \chi_\theta) \tag{4.215}$$

where

$$D = \frac{Eh}{(1 - v^2)} \tag{3.63}$$

Substituting Eqs. 4.201 and 4.213 into 4.214 and 4.215,

$$M_\theta = -D\left[\left(\frac{v}{r_1} + \frac{\partial w}{r_1 \partial \phi} \right) \frac{\cos \phi}{r_o} + v \frac{\partial}{r_1 \partial \phi} \left(\frac{v}{r_1} + \frac{\partial w}{r_1 \partial \phi} \right) \right] \tag{4.216}$$

$$M_\phi = -D\left[\frac{\partial}{r_1 \partial \phi} \left(\frac{v}{r_1} + \frac{\partial w}{r_1 \partial \phi} \right) + v \left(\frac{v}{r_1} + \frac{\partial w}{r_1 \partial \phi} \right) \frac{\cos \phi}{r_o} \right] \tag{4.217}$$

Equations 4.216 and 4.217 will be used at a later stage to solve for the stress N_ϕ.

The equations for the deformations of the dome are taken from cylindrical shell theory. To begin, the deformation equations are solved at the base of the cylindrical shell where $y = 0$. The equations appear differently from those listed in Chap. 3, but they are equivalent.

Translations due to out-of-plane shears

$$w_{y=0} = D_{11}Q_o = \frac{2\beta r^2 Q_o}{Eh} \tag{3.94}$$

Translations due to bending moments

$$w_{y=0} = D_{12}M_o = \frac{2\beta^2 r^2 M_o}{Eh} \tag{3.95}$$

Rotations due to out-of-plane shears

$$\frac{dw}{dy}_{y=0} = D_{21}Q_o = \frac{2\beta^2 r^2 Q_o}{Eh} \tag{3.97}$$

Rotations due to bending moments

$$\frac{dw}{dy}_{y=0} = D_{22}M_o = \frac{4\beta^3 r^2 M_o}{Eh} \tag{3.98}$$

where

$$\beta^4 = \frac{Eh}{4r^2 D} \tag{3.77}$$

The above equations are applied to circular domes, albeit a few modifications to reflect the changes in geometry (Fig. 4.35).

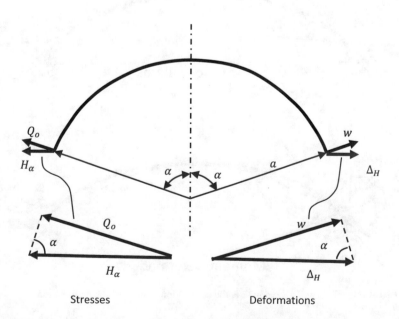

Stresses   Deformations

Fig. 4.35 Relationships between shears and deformations at the base

$M_0 = M_\alpha$ (bending moment at the base)
$Q_o = H_\alpha \sin \alpha$
   (out-of-plane shear at the base)
$\Delta_H = w \sin \alpha$
   (horizontal deformation)
$r = a$
$\phi = \alpha$ (half angle of the dome, from the vertical to the base)

The horizontal deformations and slope at the base of the shell are therefore defined in terms of the dome geometry. The sign convention is taken from Fig. 4.36.

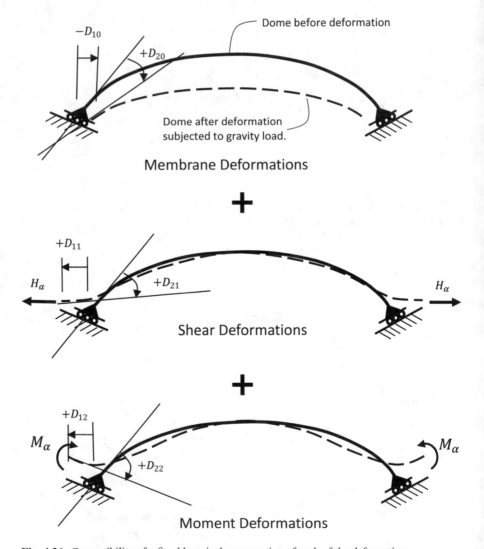

Fig. 4.36 Compatibility of a fixed base is the summation of each of the deformations

$$\Delta_H = D_{11}H_\alpha = \frac{2\beta a^2 H_\alpha \sin^2 \alpha}{Eh} \tag{4.218}$$

$$\Delta_H = D_{12}M_\alpha = \frac{2\beta^2 a^2 M_\alpha \sin \alpha}{Eh} \tag{4.219}$$

$$\Delta_\alpha = D_{21}H_\alpha = \frac{2\beta^2 a^2 H_\alpha \sin \alpha}{Eh} \tag{4.220}$$

$$\Delta_\alpha = D_{22}M_\alpha = \frac{4\beta^3 a^2 M_\alpha}{Eh} \tag{4.221}$$

Defining the deformation symbols,

First subscript	Due to	Second subscript
1 – Horizontal translation		0 – Applied loads
2 – Rotations		1 – Applied shear
		2 – Applied moment

Thus, $D_{10} = $ a horizontal translation due to applied loads.

4.6 Compatibility Equations

A solution is found by forming compatibility equations. This is done by summing the displacements and rotations at the base of the shell. Fundamentally, compatibility at the base requires that the deformations match the boundary conditions. If on a sliding foundation, the shell will deform freely and the magnitude of the deformations will be those determined by the membrane solution. If the base is fixed, a shear and a bending moment are required to correct the deformations (see Fig. 4.36). Furthermore, if the base is pinned, only a shear force is required to correct the membrane deformations. The compatibility concept is similar to the solution of cylindrical shells. However, the shapes of the dome deformations are based on the internal angle ϕ (see Fig. 4.37).

Fixed Base

The compatibility equations are therefore the sum of the horizontal deformations and slope, which are caused by a shear and moment. See Fig. 4.36 for sign of each term.

$$\sum \Delta_H = -D_{10} + D_{11}H_\alpha - D_{12}M_\alpha = 0 \tag{4.222}$$

$$\sum \Delta_\phi = +D_{20} + D_{21}H_\alpha - D_{22}M_\alpha = 0 \tag{4.223}$$

The two compatibility equations have two unknowns (Q_α and M_α), and therefore a solution is possible. Setting the compatibility equations in matrix form,

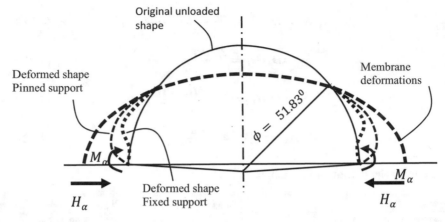

Deformations for a high dome

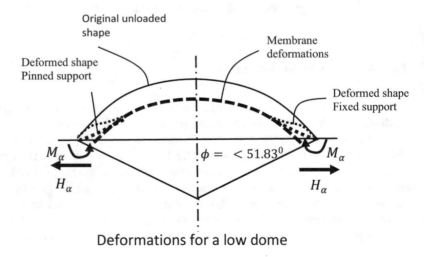

Deformations for a low dome

Fig. 4.37 Types of dome deformations, depending on angle ϕ

$$\begin{bmatrix} D_{11} & D_{12} \\ D_{21} & D_{22} \end{bmatrix} \begin{Bmatrix} H_\alpha \\ -M_\alpha \end{Bmatrix} = \begin{Bmatrix} D_{10} \\ -D_{20} \end{Bmatrix} \tag{4.224}$$

The shear and moment is solved by inverting the 2×2 matrix.

$$\begin{Bmatrix} H_\alpha \\ -M_\alpha \end{Bmatrix} = \begin{bmatrix} D_{11} & D_{12} \\ D_{21} & D_{22} \end{bmatrix}^{-1} \begin{Bmatrix} D_{10} \\ -D_{20} \end{Bmatrix} \tag{4.225}$$

$$\left\{ \begin{array}{c} H_\alpha \\ -M_\alpha \end{array} \right\} = \frac{1}{(D_{11})D_{22} - (D_{12})D_{21}} \left[\begin{array}{cc} D_{22} & -D_{12} \\ -D_{21} & D_{11} \end{array} \right] \left\{ \begin{array}{c} D_{10} \\ -D_{20} \end{array} \right\} \qquad (4.226)$$

Pinned Base

For the case of a pinned base, the concept is the same and Fig. 4.36 applies, except that the bending moment is equal to zero. Only one compatibility equation is applicable.

$$-D_{10} + D_{11}H_\alpha = 0 \qquad (4.227)$$

The unknown shear is solved directly.

$$H_\alpha = \frac{D_{10}}{D_{11}} \qquad (4.228)$$

$$M_\alpha = 0$$

Sliding Base

For the case of a sliding bearing, $M_\alpha = H_\alpha = 0$. The only equations that are applicable are the membrane equations.

Deformation and Stress Equations

The deformations along the length of the shell are from cylindrical shell theory.

$$w = e^{-\beta y}(C_3 \cos \beta y + C_4 \sin \beta y) \qquad (3.80)$$

$$\frac{\partial w}{\partial y} = \beta e^{-\beta y}[-C_3(\sin \beta y + \cos \beta y) + C_4(\cos \beta y - \sin \beta y)] \qquad (3.81)$$

$$\frac{\partial^2 w}{\partial y^2} = 2\beta^2 e^{-\beta y}[C_3 \sin \beta y - C_4 \cos \beta y] \qquad (3.82)$$

$$\frac{\partial^3 w}{\partial y^3} = 2\beta^3 e^{-\beta y}[C_3(\cos \beta y - \sin \beta y) + C_4(\sin \beta y + \cos \beta y)] \qquad (3.83)$$

where

$$C_3 = \frac{-1}{2\beta^3 D}(Q_o + \beta M_o) \qquad (3.90)$$

$$C_4 = \frac{M_o}{2D\beta^2} \qquad (3.86)$$

Referring to Figs. 4.35 and 4.38, the cylindrical shell equations are modified to fit the case of a circular dome.

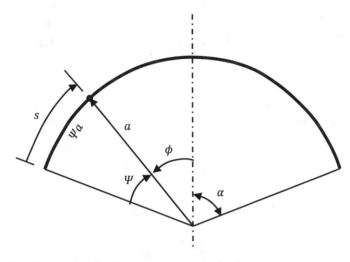

Fig. 4.38 Variables use to define the geometry of a circular dome

$$Q_o = H_\alpha \sin \alpha$$

$$M_o = M_\alpha$$

$$y = \Psi a = s$$

Thus, for the case of a circular dome, the deformation equations are defined using the above substitutions.

$$w = e^{-\beta\psi a}(C_3 \cos \beta\psi a + C_4 \sin \beta\psi a) \tag{4.229}$$

$$\frac{\partial w}{\partial s} = \beta e^{-\beta\psi a}[-C_3(\sin \beta\psi a + \cos \beta\psi a) + C_4(\cos \beta\psi a - \sin \beta\psi a)] \tag{4.230}$$

$$\frac{\partial^2 w}{\partial s^2} = 2\beta^2 e^{-\beta\psi a}[C_3 \sin \beta\psi a - C_4 \cos \beta\psi a] \tag{4.231}$$

$$\frac{\partial^3 w}{\partial s^3} = 2\beta^3 e^{-\beta\psi a}[C_3(\cos \beta\psi a - \sin \beta\psi a) + C_4(\sin \beta\psi a + \cos \beta\psi a)] \tag{4.232}$$

The solution for C_3 and C_4:

$$C_3 = -\frac{H_\alpha \sin \alpha}{2\beta^3 D} - \frac{M_\alpha}{2\beta^2 D} \tag{4.233}$$

$$C_4 = \frac{M_\alpha}{2D\beta^2} \tag{4.234}$$

Also adapted are the cylindrical shell equations for the moments, shears, and membrane forces (Eqs. 3.64, 3.68, 3.71, 3.73). The equations are altered to suit the analysis of circular domes.

$$N_\theta = -\frac{Ehw}{a} \tag{4.235}$$

$$M_\theta = \nu M_\phi \tag{4.236}$$

$$M_\phi = -D\frac{\partial^2 w}{ds^2} \tag{4.237}$$

$$Q_\phi = -D\frac{\partial^3 w}{ds^3} \tag{4.238}$$

The Solution of the N_ϕ Term

The only stress equation not listed is the N_ϕ term, which cannot be adapted from cylindrical shell theory. In cylindrical shell theory, the walls are vertical and straight. In domes, the walls are sloped. Thus, the N_ϕ term from cylindrical shell theory cannot be applied to dome theory.

From Eqs. 4.192 to 4.194 (equilibrium equations),

$$\frac{\partial(N_\phi r_o)}{\partial\phi} - Q_\phi r_o - N_\theta r_1 \cos\phi + P_\phi r_1 r_o = 0 \tag{4.192}$$

$$\frac{\partial(Q_\phi r_o)}{\partial\phi} - N_\phi\, r_o + N_\theta r_1 \sin\phi + P_z r_1 r_o = 0 \tag{4.193}$$

$$-\frac{\partial(M_\phi r_o)}{\partial\phi} + Q_\phi r_o r_1 + M_\theta r_1 \cos\phi = 0 \tag{4.194}$$

Substituting the following parameters,

$$P_\phi = 0$$

$$r_1 = r_2 = a$$

$$r_o = a\sin\phi$$

Simplifying by these substitutions and dividing by a.

$$\frac{\partial(N_\phi \sin\phi)}{\partial\phi} - Q_\phi \sin\phi - N_\theta \cos\phi = 0 \tag{4.239}$$

$$\frac{\partial(Q_\phi \sin\phi)}{\partial\phi} - N_\phi \sin\phi + N_\theta \sin\phi + P_z a \sin\phi = 0 \tag{4.240}$$

$$\frac{\partial \left(M_\phi \sin \phi \right)}{\partial \phi} - Q_\phi a \, \sin \phi - M_\theta \cos \phi = 0 \tag{4.241}$$

Furthermore, set $r_1 = a, \delta = \left(\frac{v}{r_1} + \frac{\partial w}{r_1 \partial \phi} \right) = \frac{\left(v + \frac{\partial w}{\partial \phi} \right)}{a}$ and replace $r_o = a \sin \phi$ of Eqs. 4.216 and 4.217,

$$M_\theta = -D \left[(\delta) \frac{\cos \phi}{a \sin \phi} + v \frac{\partial}{a \partial \phi} (\delta) \right]$$

$$M_\phi = -D \left[\frac{\partial}{a \partial \phi} (\delta) + v(\delta) \frac{\cos \phi}{a \sin \phi} \right]$$

or

$$M_\theta = -\frac{D}{a} \left[\delta \cot \phi + v \frac{\partial \delta}{\partial \phi} \right] \tag{4.242}$$

$$M_\phi = -\frac{D}{a} \left[\frac{\partial \delta}{\partial \phi} + v \delta \cot \phi \right] \tag{4.243}$$

Substitute M_θ and M_ϕ into the equilibrium Eq. 4.241.

$$Q_\phi a \, \sin \phi = -\frac{D}{a} \frac{\partial \left[\left(\frac{\partial \delta}{\partial \phi} + v \delta \cot \phi \right) \sin \phi \right]}{\partial \phi} + \frac{D}{a}$$

$$\times \left(\delta \cot \phi + v \frac{\partial \delta}{\partial \phi} \right) \cos \phi \tag{4.244}$$

Rearrange and divide by $D/_r$.

$$-\frac{Q_\phi \, a^2 \sin \phi}{D} = \frac{\partial \left[\left(\frac{\partial \delta}{\partial \phi} + v \delta \cot \phi \right) \sin \phi \right]}{\partial \phi} - \left(\delta \cot \phi + v \frac{\partial \delta}{\partial \phi} \right) \cos \phi \tag{4.245}$$

Taking the derivative and expanding the equation,

$$\frac{d^2 \delta}{d \phi^2} \sin \phi + v \frac{d \delta}{d \phi} \cos \phi - \delta \cot \phi \cos \phi - v \frac{d \delta}{d \phi} \cos \phi = -\frac{Q_\phi \, a^2 \sin \phi}{D} \tag{4.246}$$

Divide by $\sin \phi$ and eliminate the second and seventh terms.

$$\frac{d^2\delta}{d\phi^2} + -\nu\,\delta\csc^2\phi + \frac{d\delta}{d\phi}\cot\phi + \nu\delta\cot^2\phi - \delta\cot^2\phi = -\frac{Q_\phi\,a^2}{D} \qquad (4.247)$$

Using Eqs. 4.239 and 4.240, isolate the N_θ term.

$$N_\theta\cos\phi = \frac{\partial(N_\phi\sin\phi)}{\partial\phi} - Q_\phi\sin\phi \qquad (4.248)$$

$$N_\theta\sin\phi = -\frac{\partial(Q_\phi\sin\phi)}{\partial\phi} + N_\phi\sin\phi - P_z a\sin\phi = 0 \qquad (4.249)$$

Divide 4.248 by $\cos\phi$ and 4.249 by $\sin\phi$.

$$N_\theta = \frac{\partial(N_\phi\sin\phi)}{\partial\phi}\sec\phi - Q_\phi\tan\phi \qquad (4.250)$$

$$N_\theta = -\frac{\partial(Q_\phi\sin\phi)}{\partial\phi}\csc\phi + N_\phi - P_z a \qquad (4.251)$$

Set the two equations equal to each other to eliminate N_θ.

$$\frac{\partial(N_\phi\sin\phi)}{\partial\phi}\sec\phi - Q_\phi\tan\phi = -\frac{\partial(Q_\phi\sin\phi)}{\partial\phi}\csc\phi + N_\phi - P_z a$$

Rearrange and multiply each side by $\sin\phi\cos\phi$.

$$\frac{\partial(N_\phi\sin\phi)}{\partial\phi}\sin\phi - Q_\phi\sin^2\phi + \frac{\partial(Q_\phi\sin\phi)}{\partial\phi}\cos\phi - N_\phi\sin\phi\cos\phi$$
$$+ P_z a\,\sin\phi\cos\phi$$
$$= 0 \qquad (4.252)$$

Using the product rule for differentiation to combine the terms of Eq. 4.252,

$$(fg)' = f'g + fg'$$

Equation 4.252 is expressed in the following form:

$$\frac{d[(N_\phi\sin\phi)\sin\phi]}{d\phi} + \frac{d[(Q_\phi\sin\phi)\cos\phi]}{d\phi} + \frac{d\left[P_z a\left(\frac{\sin^2\phi}{2}\right)\right]}{d\phi} = 0 \qquad (4.253)$$

Integrating the above equation,

$$N_\phi \sin^2\phi + Q_\phi \sin\phi \cos\phi + \frac{P_z a \sin^2\phi}{2} = 0 \tag{4.254}$$

Divide by $\sin^2\phi$ and rearrange.

$$N_\phi = -Q_\phi \cot\phi - \frac{P_z a}{2} \tag{4.255}$$

The term $\frac{P_z a}{2}$ is the membrane solution for external pressure, which is considered separately. The above equation therefore reduces to the following form:

$$N_\phi = -Q_\phi \cot\phi \tag{4.256}$$

Since Q_ϕ was previously determined,

$$Q_\phi = -D\frac{\partial^3 w}{ds^3} \tag{4.238}$$

where

$$\frac{\partial^3 w}{\partial s^3} = 2\beta^3 e^{-\beta\psi a}[C_3(\cos\beta\psi a - \sin\beta\psi a) + C_4(\sin\beta\psi a + \cos\beta\psi a)] \tag{4.232}$$

and

$$C_3 = -\frac{H_a \sin\alpha}{2\beta^3 D} - \frac{M_a}{2\beta^2 D} \tag{4.233}$$

$$C_4 = \frac{M_a}{2D\beta^2} \tag{4.234}$$

Incorporating 4.232, 4.233, and 4.234 into 4.256,

$$N_\phi = D\left\{2\beta^3 e^{-\beta\psi a}\left[\left(-\frac{H_a \sin\alpha}{2\beta^3 D} - \frac{M_a}{2\beta^2 D}\right)(\cos\beta\psi a - \sin\beta\psi a)? + \left(\frac{M_a}{2D\beta^2}\right)(\sin\beta\psi a + \cos\beta\psi a)\right]\right\}\cot\phi \tag{4.257}$$

Removing $2D$ and simplifying,

$$N_\phi = \beta^3 e^{-\beta\psi a}\left[\left(-\frac{H_a \sin\alpha}{\beta^3} - \frac{M_a}{\beta^2}\right)(\cos\beta\psi a - \sin\beta\psi a)? + \left(\frac{M_a}{\beta^2}\right)(\sin\beta\psi a + \cos\beta\psi a)\right]\cot\phi \tag{4.258}$$

$$N_\phi = \beta^3 e^{-\beta\psi a}\left[\left(-\frac{H_\alpha \sin\alpha}{\beta^3}\right)(\cos\beta\psi a - \sin\beta\psi a) + \left(\frac{M_\alpha}{\beta^2}\right)(-\cos\beta\psi a + \sin\beta\psi a)\right.$$

$$\left. \times +\left(\frac{M_\alpha}{\beta^2}\right)(\sin\beta\psi a + \cos\beta\psi a)\right]\cot\phi$$

$$(4.259)$$

$$N_\phi = \beta^3 e^{-\beta\psi a}\left[\left(-\frac{H_\alpha \sin\alpha}{\beta^3}\right)(\cos\beta\psi a - \sin\beta\psi a)\right.$$

$$\left. \times +\left(\frac{M_\alpha}{\beta^2}\right)(-\cos\beta\psi a + \sin\beta\psi a + \sin\beta\psi a + \cos\beta\psi a)\right]\cot\phi \quad (4.260)$$

Finally, the stress N_ϕ is determined.

$$N_\phi = e^{-\beta\psi a}[(-H_\alpha \sin\alpha)(\cos\beta\psi a - \sin\beta\psi a) + (M_\alpha\beta)(2\sin\beta\psi a)]\cot\phi$$

$$(4.261)$$

4.7 Steps in Solving the Deformations and Stresses in the Shell

The solution of the circular dome is similar to the cylindrical shell—the order of the solution follows an identical pattern. Similar to Chap. 3, the equations are reproduced for convenience, and the steps are given to allow the reader to follow a "cook book" approach to analysis of domes.

1. **Solve for base reactions H_α and M_α.**

Initial parameters		
	Diameter (L)	
	Height (H)	
	Thickness (h)	
	Young's modulus (E)	
	Poisson's ratio (ν)	
	$a = \frac{(H)^2 + \left(\frac{L}{2}\right)^2}{2(H)}$ (radius)	4.262
	$\alpha = 180 - 2\,\tan^{-1}\left(\frac{L}{2H}\right)$ (half angle)	4.263
	$I = \frac{h^3}{12}$ (moment of inertia per unit width of shell)	4.264
	$D = \frac{EI}{(1-\nu^2)}$	3.63
	$\beta^4 = \frac{Eh}{4a^2D} = \frac{3(1-\nu^2)}{a^2h^2}$	3.77
	$\beta = \sqrt[4]{\beta^4}$	
	$\beta^2 = \left(\sqrt[4]{\beta^4}\right)^2$	
	$\beta^3 = \left(\sqrt[4]{\beta^4}\right)^3$	

(continued)

Membrane deformations at the base ($\phi = \alpha$)		
Uniform distributed load (gravity)	$D_{10} = \frac{a^2 q}{Eh} \left(\frac{1+v}{1+\cos \alpha} - \cos \alpha \right) \sin \alpha$	4.138
	$D_{20} = \frac{aq}{Eh} \left(2 + v \right) \sin \alpha$	4.145
or		
Uniform horizontal load	$D_{10} = \frac{a^2 P}{2Eh} \sin \alpha \; \left(- \cos 2\alpha + v \right)$	4.147
	$D_{20} = - \frac{ap}{2Eh} \left(3 + v \right) \sin 2\alpha$	4.157
or		
Uniform pressure	$D_{10} = \frac{a^2 P_z}{2Eh} \sin \alpha \; \left(-1 + v \right)$	4.159
	$D_{20} = 0$	4.162
Deformations at the base from boundary effects		
	$D_{11} H_\alpha = \frac{2\beta a^2 \sin^2 \alpha}{Eh} H_\alpha$	4.218
	$D_{12} M_\alpha = \frac{2\beta^2 a^2 \sin \alpha}{Eh} M_\alpha$	4.219
	$D_{21} H_\alpha = \frac{2\beta^2 a^2 \sin \alpha}{Eh} H_\alpha$	4.220
	$D_{22} M_\alpha = \frac{4\beta^3 a^2}{Eh} M_\alpha$	4.221
Compatibility equations		
Fixed base	$\begin{Bmatrix} H_\alpha \\ M_\alpha \end{Bmatrix} = \frac{1}{(D_{11})D_{22} - (D_{12})D_{21}} \begin{bmatrix} D_{22} & -D_{12} \\ -D_{21} & D_{11} \end{bmatrix} \begin{Bmatrix} -D_{10} \\ -D_{20} \end{Bmatrix}$ *Signs changed to ease programing of equations	4.226*
or		
Pinned base	$H_\alpha = \frac{D_{10}}{D_{11}}$ $M_\alpha = 0$	4.228
or		
Sliding base	$H_\alpha = 0$ $M_\alpha = 0$	

2. Solve for the constants C_3 and C_4.

$C_3 = - \frac{H_\alpha \sin \alpha}{2\beta^3 D} - \frac{M_\alpha}{2\beta^2 D}$	4.233
$C_4 = \frac{M_\alpha}{2D\beta^2}$	4.234

3. Solve for the membrane deformations in the shell at increments of ϕ.

Uniform distributed load (gravity)	$\Delta'_H = \frac{a^2 q}{Eh} \left(\frac{1+v}{1+\cos \phi} - \cos \phi \right) \sin \phi$	4.138
	$\Delta'_\phi = \frac{aq}{Eh} \left(2 + v \right) \sin \phi$	4.145
or		
Uniform horizontal load	$\Delta'_H = \frac{a^2 P}{2Eh} \sin \phi \; \left(- \cos 2\phi + v \right)$	4.147
	$\Delta'_\phi = - \frac{ap}{2Eh} \left(3 + v \right) \sin 2\phi$	4.157
or		
Uniform pressure	$\Delta'_H = \frac{a^2 P_z}{2Eh} \sin \phi \; \left(-1 + v \right)$	4.159
	$\Delta'_\phi = 0$	4.162

4. Solve for the deformations in the shell due to boundary effects at increments of ϕ.

$w = -e^{-\beta\psi a}(C_3 \cos \beta\psi a + C_4 \sin \beta\psi a)$	4.229*
$\Delta_\phi = \frac{\partial w}{\partial s} = \beta e^{-\beta\psi a}[-C_3(\sin \beta\psi a + \cos \beta\psi a) + C_4(\cos \beta\psi a - \sin \beta\psi a)]$	4.230
$\frac{\partial^2 w}{\partial s^2} = -2\beta^2 e^{-\beta\psi a}[C_3 \sin \beta\psi a - C_4 \cos \beta\psi a]$	4.231*
$\frac{\partial^3 w}{\partial s^3} = 2\beta^3 e^{-\beta\psi a}[C_3(\cos \beta\psi a - \sin \beta\psi a) + C_4(\sin \beta\psi a + \cos \beta\psi a)]$	4.232
$\Delta_H = w \sin \phi$	4.265
*Signs changed to ease programing of equations	

5. Solve for the total deformations.

$\Delta(total) = \Delta'_H + \Delta_H$	4.266
$\Delta_\phi(total) = \Delta'_\phi + \Delta_\phi$	4.267

6. Solve for membrane and boundary effects stresses in increments of ϕ.

Membrane stresses		
Uniform distributed load (gravity)	$N'_\phi = -\frac{qa}{(1+\cos\phi)}$	4.50
	$N'_\theta = aq\left(\frac{1}{1+\cos\phi} - \cos\phi\right)$	4.53
or		
Uniform horizontal load	$N'_\phi = -\frac{Pa}{2}$	4.62
	$N'_\theta = \frac{-Pa}{2} \cos 2\phi$	4.68
or		
Uniform pressure	$N'_\phi = \frac{-Pa}{2}$	4.74
	$N'_\theta = -\frac{Pa}{2}$	4.76
Stresses from boundary effects		
	$N_\theta = \frac{Ehw}{a}$	4.235*
	$M_\phi = D\frac{\partial^2 w}{ds^2}$	4.237*
	$M_\theta = \nu M_\phi$	4.236
	$Q_\phi = D\frac{\partial^3 w}{ds^3}$	4.238*
	$N_\phi = e^{-\beta\psi a}[(-H_\alpha \sin \alpha)(\cos\beta\psi a - \sin \beta\psi a) + (M_\alpha\beta)(2 \sin \beta\psi a)] \cot \phi$	4.261
	*Signs changed to ease programing of equations	
Total stresses		
	$N_\theta(total) = N'_\theta + N_\theta$	4.267
	$N_\phi(total) = N'_\phi + N_\phi$	4.268

4.8 Worked Example

A 100 mm-thick reinforced concrete dome is 2500 mm high and a span of 6000 mm. The basic geometrical configuration is given in Fig. 4.39. The basic parameters that will be used in the design are given below:

Thickness = 100 mm
Plaster thickness (inside and out) = 20 mm
Concrete strength = 30 MPa
$E = 28,000$ MPa (Young's modulus)
$\nu = 0.15$ (Poisson's ratio)
Foundation size = 1000 mm × 300 mm

Assume a nominal live load for maintenance = 0.5 kN/m^2

1. **Solve for base reactions H and M_α.**

Geometry of the dome and parameters

$$a = \frac{(H)^2 + \left(\frac{L}{2}\right)^2}{2(H)} = \frac{(2500)^2 + \left(\frac{6000}{2}\right)^2}{2(2500)} = 3050 \text{ mm} \quad (4.262)$$

$$\alpha = 180 - 2 \tan^{-1}\left(\frac{L}{2H}\right) = 180 - 2 \tan^{-1}\left[\frac{6000}{2(2500)}\right]$$

$$= 79.61° \text{ or } 1.389 \text{ rad.} \quad (4.263)$$

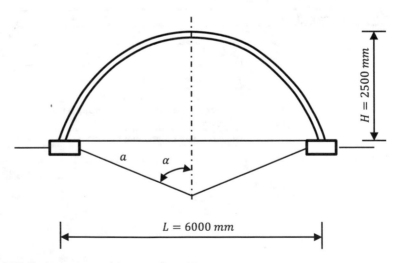

Fig. 4.39 Basic geometry of the example problem

$$I = \frac{h^3}{12} = \frac{0.1^3}{12}$$

$$= 8.33 \times 10^{-5} \text{ m}^4 \text{ (moment of inertia per meter width of shell)} \qquad (4.264)$$

$$D = \frac{(28 \times 10^6) 8.33 \times 10^{-5}}{(1 - 0.15^2)} = 2387.04 \qquad (3.63)$$

$$\beta^4 = \frac{Eh}{4r^2 D} = \frac{28 \times 10^6 (0.1)}{4 (3.05)^2 (2387.04)} = 31.52 \qquad (3.77)$$

$$\beta = \sqrt[4]{\beta^4} = \sqrt[4]{31.52} = 2.37$$

$$\beta^2 = 2.37^2 = 5.61$$

$$\beta^3 = 5.61^3 = 13.30$$

Loading

Using a dead load factor of 1.2 and a live load factor of 1.6 for limit state design,

$$q = 1.2[24(0.1) + 21(0.04)] + 1.6(0.5) = 4.69 \text{ kN/m}^2$$

The unit weight of concrete is assumed to be 24 kN/m^3 and plaster 21 kN/m^3.

Membrane deformations at base

$$D_{10} = \frac{a^2 q}{Eh} \left(\frac{1+v}{1+\cos \alpha} - \cos \alpha \right) \sin \alpha$$

$$= \frac{3.05^2 (4.69)}{28 \times 10^6 (0.1)} \left(\frac{1 + 0.15}{1 + \cos 1.389} - \cos 1.389 \right) \sin 1.389$$

$$= 1.22 \times 10^{-5} \qquad (4.138)$$

$$D_{20} = \frac{aq}{Eh} (2+v) \sin \alpha = \frac{3.05 \, (4.69)}{28 \times 10^6 (0.1)} (2 + 0.15) \sin 1.389$$

$$= 1.080 \times 10^{-5} \qquad (4.145)$$

Deformations at the base due to boundary effects

$$D_{11} H_\alpha = \frac{2\beta a^2 \sin^2 \alpha}{Eh} H_\alpha = \frac{2 \, (2.37) 3.05^2 (\sin 1.389)^2}{28 \times 10^6 \, (0.1)} H_\alpha$$

$$= 1.52 \times 10^{-5} H_\alpha \qquad (4.218)$$

$$D_{12}M_\alpha = \frac{2\beta^2 a^2 \sin\alpha}{Eh} M_\alpha = \frac{2(2.37)^2 (3.05)^2 \sin 1.389}{28 \times 10^6 (0.1)} M_\alpha$$

$$= 3.67 \times 10^{-5} M_\alpha \tag{4.219}$$

$$D_{21}H_\alpha = \frac{2\beta^2 a^2 \sin\alpha}{Eh} H_\alpha = 3.67 \times 10^{-5} H_\alpha \tag{4.220}$$

$$D_{22}M_\alpha = \frac{4\beta^3 a^2}{Eh} M_\alpha = \frac{4(2.37)^3 (3.05)^2}{28 \times 10^6 (0.1)} M_\alpha = 17.69 \times 10^{-5} M_\alpha \tag{4.221}$$

Compatibility equations

$$\left\{ \begin{matrix} H_\alpha \\ M_\alpha \end{matrix} \right\} = \frac{1}{(D_{11})D_{22} - (D_{12})D_{21}} \begin{bmatrix} D_{22} & -D_{12} \\ -D_{21} & D_{11} \end{bmatrix} \left\{ \begin{matrix} -D_{10} \\ -D_{20} \end{matrix} \right\} \tag{4.226}$$

$$\left\{ \begin{matrix} H_\alpha \\ M_\alpha \end{matrix} \right\} = \frac{1}{1.52(17.69) - 3.67(3.67)} \begin{bmatrix} 17.69 & -3.67 \\ -3.67 & 1.52 \end{bmatrix} \left\{ \begin{matrix} -1.22 \\ -1.08 \end{matrix} \right\} \times 10^{-5}$$

$$\left\{ \begin{matrix} H_\alpha \\ M_\alpha \end{matrix} \right\} = \begin{bmatrix} 1.31 & -0.27 \\ -0.27 & 0.11 \end{bmatrix} \left\{ \begin{matrix} -1.22 \\ -1.08 \end{matrix} \right\} \times 10^{-5} = \left\{ \begin{matrix} -1.30 \\ 0.21 \end{matrix} \right\}$$

Therefore,

$$H_\alpha = -1.30 \text{ kN/m}$$

$$M_\alpha = 0.21 \text{ kN.m/m}$$

The negative shear indicates that the direction of H_α is opposite to the assumed direction (see Figs. 4.36 and 4.37).

2. **Solve for the constants C_3 and C_4.**

$$C_3 = -\frac{H_\alpha \sin\alpha}{2\beta^3 D} - \frac{M_\alpha}{2\beta^2 D} = -\frac{-1.30 \sin 1.389}{2(2.37)^3 \, 2387.04} - \frac{0.21}{2(2.37)^2 \, 2387.04}$$

$$= 1.24 \times 10^{-5} \tag{4.233}$$

$$C_4 = \frac{M_\alpha}{2D\beta^2} = \frac{0.21}{2(2387.04)(2.37)^2} = 7.81 \times 10^{-6} \tag{4.234}$$

3. **Solve for the membrane deformations in the shell in increments of ϕ.**

Uniform distributed load (gravity load)

$$\Delta'_H = \frac{a^2 q}{Eh}\left(\frac{1+v}{1+\cos\phi} - \cos\phi\right)\sin\phi$$

$$= \frac{3.05^2(4.69)}{28\times10^6(0.1)}\left(\frac{1+0.15}{1+\cos\phi} - \cos\phi\right)\sin\phi \qquad (4.138)$$

$$\Delta'_\phi = \frac{aq}{Eh}(2+v)\sin\phi = \frac{3.05(4.69)}{28\times10^6(0.1)}(2+0.15)\sin\phi \qquad (4.145)$$

The above Eqs. (4.133 and 4.140) are in terms of the angle ϕ, since this equation is solved in incremental angles from the base to the apex of the shell. Subsequent equations will likewise be expressed in term of the angle ϕ.

4. **Solve for the deformations in the shell due to boundary effects at increments of ϕ.**

$$w = -e^{-\beta\psi a}(C_3\cos\beta\psi a + C_4\sin\beta\psi a)$$

$$= -e^{-2.37\Psi 3.05}\left[1.24(10)^{-5}\cos 2.37\Psi 3.05 + 7.81(10)^{-6}\sin 2.37\Psi 3.05\right]$$

$$(4.229)$$

where

$$\Delta_H = w\sin\phi$$

$$\Delta_\phi = \frac{\partial w}{\partial s} = \beta e^{-\beta\psi a}[-C_3(\sin\beta\psi a + \cos\beta\psi a) + C_4(\cos\beta\psi a - \sin\beta\psi a)] =$$

$$2.37e^{-2.37\Psi 3.05}[-1.24(10)^{-5}(\sin 2.37\Psi 3.05 + \cos 2.37\Psi 3.05)+$$

$$7.81\times10^{-6}(\cos 2.37\Psi 3.05 - \sin 2.37\Psi 3.05)]$$

$$(4.230)$$

$$\frac{\partial^2 w}{\partial s^2} = -2\beta^2 e^{-\beta\psi a}[C_3\sin\beta\psi a - C_4\cos\beta\psi a] =$$

$$-2(2.37)^2 e^{-2.37\Psi 3.05}\left[1.24(10)^{-5}\sin 2.37\Psi 3.05 + 7.81(10)^{-6}\cos 2.37\Psi 3.05\right]$$

$$(4.231)$$

$$\frac{\partial^3 w}{\partial s^3} = 2\beta^3 e^{-\beta\psi a}[C_3(\cos\beta\psi a - \sin\beta\psi a) + C_4(\sin\beta\psi a + \cos\beta\psi a)] =$$
$$2(2.37)^3 e^{-2.37\Psi 3.05}[1.24(10)^{-5}(\cos 2.37\Psi 3.05 - \sin 2.37\Psi 3.05) -$$
$$7.81(10)^{-6}(\sin 2.37\Psi 3.05 + \cos 2.37\Psi 3.05)] \tag{4.232}$$

5. Solve for the total deformations at increments of ϕ.

$$\Delta(total) = \Delta'_H + \Delta_H \tag{4.265}$$

$$\Delta_\phi(total) = \Delta'_\phi + \Delta_\phi \tag{4.266}$$

6. Solve for membrane stresses, moments, and shear at increments of ϕ.

Uniform distributed load (gravity load)

$$N'_\phi = -\frac{qa}{(1+\cos\phi)} = -\frac{4.69(3.05)}{(1+\cos\phi)} \tag{4.50}$$

$$N'_\theta = aq\left(\frac{1}{1+\cos\phi} - \cos\phi\right) = 4.69(3.05)\left(\frac{1}{1+\cos\phi} - \cos\phi\right) \tag{4.53}$$

Stresses from boundary effects

$$N_\theta = \frac{Ehw}{a} = \frac{28(10)^6(0.1)}{3.05}$$
$$\left\{-e^{-2.37\Psi 3.05}\left[1.24(10)^{-5}\cos 2.37\Psi 3.05 + 7.81(10)^{-6}\sin 2.37\Psi 3.05\right]\right\} \tag{4.235}$$

$$M_\phi = D\frac{\partial^2 w}{ds^2} = 2387.04$$
$$\left\{-2(2.37)^2 e^{-2.37\Psi 3.05}\left[1.24(10)^{-5}\sin 2.37\Psi 3.05 - 7.81(10)^{-6}\cos 2.37\Psi 3.05\right]\right\} \tag{4.237}$$

$$M_\theta = \nu M_\phi = 0.15 M_\phi \tag{4.236}$$

$$Q_\phi = D\frac{\partial^3 w}{ds^3} =$$
$$2387.04\{-2(2.37)^3 e^{-2.37\Psi 3.05}[1.24(10)^{-5}(\cos 2.37\Psi 3.05 - \sin 2.37\Psi 3.05) +$$
$$7.81 \times 10^{-6}(\sin 2.37\Psi 3.05 + \cos 2.37\Psi 3.05)]\} \tag{4.238}$$

$$N_\phi = e^{-\beta\psi a}[(-H_\alpha \sin\phi)(\cos\beta\psi a - \sin\beta\psi a)+$$
$$(M_\alpha\beta)(2 \sin\beta\psi a)]\cot\phi = e^{-2.37\Psi 3.05}[(-1.30 \sin\phi)(\cos 2.37\Psi 3.05 - \sin 2.37\Psi 3.05)$$
$$+0.21(2.37)(2 \sin 2.37\Psi 3.05)]\cot\phi$$

$$(4.261)$$

Total stress

$$N_\theta(total) = N'_\theta + N_\theta \tag{4.267}$$

$$N_\phi(total) = N'_\phi + N_\phi \tag{4.268}$$

The results of the analysis are given in Table 4.1.

Because the shell is fixed, the horizontal deflections and the slope must be equal to zero at the base. The symmetry of loads requires that the horizontal deflections at the apex must also be zero. These requirements are seen in Table 1, which is an indicator that calculations were applied correctly. As further indicators, the bending moments and shears must be localized at the base of the shell. In this particular case, the moments and shears influence about a third of the shell. The distance of influence, however, is directly related to the stiffness of the material. If the shell was constructed of a thinner material (e.g., steel), the influence of the boundary effects reduce significantly and may drop to a tenth of the shell height. The hoop stress are also calculated at $\phi = 51.83°$ ($\Psi = 27.78°$), which produced the correct value of zero stress.

Tables 4.1 and 4.2 are expressed graphically in Figs. 4.40, 4.41, 4.42, 4.43, and 4.44.

What should be noted is that the deformations are very small, which is typical of shell structures. The shell is also primarily subjected to membrane stresses, which enables an economy of materials. Near the base, however, bending moments exist and therefore the shell will most likely require a localized thickening and additional reinforcement. Fig. 4.45 is a suggested reinforcement layout for the region influenced by bending and shears. As illustrated, two layers of reinforcement is required to resist bending, and the shell has a localized thickening. In the rest of the shell, only one layer of reinforcement is required since this region is only subjected to membrane stresses. It should also be mentioned that the minimum thickness of a concrete shell should not be less than 75 mm. This will permit a layer of 12 mm mesh and 25 mm of cover on the inside and out.

Table 4.1 Horizontal and slope deformations in the shell

ψ (deg)	ϕ (rad)	Δ'_H (mm)	Δ_H (mm)	$\Delta_H(total)$ (mm)	Δ'_ϕ	Δ_ϕ	$\Delta_\phi(total)$	$\frac{\partial_2 w^2}{\partial^2 y}$	$\frac{\partial_2 w^3}{\partial^3 y}$
0	1.389	1.22E-02	−1.22E-02	0.00E+00	1.08E-05	−1.08E-05	0.00E+00	8.77E-05	5.37E-04
1	1.372	1.17E-02	−1.15E-02	1.98E-04	1.08E-05	−1.48E-05	−3.98E-06	6.13E-05	4.56E-04
2	1.355	1.11E-02	−1.06E-02	5.73E-04	1.07E-05	−1.74E-05	−6.68E-06	3.91E-05	3.81E-04
5	1.302	9.67E-03	−7.49E-03	2.18E-03	1.06E-05	−1.97E-05	−9.06E-06	−5.89E-06	1.93E-04
10	1.215	7.37E-03	−2.98E-03	4.39E-03	1.03E-05	−1.38E-05	−3.54E-06	−2.99E-05	1.36E-05
20	1.040	3.47E-03	3.84E-04	3.85E-03	9.47E-06	−1.52E-06	7.95E-06	−1.22E-05	−4.07E-05
27.78	0.905	1.14E-03	3.39E-04	1.48E-03	8.64E-06	8.14E-07	9.45E-06	−9.85E-07	−1.38E-05
30	0.866	5.92E-04	2.53E-04	8.44E-04	8.37E-06	8.48E-07	9.21E-06	2.96E-07	−8.12E-06
40	0.691	−1.20E-03	1.37E-05	−1.19E-03	7.00E-06	2.68E-07	7.27E-06	1.03E-06	1.87E-06
50	0.517	−1.96E-03	−1.13E-05	−1.97E-03	5.43E-06	−2.18E-08	5.41E-06	1.54E-07	9.74E-07
60	0.342	−1.83E-03	−1.90E-06	−1.83E-03	3.69E-06	−2.53E-08	3.66E-06	−5.61E-08	1.81E-08
70	0.168	−1.06E-03	1.43E-07	−1.06E-03	1.83E-06	−2.62E-09	1.83E-06	−2.20E-08	−7.50E-08
79.61	0.000	0.00E+00	0.00E+00	0.00E+00	0.00E+00	1.59E-09	1.59E-09	3.85E-10	−1.60E-08

Table 4.2 Stresses, moments, and shears in the shell

ψ	N'_ϕ (kN/m)	N_ϕ (kN/m)	N_ϕ(tot) (kN/m)	N'_θ (kN/m)	N_θ (kN/m)	N_θ(tot) (kN/m)	M_ϕ (kN. m/m)	M_θ (kN. m/m)	Q_ϕ (kN/m)
0	−12.12	0.24	−11.88	9.54	−11.36	−1.82	0.21	0.03	1.28
1	−11.95	0.22	−11.73	9.12	−10.73	−1.61	0.15	0.02	1.09
2	−11.78	0.20	−11.58	8.71	−9.94	−1.23	0.09	0.01	0.91
5	−11.30	0.13	−11.18	7.51	−7.13	0.38	−0.01	0.00	0.46
10	−10.61	0.01	−10.60	5.63	−2.91	2.71	−0.07	−0.01	0.03
20	−9.50	−0.06	−9.56	2.26	0.41	2.67	−0.03	0.00	−0.10
27.78	−8.84	−0.03	−8.87	0.00	0.40	0.40	0.00	0.00	−0.03
30	−8.68	−0.02	−8.70	−0.59	0.30	−0.28	0.00	0.00	−0.02
40	−8.08	0.01	−8.07	−2.94	0.02	−2.92	0.00	0.00	0.00
50	−7.65	0.00	−7.65	−4.78	−0.02	−4.81	0.00	0.00	0.00
60	−7.37	0.00	−7.37	−6.11	−0.01	−6.11	0.00	0.00	0.00
70	−7.20	0.00	−7.20	−6.90	0.00	−6.90	0.00	0.00	0.00
79.61	−7.15	0.00	−7.15	−7.15	0.00	−7.15	0.00	0.00	0.00

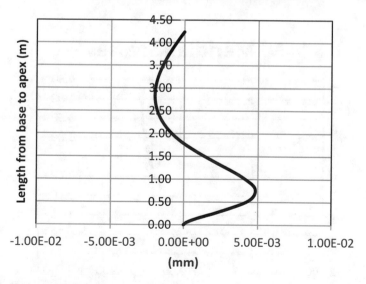

Fig. 4.40 Horizontal deformations in a dome

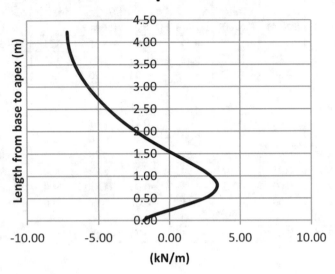

Fig. 4.41 Hoop stresses in the dome

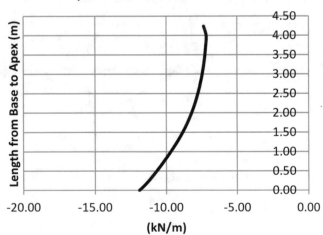

Fig. 4.42 Meridian stresses along the meridian

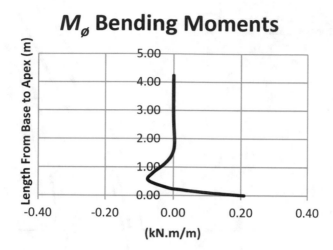

Fig. 4.43 Bending moments along the meridian

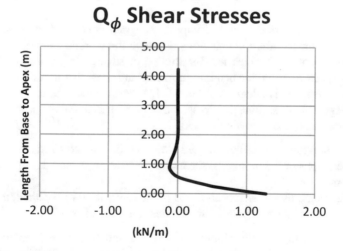

Fig. 4.44 Out-of-plane shears along the meridian

4.9 Exercises

4.9.1 Using a spreadsheet or alternative programing language, program the membrane theory. Using the membrane theory, solve for the membrane stresses (N'_ϕ and N'_Θ) and the deformations (Δ'_H and Δ'_ϕ) for the worked example in Sect. 4.8, in increments of five degrees. Furthermore, solve for the stresses and deformations at one degree increments for the first five degrees.

Fig. 4.45 Suggested layout
of reinforcement at the base
of the shell

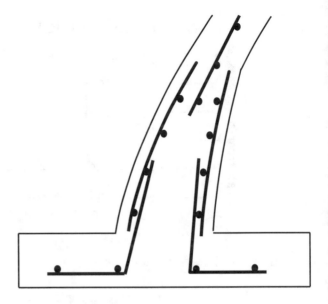

4.9.2 Using the spreadsheet or computer program developed in question 4.9.1, include the boundary solution equations. Resolve the worked example in Sect. 4.8 for a fixed support. Determine the distribution of stresses (N_ϕ and N_Θ), the deformations (Δ_H and Δ_ϕ), and the bending moments and shear distribution (M_ϕ, M_Θ and Q_ϕ), in increments of five degrees. As in 4.9.1, solve at one degree increments for the first five degrees to determine the stress, moment, and shear curves near the support. Furthermore, make the following checks:

(a) Are the horizontal deformations and slopes at the base equal to zero?
(b) Do the graphs of the moments, shears, in-plane stresses, and deformations have the correct form?
(c) Are the moments and shears confined to the lower third of the shell, and are the peak deformations and hoops stresses near the base of the shell?

4.9.3 Considering the distribution of stresses and forces determined in 4.9.2, demonstrate how the shell will potentially crack in hoop and bending stresses, if constructed of reinforced concrete. Use a sketch to illustrate the location and orientation of cracks.

4.9.4 Resolve for the worked example in Sect. 4.8 for the case of a pinned support. Determine the distribution of stresses (N_ϕ and N_Θ), deformations (w and ϕ_y), and the bending moments and shears (M_ϕ, M_Θ and Q_ϕ) in increments of five degrees. As previously, solve for the stresses and deformations at one degree increments for the first five degrees.

4.9.5 Reconsider the worked example Sect. 4.8 with a 1 m wide hole (skylight or oculus), located at the top of the dome, and concentric. Assume a 300 mm × 300 mm ring beam is located around the edge of the hole. Determine the stress distribution N_ϕ

and N_Θ in the walls of the dome in increments of five degrees. Only consider membrane stresses.

To solve this problem, the meridian and hoop stress equations must be resolved by subtracting the weight of the removed cap (the removed portion above the hole of the dome) from the membrane equations. Provide the derivation of the revised membrane equations.

4.9.6 Using the worked example Sect. 4.8, solve for the ring force at the base of the dome for H values of 3, 2.5, 2, 1.5, 1, and 0.5 m. Determine the H values for the maximum and minimum ring tensions. Compare the ring tensions and draw a conclusion about the effect of changing the height of the dome.

References

Billington, D. P., 1982. *Thin Shell Concrete Structures.* 2 ed. New York: McGraw-Hill.

Chatterjee, B. K., 1971. *Theory and Design of Concrete Structures.* Calcutta: Oxford and IBH Publishing Co.

Flugge, W., 1960. *Stresses in Shells.* Berlin: Springer-Verag.

Higdon, A., Ohlsen, E. H., Stiles, W. B. & Weese, J. A., 1976. *Mechanics of Materials.* New York: John Wiley and Sons.

Roy, R., 2004. *Fibre Reinforced Shotcrete Domes,* MSc thesis. Johannesburg: University of the Witwatersrand.

Szilard, R., 1974. *Theory and Analysis of Plates.* New Jersey: Prentice-Hall..

Timoshenko, S., 1959. *Theory of Plates and Shells.* 2 ed. New York: McGraw-Hill Book Co..

Torroja, E., 1958. *Philosophy of Structures.* London: Cambridge University Press.

Wilson, A., 2005. *Practical Design of Concrete Shells.* Texas: Monolithic Dome Institute.

Wood, R. H., 1961. *Plastic and Elastic Design of Slabs and Plates.* London: Thames and Hudson.

Chapter 5
Derivatives of Dome Theory: The Conoidal, Elliptical, and Conical Domes and the Hyperbolic Shell

Mitchell Gohnert

Abstract The membrane theories are presented for the conical inverted cone, the conical dome, the elliptical dome, the conoidal dome, and the hyperbolic tower. Boundary effects, or support conditions, are not derived but considered. The membrane results for each of the shell domes are compared to see the limitations and efficiency of stress flow. A full derivation of the membrane theories is provided.

5.1 Introduction

Dome theory, presented in Chap. 4, is derived in a general form. In other words, the theory is generalized to allow the equations to be applied to other shell shapes, such as the conoidal, elliptical, hyperbolic, and conical shells. These shells are therefore derivatives of circular dome theory. The differences between the shell types is the change in the principal radii of curvature, defined as r_1 and r_2. Furthermore, the curvature defines the classification of shell, which is referred to as Gaussian curvature (Wilson 2005). Shell classifications for various shells are listed in Table 5.1. Positive Gaussian curvature is where the curvatures have the same signs and negative Gaussian curvature is where the curvatures have opposite signs. A third classification is where the shell that has zero Gaussian curvature. These shells have curvature, but in a single direction.

The derivations in this chapter covers only the membrane theory. Thus, understanding the classification of shells is vital to ascertain the influence of the boundary effects. As previously defined, boundary effects are bending moments and shears that occur at the base of the shell. Depending on the classification of the shell, the boundary effects will have a greater or lesser influence. Shells with positive curvature tend to have a lesser influence; bending will occur, but the stresses tend to die out quickly. Shells of negative Gaussian curvature, however, are shells that are the most influenced by boundary effects: bending and shears tend to travel up higher up

M. Gohnert
University of the Witwatersrand, Johannsesburg, South Africa
e-mail: mitchell.gohnert@wits.ac.za

Table 5.1 Classification of shells

Positive Gaussian curvature (synclastic shells)	Zero Gaussian curvature (single curvature shells)	Negative Gaussian curvature (anti-synclastic shells)
Circular domes	Circular cylinders	Hyperbolic shell
Elliptical domes	Conical dome	Hyperbola shell
Conoidal dome		
Catenary domes		

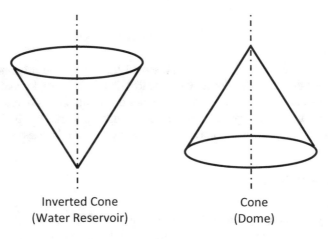

Inverted Cone
(Water Reservoir)

Cone
(Dome)

Fig. 5.1 Orientations of the conical shell

in the shell walls. Another factor, which significantly influences the extent of the boundary effects, is the stiffness of the shell. The stiffer the shell, the greater the influence of the boundary effects. Therefore, a "thin" shell will cause fewer errors in the analysis.

Since the membrane theory is only considered, the analysis is simplified (Billington 1982; Chronowicz 1959). Shell theory is, by nature, complex. Complexity is inherently dangerous, in a sense that mistakes in the analysis are probable. Any simplification is welcome, as long as the errors in the analysis are identified and catered for in the design. It is therefore essential to strengthen the boundaries to counter boundary effects, which exit, but are not analyzed.

5.2   Conical Shells

As shown in Fig. 5.1, the conical shell may be orientated in two ways. With the apex pointing down, the shell is used as a water reservoir; the shape is also referred to as an inverted cone. With the apex pointing up, the shell is used as a dome, or roof covering. Both of these shells are derived in this chapter (Billington 1982).

5.2.1 Conical Water Reservoir

The differential element of the conical shell is defined in Fig. 5.2.

Before the equilibrium equations are developed, the area of the differential element must be defined. The geometry is given in Fig. 5.3.

The area is determined by taking the average of the top and bottom lengths and multiplying by the height.

$$\left[\frac{\left(r_o + \frac{\partial r_o}{\partial y}dy\right) + r_o}{2}\right] d\theta dy \tag{5.1}$$

$$\left(\frac{r_o}{2} + \frac{\partial r_o}{2\partial y}dy + \frac{r_o}{2}\right) d\theta dy \tag{5.2}$$

Since $\frac{\partial r_o}{2\partial y}dy^2 d\theta$ is a higher order term (very small compared to the other terms), the expression for the area reduces to:

$$r_o d\theta dy \tag{5.3}$$

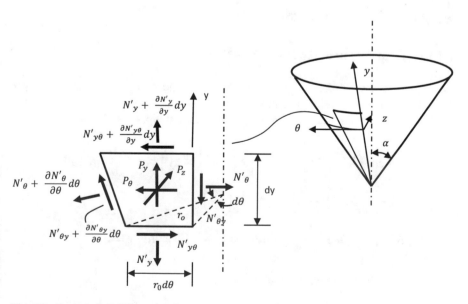

Fig. 5.2 Conical shell differential element

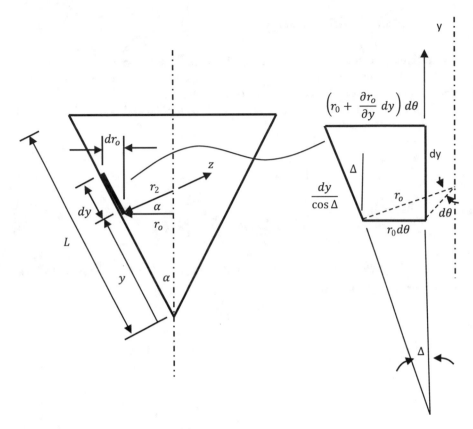

Fig. 5.3 Radial geometry of the differential element

5.2.1.1 Equilibrium in the Hoop Direction

Figure 5.4 is the differential element with only the components of stress that contribute to the equilibrium equation.

Referring to Fig. 5.4 and summing stresses in the hoop, or θ-direction:

$$
\begin{aligned}
&- N'_\theta dy \cos\left(\frac{d\theta}{2}\right) - N'_{y\theta} r_o d\theta + \left(N'_{y\theta} + \frac{\partial N'_{y\theta}}{\partial y} dy\right)\left(r_o + \frac{\partial r_o}{\partial y} dy\right) d\theta \\
&+ \left(N'_\theta + \frac{\partial N'_\theta}{\partial \theta} d\theta\right)\frac{dy}{\cos\Delta}\cos\Delta \cos\left(\frac{d\theta}{2}\right) \\
&+ \left(N'_{\theta y} + \frac{\partial N'_{\theta y}}{\partial \theta} d\theta\right)\frac{dy}{\cos\Delta}\sin\Delta \cos\left(\frac{d\theta}{2}\right) + P_\theta r_o d\theta dy = 0
\end{aligned}
\tag{5.4}
$$

Expanding the above equation and assuming,

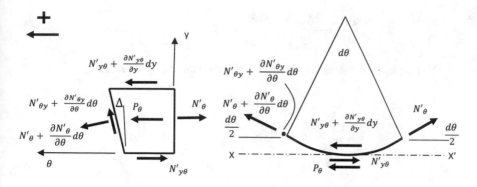

Fig. 5.4 Components of stress that contribute to the equilibrium equation

$$\cos \Delta \approx \cos\left(\frac{d\theta}{2}\right) \approx 1$$

$$\sin \Delta \approx \Delta$$

Furthermore, from Fig. 5.3,

$$\frac{dy}{\cos \Delta}\sin \Delta = \frac{\partial r_o}{\partial y}dy\,d\theta$$

$$\Delta = \frac{\partial r_o}{\partial y}d\theta$$

$$-N'_\theta dy - N'_{y\theta}r_o d\theta + N'_{y\theta}r_o d\theta + N'_{y\theta}\frac{\partial r_o}{\partial y}dyd\theta + \frac{\partial N'_{y\theta}}{\partial y}dy\,r_o d\theta$$

$$+\frac{\partial N'_{y\theta}}{\partial y}dy\,\frac{\partial r_o}{\partial y}dyd\theta + N'_\theta dy + \frac{\partial N'_\theta}{\partial \theta}d\theta dy + N'_{\theta y}dy\,\frac{\partial r_o}{\partial y}d\theta$$

$$+\frac{\partial N'_{\theta y}}{\partial \theta}d\theta\,dy\,\frac{\partial r_o}{\partial y}d\theta + P_\theta r_o d\theta dy$$

$$= 0 \tag{5.5}$$

Simplify the equation by assuming the sixth and tenth terms are higher order terms (significantly less in magnitude that the other terms), and therefore these terms may be eliminated. Furthermore, eliminate the first, second, third and seventh terms.

$$N'_{y\theta}\frac{\partial r_o}{\partial y}dyd\theta + \frac{\partial N'_{y\theta}}{\partial y}dy\,r_o d\theta + \frac{\partial N'_\theta}{\partial \theta}d\theta dy + N'_{\theta y}dy\,\frac{\partial r_o}{\partial y}d\theta + P_\theta r_o d\theta dy = 0 \tag{5.6}$$

Divide by $d\theta dy$.

$$N'_{y\theta}\frac{\partial r_o}{\partial y} + \frac{\partial N'_{y\theta}}{\partial y}r_o + \frac{\partial N'_{\theta}}{\partial \theta} + N'_{\theta y}\frac{\partial r_o}{\partial y} + P_{\theta}r_o = 0 \qquad (5.7)$$

The first and the second terms are combined using the product rule.

$$(fg)' = f'g + fg'$$

$$\frac{\partial (N'_{y\theta}r_o)}{\partial y} + \frac{\partial N'_{\theta}}{\partial \theta} + N'_{\theta y}\frac{\partial r_o}{\partial y} + P_{\theta}r_o = 0 \qquad (5.8)$$

5.2.1.2 Equilibrium in the y-Direction (Fig. 5.5)

Summing stresses in the vertical, or y-direction,

$$N'_{y}r_o d\theta - \left(N'_{y} + \frac{\partial N'_{y}}{\partial y}dy\right)\left(r_o + \frac{\partial r_o}{\partial y}dy\right)d\theta + N'_{\theta y}dy$$

$$- \left(N'_{\theta y} + \frac{\partial N'_{\theta y}}{\partial \theta}d\theta\right)\frac{dy}{\cos \Delta}\cos \Delta + \left(N'_{\theta} + \frac{\partial N'_{\theta}}{\partial \theta}d\theta\right)\frac{dy}{\cos \Delta}\sin \Delta$$

$$- P_{y}r_o d\theta dy$$
$$= 0 \qquad (5.9)$$

Expand the above equation and assume $\cos\left(\frac{d\theta}{2}\right) \approx \cos \Delta \approx 1$, $\sin\Delta \approx \Delta$ and $\Delta = \frac{\partial r_o}{\partial y}d\theta$.

Fig. 5.5 Components in the vertical direction

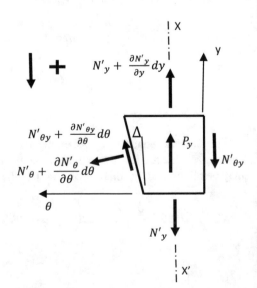

$$N'_y r_o\, d\theta - N'_y r_o d\theta - \frac{\partial N'_y}{\partial y} dy\, r_o d\theta - N'_y \frac{\partial r_o}{\partial y} dy d\theta - \frac{\partial N'_y}{\partial y} dy\, \frac{\partial r_o}{\partial y} dy d\theta$$

$$+ N'_{\theta y} dy - N'_{\theta y} dy - \frac{\partial N'_{\theta y}}{\partial \theta} d\theta dy + N'_\theta dy \frac{\partial r_o}{\partial y} d\theta + \frac{\partial N'_\theta}{\partial \theta} d\theta dy \frac{\partial r_o}{\partial y} d\theta$$

$$- P_y r_o d\theta dy$$
$$= 0 \tag{5.10}$$

Eliminate the higher order terms (fifth and tenth terms) and the first, second, sixth, and seventh terms.

$$- \frac{\partial N'_y}{\partial y} dy\, r_o d\theta - N'_y \frac{\partial r_o}{\partial y} dy d\theta - \frac{\partial N'_{\theta y}}{\partial \theta} d\theta dy + N'_\theta dy \frac{\partial r_o}{\partial y} d\theta - P_y r_o d\theta dy$$
$$= 0 \tag{5.11}$$

Divide by $dy d\theta$.

$$- \frac{\partial N'_y}{\partial y} r_0 - N'_y \frac{\partial r_o}{\partial y} - \frac{\partial N'_{\theta y}}{\partial \theta} + N'_\theta \frac{\partial r_o}{\partial y} - P_y r_o = 0 \tag{5.12}$$

Changing the sign and using the product rule to combine the first and second terms,

$$\frac{\partial \left(N'_y r_0 \right)}{\partial y} + \frac{\partial N'_{\theta y}}{\partial \theta} - N'_\theta \frac{\partial r_o}{\partial y} + P_y r_o = 0 \tag{5.13}$$

5.2.1.3   Equilibrium in the z-Direction

Referring to Fig. 5.6, only the components that contribute to the equilibrium equation are included. The N'_θ terms are in the direction of r_o.

Summing forces in the z-direction,

$$N'_\theta dy \sin\left(\frac{d\theta}{2}\right) \cos\alpha + \left(N'_\theta + \frac{\partial N'_\theta}{\partial \theta} d\theta \right) \frac{dy}{\cos\Delta} \cos\Delta \sin\left(\frac{d\theta}{2}\right) \cos\alpha$$

$$+ P_z\, r_o\, d\theta dy$$
$$= 0 \tag{5.14}$$

Expand the above equation and assume $\sin\left(\frac{d\theta}{2}\right) \approx \frac{d\theta}{2}$ and $\cos\Delta \approx 1$.

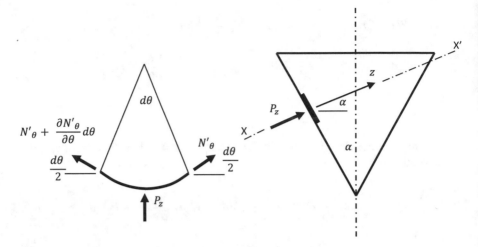

Fig. 5.6 Components of stress that contribute to equilibrium in the z-direction

$$N'_\theta dy \frac{d\theta}{2} \cos \alpha + N'_\theta dy \frac{d\theta}{2} \cos \alpha + \frac{\partial N'_\theta}{\partial \theta} d\theta dy \frac{d\theta}{2} \cos \alpha + P_z r_o d\theta dy = 0 \quad (5.15)$$

The third term is a higher order term and may be eliminated. Terms one and two are combined.

$$N'_\theta dy d\theta \cos \alpha + P_z r_o d\theta dy = 0 \quad (5.16)$$

Divide by $dyd\theta$,

$$N'_\theta \cos \alpha + P_z r_o = 0 \quad (5.17)$$

Since $r_2 \cos \alpha = r_0$,

$$r_2 = \frac{r_o}{\cos \alpha} \quad (5.18)$$

Dividing Eq. 5.17 by r_o and substituting 5.18,

$$\frac{N'_\theta}{r_2} + P_z = 0 \quad (5.19)$$

5.2.1.4 General Equations for Membrane Stresses

$$\frac{\partial \left(N'_{y\theta} r_o \right)}{\partial y} + \frac{\partial N'_\theta}{\partial \theta} + N'_{\theta y} \frac{\partial r_o}{\partial y} + P_\theta r_o = 0 \quad (5.8)$$

$$\frac{\partial \left(N'_y r_0\right)}{\partial y} + \frac{\partial N'_{\theta y}}{\partial \theta} - N'_\theta \frac{\partial r_0}{\partial y} + P_y r_0 = 0 \tag{5.13}$$

$$\frac{N'_\theta}{r_2} + P_z = 0 \tag{5.19}$$

Since,

$$r_0 = y \sin \alpha \tag{5.20}$$

From Fig. 5.3,

$$\partial y \sin \alpha = \partial r_0$$

or,

$$\frac{\partial r_0}{\partial y} = \sin \alpha \tag{5.21}$$

$$r_2 \cos \alpha = y \sin \alpha$$

$$\tan \alpha = \frac{r_2}{y}$$

or

$$y \tan \alpha = r_2 \tag{5.22}$$

Substituting 5.22 into 5.19 and rearranging,

$$N'_\theta = -P_z y \tan \alpha \tag{5.23}$$

Substituting 5.20 and 5.21 into 5.7 (equilibrium equation prior to using the product rule),

$$N'_{y\theta} \frac{\partial r_0}{\partial y} + \frac{\partial N'_{y\theta}}{\partial y} r_0 + \frac{\partial N'_\theta}{\partial \theta} + N'_{\theta y} \frac{\partial r_0}{\partial y} + P_\theta r_0 = 0 \tag{5.7}$$

$$N'_{y\theta} \sin \alpha + \frac{\partial N'_{y\theta}}{\partial y} y \sin \alpha + \frac{\partial N'_\theta}{\partial \theta} + N'_{\theta y} \sin \alpha + P_\theta y \sin \alpha = 0 \tag{5.24}$$

Combining the like terms and rearranging,

$$2N'_{y\theta} \sin \alpha + \frac{\partial N'_{y\theta}}{\partial y} y \sin \alpha = - \frac{\partial N'_\theta}{\partial \theta} - P_\theta y \sin \alpha \tag{5.25}$$

Divide by $y \sin \alpha$,

$$\frac{2N'_{y\theta}}{y} + \frac{\partial N'_{y\theta}}{\partial y} = -\frac{\partial N'_\theta}{\partial \theta}\frac{1}{y\sin\alpha} - P_\theta \tag{5.26}$$

From Eq. 5.12 (equilibrium equation prior to using the product rule),

$$-\frac{\partial N'_y}{\partial y}r_0 - N'_y\frac{\partial r_o}{\partial y} - \frac{\partial N'_{\theta y}}{\partial \theta} + N'_\theta\frac{\partial r_o}{\partial y} - P_y r_o = 0 \tag{5.12}$$

Substituting Eqs. 5.20 and 5.21 into the above equation,

$$-\frac{\partial N'_y}{\partial y}y\sin\alpha - N'_y\sin\alpha - \frac{\partial N'_{\theta y}}{\partial \theta} + N'_\theta\sin\alpha - P_y y\sin\alpha = 0 \tag{5.27}$$

Dividing by $y\sin\alpha$ and rearranging,

$$\frac{\partial N'_y}{\partial y} + \frac{N'_y}{y} = -\frac{\partial N'_{\theta y}}{\partial \theta}\frac{1}{y\sin\alpha} + \frac{N'_\theta}{y} - P_y \tag{5.28}$$

Applying the general equations, solve for the stresses in the shell due to self-weight, water, and wind loads.

5.2.1.5   Membrane Stresses for Self-Weight (Gravity) Load

Referring to Fig. 5.7, the external loading for a gravity load is defined as follows:

$$P_z = -q\sin\alpha \tag{5.29}$$

$$P_y = -q\cos\alpha \tag{5.30}$$

$$P_\theta = 0 \tag{5.31}$$

Furthermore, for symmetrical loading,

$$N'_{y\theta} = N'_{\theta y} = 0$$

Using the general Eq. 5.23 to solve for N'_θ,

$$N'_\theta = -P_z y\tan\alpha \tag{5.23}$$

Substituting 5.29,

$$N'_\theta = qy\ \sin\alpha\tan\alpha \tag{5.32}$$

Solving for N'_y from Eq. 5.28,

Fig. 5.7 Configuration for gravity loads

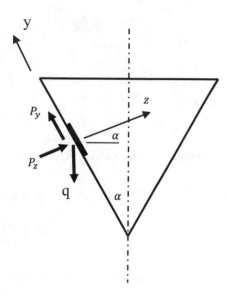

$$\frac{\partial N'_y}{\partial y} + \frac{N'_y}{y} = -\frac{\partial N'_{\theta y}}{\partial \theta}\frac{1}{y\sin\alpha} + \frac{N'_\theta}{y} - P_y \qquad (5.28)$$

The left-hand side of the 5.28 is combined by using the product rule.

$$\frac{\partial N'_y}{\partial y} + \frac{N'_y}{y} = \frac{1}{y}\frac{\partial\left(N'_y y\right)}{\partial y}$$

The first term of the right-hand side of 5.28 is eliminated, since $N'_{\theta y}$ is zero for symmetrical loads. Equation 5.28 becomes,

$$\frac{1}{y}\frac{\partial\left(N'_y y\right)}{\partial y} = \frac{N'_\theta}{y} - P_y$$

or

$$\frac{\partial\left(N'_y y\right)}{\partial y} = N'_\theta - P_y y \qquad (5.33)$$

Replacing N'_θ,

$$\frac{\partial\left(N'_y y\right)}{\partial y} = -P_z y \tan\alpha - P_y y \qquad (5.34)$$

Integrating the above expression from y to L,

$$\int \frac{\partial (N'_y y)}{\partial y} = \int_y^L \left(-P_z y \tan \alpha - P_y y \right) dy \tag{5.35}$$

$$N'_y y = \left(-P_z \tan \alpha - P_y \right) \int_y^L y \, dy \tag{5.36}$$

$$N'_y y = \left(-P_z \tan \alpha - P_y \right) \frac{y^2}{2} + C \tag{5.37}$$

Replace P_z and P_y with Eqs. 5.29 and 5.30.

$$N'_y y = -\left(-q \sin \alpha \tan \alpha - q \cos \alpha \right) \frac{y^2}{2} + C \tag{5.38}$$

$$N'_y y = \frac{q y^2}{2} \left(\frac{\sin^2 \alpha}{\cos \alpha} + \cos \alpha \right) + C \tag{5.39}$$

Manipulating the trigonometric terms,

$$\frac{\sin^2 \alpha}{\cos \alpha} + \cos \alpha = \frac{\sin^2 \alpha}{\cos \alpha} + \frac{\cos^2 \alpha}{\cos \alpha} = \frac{1}{\cos \alpha}$$

$$N'_y = \frac{q y^2}{2 y \cos \alpha} + \frac{C}{y} \tag{5.40}$$

The constant C is solved from boundary conditions. At $y = L$, $N'_y = 0$.

$$0 = \frac{q L^2}{2 L \cos \alpha} + \frac{C}{L} \tag{5.41}$$

Solving for C,

$$C = \frac{-q L^2}{2 \cos \alpha} \tag{5.42}$$

Substituting C into 5.40,

$$N'_y = \frac{q y^2}{2 y \cos \alpha} + \frac{-q L^2}{2 y \cos \alpha} \tag{5.43}$$

Simplified,

$$N'_y = \frac{q \left(y^2 - L^2 \right)}{2 y \cos \alpha} \tag{5.44}$$

5.2.1.6 Membrane Stresses for Water Load

Conical shells are used primarily as water tanks. The water pressure is always normal to the surface of the tank, since water is not capable of resisting a shear force.

$$P_z = -\gamma(L - y) \cos \alpha \tag{5.45}$$

$$P_y = 0 \tag{5.46}$$

$$P_\theta = 0 \tag{5.47}$$

where γ is the unit weight of water.

Using the general equation for N'_θ and substituting 5.45,

$$N'_\theta = -P_z y \tan \alpha \tag{5.23}$$

$$N'_\theta = \gamma(L - y) \cos \alpha y \tan \alpha \tag{5.48}$$

$$\boldsymbol{N'_\theta = \gamma y(L - y) \sin \alpha} \tag{5.49}$$

Using Eq. 5.35 and substituting equations 5.45 and 5.46,

$$\int \frac{\partial (N'_y y)}{\partial y} = \int_y^L \left(-P_z y \tan \alpha - P_y y \right) dy \tag{5.35}$$

$$N'_y y = -\int_y^L \left(-\gamma y(L - y) \cos \alpha \tan \alpha + 0 \right) dy \tag{5.50}$$

$$N'_y y = \gamma \cos \alpha \tan \alpha \int_y^L y(L - y) dy \tag{5.51}$$

$$N'_y y = \gamma \sin \alpha \left(\frac{Ly^2}{2} - \frac{y^3}{3} \right) + C \tag{5.52}$$

or

$$N'_y = \frac{\gamma \sin \alpha}{y} \left(\frac{Ly^2}{2} - \frac{y^3}{3} \right) + \frac{C}{y} \tag{5.53}$$

Solve for C with boundary conditions. At $y = L$, $N'_y = 0$

$$0 = \frac{\gamma \sin \alpha}{L} \left(\frac{L^3}{2} - \frac{L^3}{3} \right) + \frac{C}{L} \tag{5.54}$$

Manipulating 5.54,

$$0 = \frac{\gamma \sin \alpha}{L}\left(\frac{L^3}{2} - \frac{L^3}{3}\right) + \frac{C}{L} = \frac{\gamma \sin \alpha}{L}\left(\frac{3L^3 - 2L^3}{6}\right) + \frac{C}{L}$$

$$= \frac{\gamma L^3 \sin \alpha}{6L} + \frac{C}{L} \tag{5.55}$$

Solving for C,

$$C = -\frac{\gamma L^3 \sin \alpha}{6} \tag{5.56}$$

Substituting C into equation 5.53,

$$N'_y = \frac{\gamma \sin \alpha}{y}\left(\frac{Ly^2}{2} - \frac{y^3}{3}\right) - \frac{\gamma L^3 \sin \alpha}{6y} \tag{5.57}$$

Combining the terms,

$$N'_y = \frac{\gamma \sin \alpha}{2y}\left(Ly^2 - \frac{2y^3}{3} - \frac{L^3}{3}\right) \tag{5.58}$$

5.2.1.7 Membrane Stresses for Wind Load

The distribution of wind load is given by the following equation:

$$P_z = P_w A \cos \alpha \cos \theta \tag{5.59}$$

Expressed graphically in Fig. 5.8.where

P_w = the peak wind pressure at the top of the inverted cone
A = the projected area of the cone (height x base x ½)
θ = is zero at normal windward load

Using the general equation for the stress in the hoop direction (Eq. 5.23),

$$N'_\theta = -P_z y \tan \alpha \tag{5.23}$$

Substituting the distribution of wind load P_z,

$$N'_\theta = -(P_w A \cos \alpha \cos \theta)y \tan \alpha \tag{5.60}$$

Simplifying, the equation for the hoop stress is determined.

$$N'_\theta = -P_w A y \sin \alpha \cos \theta \qquad (5.61)$$

From Eq. 5.26,

$$\frac{2N'_{y\theta}}{y} + \frac{\partial N'_{y\theta}}{\partial y} = -\frac{\partial N'_\theta}{\partial \theta} \frac{1}{y \sin \alpha} - P_\theta \qquad (5.26)$$

Since,

$$\frac{2N'_{y\theta}}{y} + \frac{\partial N'_{y\theta}}{\partial y} = \frac{1}{y^2} \frac{\partial (N'_{y\theta} y^2)}{\partial y}$$

$$\frac{1}{y^2} \frac{\partial (N'_{y\theta} y^2)}{\partial y} = -\frac{\partial N'_\theta}{\partial \theta} \frac{1}{y \sin \alpha} - P_\theta \qquad (5.62)$$

Taking the derivative of Eq. 5.61 with respect to θ,

$$\frac{\partial N'_\theta}{\partial \theta} = y P_w A \sin \alpha \sin \theta \qquad (5.63)$$

Substituting into 5.63 into 5.62,

$$\frac{1}{y^2} \frac{\partial (N'_{y\theta} y^2)}{\partial y} = -y P_w A \sin \alpha \sin \theta \frac{1}{y \sin \alpha} - P_\theta \qquad (5.64)$$

Simplifying by setting $P_\theta = 0$,

Fig. 5.8 Distribution of wind pressures

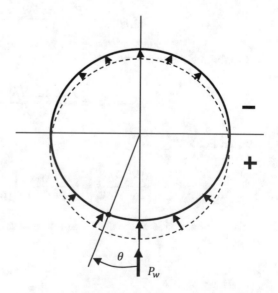

$$\frac{\partial\left(N'_{y\theta}y^2\right)}{\partial y} = -y^2\, P_w A \sin\theta \tag{5.65}$$

Integrating the expression,

$$\int \frac{\partial\left(N'_{y\theta}y^2\right)}{\partial y} = -\; P_w A \sin\theta \int y^2 dy \tag{5.66}$$

$$N'_{y\theta}y^2 = -\; \frac{P_w A \sin\theta\, y^3}{3} + C \tag{5.67}$$

or

$$N'_{y\theta} = -\; \frac{P_w A \sin\theta y}{3} + \frac{C}{y^2} \tag{5.68}$$

Solve for the integration constant C by boundary conditions. $N'_{y\theta} = 0$ at $y = L$.

$$0 = -\; \frac{P_w A \sin\theta L}{3} + \frac{C}{L^2} \tag{5.69}$$

Solving for C,

$$C = \frac{P_w A \sin\theta L^3}{3} \tag{5.70}$$

Incorporating the value of C into Eq. 5.68,

$$N'_{y\theta} = -\; \frac{P_w A \sin\theta y}{3} + \frac{P_w A \sin\theta L^3}{3y^2} \tag{5.71}$$

Combining the two terms,

$$N'_{y\theta} = \frac{P_w A \sin\theta\left(L^3 - y^3\right)}{3y^2} \tag{5.72}$$

From Eq. 5.28,

$$\frac{\partial N'_y}{\partial y} + \frac{N'_y}{y} = -\frac{\partial N'_{\theta y}}{\partial\theta}\frac{1}{y\sin\alpha} + \frac{N'_\theta}{y} - P_y \tag{5.28}$$

The left-hand side of the above equation is combined.

$$\frac{\partial N'_y}{\partial y} + \frac{N'_y}{y} = \frac{1}{y}\frac{\partial (N'_y y)}{\partial y} \qquad (5.73)$$

Therefore,

$$\frac{1}{y}\frac{\partial (N'_y y)}{\partial y} = -\frac{\partial N'_{\theta y}}{\partial \theta}\frac{1}{y \sin \alpha} + \frac{N'_\theta}{y} \qquad (5.74)$$

where $P_y = 0$.

Substituting 5.72 into 5.74,

$$\frac{1}{y}\frac{\partial (N'_y y)}{\partial y} = -\frac{\partial \left(\frac{P_w A \sin \theta (L^3 - y^3)}{3y^2}\right)}{\partial \theta}\frac{1}{y \sin \alpha} + \frac{N'_\theta}{y} \qquad (5.75)$$

$$\frac{1}{y}\frac{\partial (N'_y y)}{\partial y} = -\frac{P_w A (L^3 - y^3)}{3y^2}\frac{\partial (\sin \theta)}{\partial \theta}\frac{1}{y \sin \alpha} + \frac{N'_\theta}{y} \qquad (5.76)$$

$$\frac{\partial (N'_y y)}{\partial y} = -\frac{P_w A (L^3 - y^3) \cos \theta}{3y^2 \sin \alpha} + N'_\theta \qquad (5.77)$$

Substituting N'_θ (Eq. 5.61),

$$\frac{\partial (N'_y y)}{\partial y} = -\frac{P_w A (L^3 - y^3) \cos \theta}{3y^2 \sin \alpha} - P_w A y \sin \alpha \cos \theta \qquad (5.78)$$

Integrate 5.78,

$$\int \frac{\partial (N'_y y)}{\partial y} = -\frac{P_w A \cos \theta}{3 \sin \alpha}\int \frac{(L^3 - y^3)}{y^2}dy - P_w A \sin \alpha \cos \theta \int y dy \qquad (5.79)$$

Simplify for integration.

$$\int \frac{\partial (N'_y y)}{\partial y} = -\frac{P_w A \cos \theta}{3 \sin \alpha}\int \left(\frac{L^3}{y^2} - y\right)dy - P_w A \sin \alpha \cos \theta \int y dy \qquad (5.80)$$

$$N'_y y = -\frac{P_w A \cos \theta}{3 \sin \alpha}\left[-\frac{L^3}{y} - \frac{y^2}{2}\right] - P_w A \sin \alpha \cos \theta \frac{y^2}{2} + C \qquad (5.81)$$

Simplify,

$$N'_y = \frac{P_w A \cos \theta}{y \sin \alpha}\left(\frac{L^3}{3y} + \frac{y^2}{6} - \frac{y^2 \sin^2 \alpha}{2}\right) + \frac{C}{y} \qquad (5.82)$$

Solve for C using boundary conditions, $N'_y = 0$ at $y = L$.

$$0 = \frac{P_w A \cos \theta}{L \sin \alpha} \left(\frac{L^3}{3L} + \frac{L^2}{6} - \frac{L^2 \sin^2 \alpha}{2} \right) + \frac{C}{L} \tag{5.83}$$

Solving for C,

$$C = -\frac{P_w A \cos \theta}{\sin \alpha} \left(\frac{2L^2}{6} + \frac{L^2}{6} - \frac{L^2 \sin^2 \alpha}{2} \right) \tag{5.84}$$

$$C = \frac{P_w A \cos \theta}{\sin \alpha} \left(-\frac{L^2}{2} + \frac{L^2 \sin^2 \alpha}{2} \right) \tag{5.85}$$

$$C = \frac{P_w A L^2 \cos \theta}{2 \sin \alpha} \left(\sin^2 \alpha - 1 \right) \tag{5.86}$$

Substituting C into 5.82,

$$N'_y = \frac{P_w A \cos \theta}{y \sin \alpha} \left(\frac{L^3}{3y} + \frac{y^2}{6} - \frac{y^2 \sin^2 \alpha}{2} \right) + \frac{P_w A L^2 \cos \theta \left(\sin^2 \alpha - 1 \right)}{2 y \sin \alpha} \tag{5.87}$$

and simplify,

$$N'_y = \frac{P_w A \cos \theta}{y \sin \alpha} \left(\frac{L^3}{3y} + \frac{y^2}{6} - \frac{y^2 \sin^2 \alpha}{2} + \frac{L^2 \sin^2 \alpha}{2} - \frac{L^2}{2} \right) \tag{5.88}$$

$$N'_y = \frac{P_w A \cos \theta}{\sin \alpha} \left[\frac{L^3}{3y^2} + \frac{y}{6} - \frac{L^2}{2y} - \frac{\sin^2 \alpha (L^2 - y^2)}{2y} \right] \tag{5.89}$$

5.2.2 Conical Dome

The derivation of the conical dome is significantly simpler than the water tower. The configuration is taken from dome theory, but modified for the case of a conical dome. The derivation presented is for the case of self-weight loading (gravity load).

The value ϕ is a constant, defined as:

$$\phi = \frac{\pi}{2} - \alpha \tag{5.90}$$

Furthermore,

$$r_2 \sin \phi = r_o \tag{5.91}$$

or,

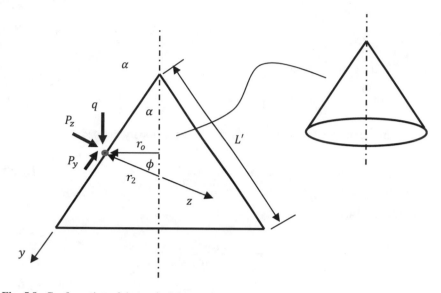

Fig. 5.9 Configuration of the conical dome

$$r_2 = \frac{r_o}{\sin \phi} \tag{5.92}$$

$$y \sin \alpha = r_o \tag{5.93}$$

Because the surface is straight and r_1 and r_2 are parallel,

$$r_1 = \infty$$

From Fig. 5.9, the pressures P_z and P_y are defined.

$$P_z = q \sin \alpha \tag{5.94}$$

$$P_y = -q \cos \alpha \tag{5.95}$$

Solving for the total load R at any point y.

$$R = \int_0^y 2\pi r_o q \, dy \tag{5.96}$$

Replace r_0,

$$R = \int_0^y 2\pi q \sin \alpha y \, dy \tag{5.97}$$

$$R = 2\pi q \, \sin\alpha \int_0^y y\,dy \tag{5.98}$$

Integrate the above equation,

$$R = \pi q y^2 \, \sin\alpha \tag{5.99}$$

From circular dome theory, the stress N'_ϕ was previously defined as Eq. 4.35.

$$N'_\phi = -\frac{R}{2\pi r_o \sin\phi} \tag{4.35}$$

Substituting Eqs. 5.99 and 5.93,

$$N'_\phi = -\frac{\pi q \, y^2 \sin\alpha}{2\pi y \sin\alpha \sin\phi} \tag{5.100}$$

$$N'_\phi = -\frac{q\,y}{2\sin\phi} \tag{5.101}$$

Since,

$$\phi = \frac{\pi}{2} - \alpha \tag{5.102}$$

which is a constant.

Using the following trigonometric identity and replacing ϕ,

$$\sin\left(\frac{\pi}{2} - \alpha\right) = \cos\alpha \tag{5.103}$$

$$N'_\phi = -\frac{q\,y}{2\cos\alpha} \tag{5.104}$$

The stress in the hoop direction was also previously defined in circular dome theory,

$$N'_\theta = \frac{R}{2\pi r_1 \sin^2\phi} - \frac{P_z r_o}{\sin\phi} \tag{4.37}$$

Since $r_1 = \infty$, the first term is approximately equal to zero and therefore may be eliminated.

$$N'_\theta = -\frac{P_z r_o}{\sin\phi} \tag{5.105}$$

Substituting Eqs. 5.93 and 5.94,

$$N'_\theta = -\frac{qy\ \sin\alpha\sin\alpha}{\sin\phi} \qquad (5.106)$$

or

$$N'_\theta = -\frac{qy\ \sin^2\alpha}{\sin\phi} \qquad (5.107)$$

Using the trigonometric identity (5.103),

$$N'_\theta = -\frac{qy\ \sin^2\alpha}{\cos\alpha} \qquad (5.108)$$

$$N'_\theta = -qy\ \sin\alpha\tan\alpha \qquad (5.109)$$

5.3 Elliptical Dome

The theory for the elliptical dome is taken directly from circular dome theory (Billington 1982; Wilson 2005; Chronowicz 1959; Flugge 1960; Beles and Soare 1976). However, the profile is distinctly different to a circular dome. The fundamental difference is that the values r_1 and r_2 differ, defining the elliptical shape. In circular dome theory, the values r_1 and r_2 are equal.

5.3.1 Total Load R

The profile of the elliptical dome follows the general equation for an ellipse.

$$\frac{x^2}{a^2} + \frac{y^2}{b^2} = 1 \qquad (5.110)$$

where a and b are the major and minor axis, respectively.
   Rearranging 5.110 and taking the square root (Fig. 5.10),

$$x = a\sqrt{1 - \frac{y^2}{b^2}} \qquad (5.111)$$

or

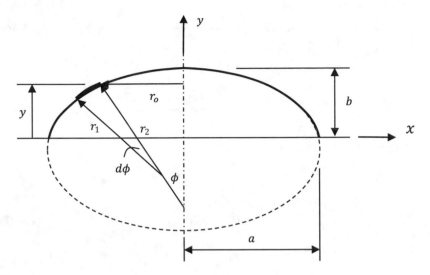

Fig. 5.10 Radial configuration of an elliptical dome

$$x = \frac{a}{b}\sqrt{b^2 - y^2} \tag{5.112}$$

Taking the derivative of 5.112 with respect to y.

$$\frac{dx}{dy} = \frac{-ay}{b\sqrt{b^2 - y^2}} \tag{5.113}$$

Taking the reciprocal of the above equation.

$$\frac{dy}{dx} = \frac{1}{\frac{dx}{dy}} = \frac{-b\sqrt{b^2 - y^2}}{ay} = -\tan\phi \tag{5.114}$$

From Fig. 5.11,

$$ds\sin\phi = dy \tag{5.115}$$

or

$$\sin\phi = \frac{dy}{ds} \tag{5.116}$$

Similarly,

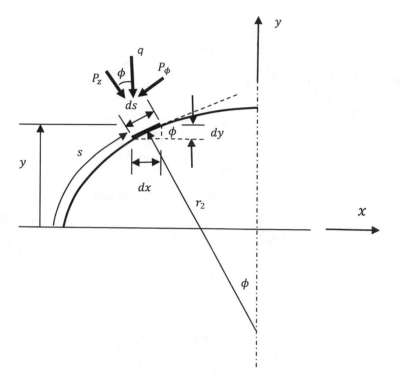

Fig. 5.11 Differential element on an elliptical shell

$$ds = \sqrt{dy^2 + dx^2} \tag{5.117}$$

Substituting 5.117 into 5.116,

$$\sin \phi = \frac{dy}{\sqrt{dy^2 + dx^2}} \tag{5.118}$$

Divide the top and bottom by dx.

$$\sin \phi = \frac{\frac{dy}{dx}}{\frac{\sqrt{dy^2 + dx^2}}{dx}} = \frac{\frac{dy}{dx}}{\sqrt{\left(\frac{dy}{dx}\right)^2 + 1^2}} \tag{5.119}$$

Since,

$$\frac{dy}{dx} = -\tan \phi \tag{5.120}$$

$$\sin \phi = \frac{-\tan \phi}{\sqrt{(\tan \phi)^2 + 1}} \tag{5.121}$$

Substituting 5.114 into 5.121,

$$\sin \phi = \frac{\frac{-b\sqrt{b^2-y^2}}{ay}}{\sqrt{\left(\frac{b\sqrt{b^2-y^2}}{ay}\right)^2 + 1}} \tag{5.122}$$

or

$$\sin \phi = \frac{-b\sqrt{b^2 - y^2}}{ay\sqrt{\left[\frac{b^2(b^2-y^2)}{a^2y^2}\right] + 1}} \tag{5.123}$$

Simplifying,

$$\sin \phi = \frac{-b\sqrt{b^2 - y^2}}{ay\sqrt{\frac{b^4-b^2y^2}{a^2y^2} + \frac{a^2y^2}{a^2y^2}}} = \frac{-b\sqrt{b^2 - y^2}}{ay\frac{1}{ay}\sqrt{b^4 - b^2y^2 + a^2y^2}}$$

$$= \frac{-b\sqrt{b^2 - y^2}}{\sqrt{b^4 + y^2(a^2 - b^2)}} \tag{5.124}$$

Similarly, the *cos* can likewise be solved.

$$\cos \phi = \frac{ay}{\sqrt{b^4 + y^2(a^2 - b^2)}} \tag{5.125}$$

Since,

$$x = r_2 \sin \phi \tag{5.126}$$

$$r_2 = \frac{x}{\sin \phi} \tag{5.127}$$

Substituting 5.112 and 5.124,

$$r_2 = \frac{\frac{a}{b}\sqrt{b^2 - y^2}}{\frac{b\sqrt{b^2-y^2}}{\sqrt{b^4+y^2(a^2-b^2)}}} \tag{5.128}$$

Since r_2 will always be positive.

$$r_2 = \frac{a}{b^2} \sqrt{b^4 + y^2 (a^2 - b^2)} \tag{5.129}$$

Similarly,

$$r_1 = \frac{\left[\sqrt{b^4 + y^2 (a^2 - b^2)}\right]^{\frac{3}{2}}}{ab^4} \tag{5.130}$$

If q represents the self-weight (refer to Fig. 5.11),

$$P_z = q \cos \phi \tag{5.131}$$

$$P_\phi = q \sin \phi \tag{5.132}$$

$$P_\theta = 0 \tag{5.133}$$

The total load R is defined as:

$$R = 2\pi q \int_y^b x \, ds \tag{5.134}$$

Since,

$$\sin \phi = \frac{dy}{ds} \tag{5.116}$$

$$\sin \phi = \frac{-\tan \phi}{\sqrt{(\tan \phi)^2 + 1}} \tag{5.121}$$

$$\frac{dy}{dx} = -\tan \phi \tag{5.120}$$

Combining these three equations,

$$\frac{dy}{ds} = \frac{-\tan \phi}{\sqrt{(\tan \phi)^2 + 1}} = \frac{\frac{dy}{dx}}{\sqrt{\left(\frac{dy}{dx}\right)^2 + 1}} \tag{5.135}$$

Taking the reciprocal of 5.135,

$$\frac{ds}{dy} = \sqrt{\left(\frac{dx}{dy}\right)^2 + 1} \tag{5.136}$$

Substituting 5.113,

$$ds = \sqrt{\left(\frac{-ay}{b\sqrt{b^2 - y^2}}\right)^2 + 1} \; dy = \sqrt{\frac{a^2 y^2}{b^2(b^2 - y^2)} + 1} \; dy \qquad (5.137)$$

Substitute 5.112 and 5.137 into 5.134.

$$R = 2\pi q \int_y^b \left[\left(\frac{a}{b}\sqrt{b^2 - y^2}\right)\sqrt{\frac{a^2 y^2}{b^2(b^2 - y^2)} + 1}\right] dy \qquad (5.138)$$

$$R = 2\pi q \int_y^b \left[\left(\frac{a}{b}\sqrt{b^2 - y^2}\right)\sqrt{\frac{a^2 y^2}{b^2(b^2 - y^2)} + \frac{b^2(b^2 - y^2)}{b^2(b^2 - y^2)}}\right] dy \qquad (5.139)$$

$$R = 2\pi q \int_y^b \left[\left(\frac{a}{b}\sqrt{b^2 - y^2}\right)\sqrt{\frac{a^2 y^2 + b^2(b^2 - y^2)}{b^2(b^2 - y^2)}}\right] dy \qquad (5.140)$$

$$R = 2\pi q \int_y^b \left[\left(\frac{a}{b}\sqrt{b^2 - y^2}\right)\frac{\sqrt{a^2 y^2 + b^2(b^2 - y^2)}}{\sqrt{b^2}\sqrt{(b^2 - y^2)}}\right] dy \qquad (5.141)$$

$$R = 2\pi q \int_y^b \left[\frac{a}{b^2}\sqrt{a^2 y^2 + b^2(b^2 - y^2)}\right] dy \qquad (5.142)$$

$$R = 2\pi q \frac{a}{b^2} \int_y^b \sqrt{a^2 y^2 + b^2(b^2 - y^2)} \, dy \qquad (5.143)$$

Reducing the above equation to a simpler form,

$$\int \sqrt{my^2 + n} \, dy \qquad (5.144)$$

Integrating from 0 to b, the total load R is solved.

$$R = 2\pi q a^2$$

$$\left[\frac{1}{2} - \frac{y}{2ab^2}\sqrt{b^4 + y^2(a^2 - b^2)} + \frac{b^2}{2a\sqrt{a^2 - b^2}}\log\frac{b\left(a + \sqrt{a^2 - b^2}\right)}{y\sqrt{a^2 - b^2} + \sqrt{b^4 + y^2(a^2 - b^2)}}\right]$$

$$(5.145)$$

or, set the value in brackets as C.

$$C = \frac{1}{2} - \frac{y}{2ab^2} \sqrt{b^4 + y^2\left(a^2 - b^2\right)}$$

$$+ \frac{b^2}{2a\sqrt{a^2 - b^2}} \log \frac{b\left(a + \sqrt{a^2 - b^2}\right)}{y\sqrt{a^2 - b^2} + \sqrt{b^4 + y^2\left(a^2 - b^2\right)}} \qquad (5.146)$$

R is therefore,

$$R = 2\pi q a^2 C \qquad (5.147)$$

5.3.2   Membrane Stresses

The meridian or vertical stresses are solved by the general equation from cylindrical dome theory.

$$N'_\phi = - \frac{R}{2\pi r_o \sin \phi} \qquad (4.35)$$

Substituting 5.147,

$$N'_\phi = - \frac{2\pi q a^2 C}{2\pi r_o \sin \phi} \qquad (5.148)$$

$$N'_\phi = - \frac{q a^2 C}{r_o \sin \phi} \qquad (5.149)$$

From Fig. 5.10,

$$r_o = r_2 \sin \phi \qquad (5.150)$$

And substituting 5.150 and 5.129,

$$N'_\phi = - \frac{q a^2 C}{\left(\frac{a}{b^2}\right)\sqrt{b^4 + y^2\left(a^2 - b^2\right)} \sin^2 \phi} \qquad (5.151)$$

Squaring Eq. 5.124,

$$\sin^2\phi = \left[\frac{-b\sqrt{b^2-y^2}}{\sqrt{b^4+y^2(a^2-b^2)}}\right]^2 = \frac{b^2(b^2-y^2)}{b^4+y^2(a^2-b^2)} \tag{5.152}$$

Substituting 5.152 into 5.151.

$$N'_\phi = -\frac{qab^2C\left[b^4+y^2(a^2-b^2)\right]}{\sqrt{b^4+y^2(a^2-b^2)}\left[b^2(b^2-y^2)\right]} \tag{5.153}$$

Since,

$$\frac{A}{\sqrt{A}} = A - \sqrt{A} = \sqrt{A} \tag{5.154}$$

$$N'_\phi = -\frac{qaC\sqrt{b^4+y^2(a^2-b^2)}}{(b^2-y^2)} \tag{5.155}$$

A more common form of 5.155,

$$N'_\phi = -\frac{qa^2}{b}\frac{Cb^2}{(b^2-y^2)}\frac{\sqrt{b^4+y^2(a^2-b^2)}}{ab} \tag{5.156}$$

If we say,

$$Q = \frac{\sqrt{b^4+y^2(a^2-b^2)}}{ab} \tag{5.157}$$

and

$$\frac{b^2}{b^2-y^2} = \frac{b^2}{b^2\left(1-\frac{y^2}{b^2}\right)} = \frac{1}{\left(1-\frac{y^2}{b^2}\right)} \tag{5.158}$$

The stress in the meridian, or vertical direction is solved.

$$N'_\phi = -\frac{qa^2}{b}\frac{CQ}{\left(1-\frac{y^2}{b^2}\right)} \tag{5.159}$$

The general Eq. 4.37 from cylindrical dome theory is used to solve for the hoop stresses.

$$N'_\theta = \frac{R}{2\pi r_1 \sin^2 \phi} - \frac{P_z r_o}{\sin \phi} \qquad (4.37)$$

Since,

$$N'_\phi = -\frac{R}{2\pi r_o \sin \phi} \qquad (4.35)$$

$$r_o = r_2 \sin \phi \qquad (5.150)$$

Combining the two above equations,

$$N'_\phi = -\frac{R}{2\pi r_2 \sin^2 \phi} \qquad (5.160)$$

or

$$R = 2\pi r_2 \sin^2 \phi \, N'_\phi \qquad (5.161)$$

also

$$P_z = q \cos \phi \qquad (5.131)$$

Substituting 5.161, 5.149 and 5.131 into Eq. 4.37,

$$N'_\theta = -\frac{r_2 N'_\phi}{r_1} - q \cos \phi r_2 \qquad (5.162)$$

Solving for the ratio of r_2 and r_1,

$$\frac{r_2}{r_1} = \frac{\frac{a}{b^2} \sqrt{b^4 + y^2 (a^2 - b^2)}}{\frac{\left[\sqrt{b^4 + y^2 (a^2 - b^2)} \right]^{\frac{3}{2}}}{ab^4}} \qquad (5.163)$$

Mathematically,

$$\frac{\sqrt{A}}{\left(\sqrt{A}\right)^3} = \sqrt{A} - \left(\sqrt{A}\right)^3 = A^{\frac{1}{2}} - A^{\frac{3}{2}} = \frac{1}{A}$$

Therefore,

$$\frac{r_2}{r_1} = \frac{a^2 b^2}{b^4 + y^2 (a^2 - b^2)} \tag{5.164}$$

Since,

$$Q = \frac{\sqrt{b^4 + y^2 (a^2 - b^2)}}{ab} \tag{5.157}$$

$$Q^2 = \frac{b^4 + y^2 (a^2 - b^2)}{a^2 b^2} \tag{5.165}$$

Equation 5.164 therefore becomes,

$$\frac{r_2}{r_1} = \frac{1}{Q^2} \tag{5.166}$$

Substituting the ratio $\frac{r_2}{r_1}$ and r_2 (Eq. 5.129) into Eq. 5.162,

$$N'_\theta = \frac{qa^2 CQ}{Q^2 b \left(1 - \frac{y^2}{b^2}\right)} - \frac{aq \cos \phi \sqrt{b^4 + y^2 (a^2 - b^2)}}{b^2} \tag{5.167}$$

Replacing $\cos\phi$ with 5.125,

$$N'_\theta = \frac{qa^2 C}{Qb \left(1 - \frac{y^2}{b^2}\right)} - \frac{a^2 yq}{b^2} \tag{5.168}$$

Simplifying the above expression,

$$N'_\theta = \frac{-qa^2}{b} \left[\frac{y}{b} - \frac{C}{Q\left(1 - \frac{y^2}{b^2}\right)}\right] \tag{5.169}$$

or,

$$N'_\theta = \frac{-qa^2}{b} \left[\frac{y}{b} - \frac{Cb^2}{Q(b^2 - y^2)}\right] \tag{5.170}$$

5.4 Conoidal Dome

The conoidal dome is the pointed dome, and the shape is achieved by overlapping two circular shapes (Billington 1982; Timoshenko 1959). The general equations derived for the circular dome are applied to the conoidal dome (Figs. 5.12 and 5.13).

Defining the geometries of the conoidal dome,

$$r' = a \sin \phi_o \tag{5.171}$$

$$r_o = a \sin \phi - r' \tag{5.172}$$

$$r_o = r_2 \sin \phi \tag{5.173}$$

Setting the above two equations equal (Fig. 5.14),

$$r_2 \sin \phi = a \sin \phi - r' \tag{5.174}$$

$$r_2 = \frac{a \sin \phi - r'}{\sin \phi} \tag{5.175}$$

$$r_2 = a - \frac{r'}{\sin \phi} \tag{5.176}$$

$$P_\theta = 0 \tag{5.177}$$

$$P_z = q \cos \phi \tag{5.178}$$

$$P_\phi = q \sin \phi \tag{5.179}$$

The load of a differential ring is given in Eq. 5.180.

$$R = q \, (2\pi r_o) a \, d\phi \tag{5.180}$$

The total load is determined by summing all of the differential rings.

$$R = q2\pi a \int_{\phi_o}^{\phi} r_o d\phi \tag{5.181}$$

Substituting the value of r_o (Eq. 5.172).

$$R = q2\pi a \int_{\phi_o}^{\phi} (a \sin \phi - r')d\phi \tag{5.182}$$

Integrating,

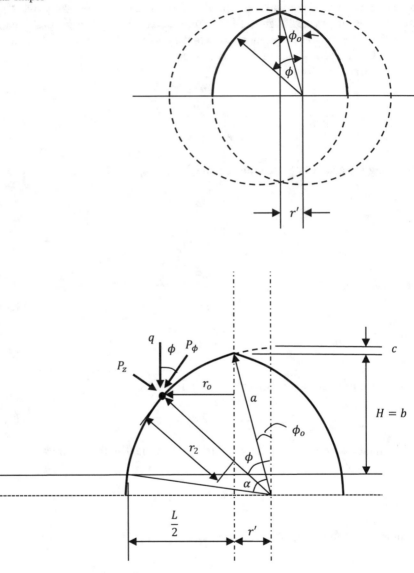

Fig. 5.12 Conoidal dome formed by two off-set circular shapes

Fig. 5.13 The radial geometry of the conoidal dome

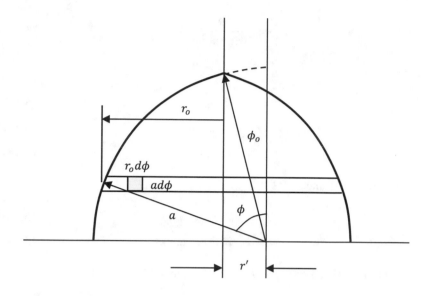

Fig. 5.14 Differential ring to solve for the total load R

$$R = q2\pi a^2(\cos\phi_o - \cos\phi) - q2\pi ar'(\phi - \phi_o) \qquad (5.183)$$

Use the general equation for a circular dome to solve for the meridian, or vertical stresses,

$$N'_\phi = -\frac{R}{2\pi r_o \sin\phi} \qquad (4.35)$$

Substituting Eqs. 5.172 and 5.183,

$$N'_\phi = -\frac{q2\pi a^2(\cos\phi_o - \cos\phi) - q2\pi ar'(\phi - \phi_o)}{2\pi(a\sin\phi - r')\sin\phi} \qquad (5.184)$$

Since,

$$r' = a\sin\phi_o \qquad (5.171)$$

$$N'_\phi = -\frac{q2\pi a^2(\cos\phi_o - \cos\phi) - q2\pi a^2\sin\phi_o(\phi - \phi_o)}{2\pi a(\sin\phi - \sin\phi_o)\sin\phi} \qquad (5.185)$$

$$N'_\phi = \frac{qa[(\cos\phi - \cos\phi_o) + \sin\phi_o(\phi - \phi_o)]}{(\sin\phi - \sin\phi_o)\sin\phi} \qquad (5.186)$$

From the general Eq. 4.37,

$$N'_\theta = \frac{R}{2\pi r_1 \sin^2 \phi} - \frac{P_z r_o}{\sin \phi} \tag{4.37}$$

Since,

$$r_1 = a \tag{5.187}$$

$$r_o = a \sin \phi - r' \tag{5.172}$$

$$r' = a \sin \phi_0 \tag{5.171}$$

Combining 5.171 and 5.172,

$$r_o = a \sin \phi - a \sin \phi_0 \tag{5.188}$$

$$r_o = a(\sin \phi - \sin \phi_0) \tag{5.189}$$

Substituting 5.178, 5.183, 5.187 and 5.188 into 4.37,

$$N'_\theta = \frac{q2\pi a^2(\cos \phi_o - \cos \phi) - q2\pi a^2 \sin \phi_0 (\phi - \phi_o)}{2\pi a \sin^2 \phi}$$
$$- \frac{(q \cos \phi)a(\sin \phi - \sin \phi_0)}{\sin \phi} \tag{5.190}$$

Simplifying the above equation,

$$N'_\theta = \frac{qa}{\sin^2 \phi}$$
$$\times [(\cos \phi_o - \cos \phi) - \sin \phi_0 (\phi - \phi_o) - \cos \phi \sin \phi (\sin \phi - \sin \phi_o)] \tag{5.191}$$

or

$$N'_\theta = - \frac{qa}{\sin^2 \phi}$$
$$\times [\cos \phi \sin \phi (\sin \phi - \sin \phi_o) + \sin \phi_0 (\phi - \phi_o) - (\cos \phi_o - \cos \phi)] \tag{5.192}$$

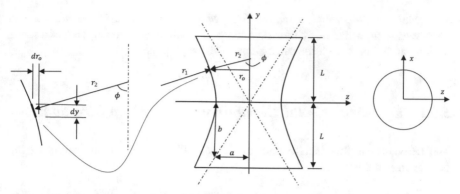

Fig. 5.15 Radial geometry of the hyperbolic shell

5.5 Hyperbolic Shell

The hyperbolic shell is used almost exclusively as a cooling tower, a structure that is very common in the production of energy. Thus, this shell is associated with nuclear and coal-fired power plants. Its popularity is due to the natural draught that occurs in the shell, used to cool water.

The hyperbolic shell equations are similar to the general equations of the circular dome, with exception of the equilibrium equation in the z-direction (the shapes differ in the radial direction) (Billington 1982; Flugge 1960; Beles and Soare 1976). The geometry of the shell is shown in Fig. 5.15.

The equation for a hyperbolic shape is given in 5.193.

$$\frac{x^2 + z^2}{a^2} - \frac{y^2}{b^2} = 1 \tag{5.193}$$

$$x^2 + z^2 = r_o^2 \tag{5.194}$$

$$r_o = r_2 \sin \phi \tag{5.195}$$

If $z = a$,

$$\frac{x^2 + a^2}{a^2} - \frac{y^2}{b^2} = 1 \tag{5.196}$$

$$\frac{x^2}{a^2} + \frac{a^2}{a^2} - \frac{y^2}{b^2} = 1 \tag{5.197}$$

Solving for y, the equation is linear.

$$y = \frac{b}{a}x \tag{5.198}$$

Equation 5.198 is the asymptote line, which passes through the origin. It is interesting to note that the cylindrical hyperbolic shell may be constructed of a series of non-parallel lines with the same slope of the asymptote.

5.5.1 Equilibrium in the Circumferential, or Hoop Direction

The derivation is identical to circular dome, and therefore a replication in the derivation is not needed.

$$N'_{\theta\varnothing}\frac{\partial r_o}{\partial\varnothing} + \frac{\partial N'_\theta}{\partial\theta}r_1 + \frac{\partial(N'_{\varnothing\theta}r_o)}{\partial\varnothing} + P_\theta r_o r_1 = 0 \qquad (4.8)$$

5.5.2 Equilibrium in the Vertical, or y-Direction

Similar to the hoop direction, the derivation in the y-direction is identical to the derivation in the meridian direction of the circular dome.

$$\frac{\partial(N'_\phi r_o)}{\partial\phi} + \frac{\partial N'_{\theta\phi}}{\partial\theta}r_1 - N'_\theta\frac{\partial r_o}{\partial\phi} + P_\phi r_1 r_o = 0 \qquad (4.12)$$

5.5.3 Equilibrium in the Radial, or z-Direction

The equilibrium equation in the z-direction is identical to the equilibrium equation for the circular dome. The circular shell curves inward, while the hyperbolic shell curves outward. This difference in curvature, however, is accommodated by the values of r_1 and r_2, defined by the geometry of the shell.

$$\frac{N'_\phi}{r_1} + \frac{N'_\theta}{r_2} + P_z = 0 \qquad (4.18)$$

5.5.4 Defining the Principle Radii r₁ and r₂

The equation for curvature for a hyperboloid is defined as:

$$\frac{1}{r_1} = -\frac{\left(\frac{d^2 r_o}{dy^2}\right)}{\left[1 + \left(\frac{dr_o}{dy}\right)^2\right]^{3/2}} \tag{5.199}$$

From Fig. 5.15,

$$r_2 \sin \phi = r_o \tag{5.200}$$

or

$$r_2 = \frac{r_o}{\sin \phi} \tag{5.201}$$

Since the shape of the hyperbolic shell is defined as:

$$\frac{x^2 + z^2}{a^2} - \frac{y^2}{b^2} = 1 \tag{5.202}$$

We simplify by noting that:

$$x^2 + z^2 = r_o^2 \tag{5.203}$$

Equation 5.193 becomes

$$\frac{r_o^2}{a^2} - \frac{y^2}{b^2} = 1 \tag{5.204}$$

Rearranging and simplifying:

$$\frac{r_o^2}{a^2} = 1 + \frac{y^2}{b^2} \tag{5.205}$$

$$r_o^2 = a^2 \left(1 + \frac{y^2}{b^2}\right) \tag{5.206}$$

$$r_o^2 = \frac{a^2 b^2}{b^2} \left(1 + \frac{y^2}{b^2}\right) \tag{5.207}$$

$$r_o^2 = \frac{a^2}{b^2} \left(b^2 + y^2\right) \tag{5.208}$$

Square root each side of the equation.

$$r_0 = \frac{a}{b} \sqrt{\left(b^2 + y^2\right)} \tag{5.209}$$

Taking the derivative of 5.209 with respect to y,

$$\frac{dr_0}{dy} = \frac{a}{b} \frac{y}{\sqrt{(b^2 + y^2)}}$$

(5.210)

Manipulating the equation,

$$\frac{dr_0}{dy} = \frac{ar_o}{br_o} \frac{y}{\sqrt{(b^2 + y^2)}}$$

(5.211)

Replacing r_o in the numerator with 5.209,

$$\frac{dr_0}{dy} = \frac{a\left(\frac{a}{b}\sqrt{(b^2 + y^2)}\right)}{br_o} \frac{y}{\sqrt{(b^2 + y^2)}}$$

(5.212)

$$\frac{dr_0}{dy} = \frac{a^2 y}{b^2 r_o}$$

(5.213)

Taking the second derivative,

$$\left(\frac{f}{g}\right)' = \frac{f'}{g} - \frac{fg'}{g^2}$$

(5.214)

$$\frac{d^2 r_o}{dy^2} = \frac{a^2}{b^2} \left[\frac{1}{\frac{a}{b}(b^2 + y^2)^{1/2}} - \frac{y}{\frac{a^2}{b^2}(b^2 + y^2)} \frac{a}{b}(b^2 + y^2)^{-1/2} y \right]$$

(5.215)

$$\frac{d^2 r_o}{dy^2} = \frac{a^2}{b^2} \left[\frac{b}{a(b^2 + y^2)^{1/2}} - \frac{by^2}{a(b^2 + y^2)^{3/2}} \right]$$

(5.216)

$$\frac{d^2 r_o}{dy^2} = \frac{a}{b} \left[\frac{1}{(b^2 + y^2)^{1/2}} - \frac{y^2}{(b^2 + y^2)^{3/2}} \right]$$

(5.217)

$$\frac{d^2 r_o}{dy^2} = \frac{a}{b} \left[\frac{(b^2 + y^2) - y^2}{(b^2 + y^2)^{3/2}} \right]$$

(5.218)

$$\frac{d^2 r_0}{dy^2} = \frac{ab}{(b^2 + y^2)^{3/2}}$$

(5.219)

Substituting 5.213 and 5.219 into 5.199,

$$\frac{1}{r_1} = -\frac{\frac{ab}{(b^2 + y^2)^{3/2}}}{\left[1 + \left(\frac{a^2 y}{b^2 r_o}\right)^2\right]^{3/2}}$$

(5.220)

$$\frac{1}{r_1} = -\frac{ab}{\left(b^2 + y^2\right)^{3/2}\left[1 + \left(\frac{a^2 y}{b^2 r_o}\right)^2\right]^{3/2}} \tag{5.221}$$

Taking the reciprocal,

$$r_1 = -\frac{\left(b^2 + y^2\right)^{3/2}\left[1 + \left(\frac{a^2 y}{b^2 r_o}\right)^2\right]^{3/2}}{ab} \tag{5.222}$$

or,

$$r_1 = -\frac{\left(b^2 + y^2\right)^{3/2}\left[1 + \left(\frac{a^4 y^2}{b^4 r_o^2}\right)\right]^{3/2}}{ab} \tag{5.223}$$

Since,

$$r_0 = \frac{a}{b}\sqrt{\left(b^2 + y^2\right)} = \frac{a}{b}\left(b^2 + y^2\right)^{1/2} \tag{5.209}$$

The term in the numerator of 5.223,

$$\left(b^2 + y^2\right)^{3/2} = \left[\left(b^2 + y^2\right)^{1/2}\right]^3 \tag{5.224}$$

By manipulation,

$$\left(b^2 + y^2\right)^{3/2} = \frac{b^3}{a^3}\left[\frac{a}{b}\left(b^2 + y^2\right)^{1/2}\right]^3 = \frac{b^3}{a^3}[r_o]^3 = \frac{b^3 r_o^3}{a^3} \tag{5.225}$$

Therefore,

$$r_1 = -\frac{\frac{b^3 r_o^3}{a^3}\left[1 + \left(\frac{a^4 y^2}{b^4 r_o^2}\right)\right]^{3/2}}{ab} \tag{5.226}$$

$$r_1 = -\frac{b^2 r_o^3}{a^4}\left[\left(1 + \frac{a^4 y^2}{b^4 r_o^2}\right)^{1/2}\right]^3 \tag{5.227}$$

$$r_1 = -\frac{b^2}{a^4}\left[r_o\left(\frac{a^4}{a^4} + \frac{a^4 y^2}{b^4 r_o^2}\right)^{1/2}\right]^3 \tag{5.228}$$

$$r_1 = -b^2 a^2 \left(\frac{r_o^2}{a^4} + \frac{y^2}{b^4} \right)^{3/2} \tag{5.229}$$

$$r_1 = -b^2 a^2 \left[\frac{1}{a^2} \left(r_o^2 + \frac{y^2 a^4}{b^4} \right) \right]^{3/2} \tag{5.230}$$

$$r_1 = -b^2 a^2 \left[\frac{1}{a^2} \left(r_o^2 + \frac{y^2 a^4}{b^4} \right)^{1/2} \right]^3 \tag{5.231}$$

$$r_1 = -\frac{b^2}{a^4} \left[\left(r_o^2 + \frac{y^2 a^4}{b^4} \right)^{1/2} \right]^3 \tag{5.232}$$

Now solving for r_2.

$$r_2 = \frac{r_o}{\sin \phi} \tag{5.201}$$

From the trigonometric identity,

$$\sin \phi = \frac{1}{\sqrt{1 + \cot^2 \phi}} \tag{5.233}$$

and from Fig. 5.15,

$$\cot \phi = \frac{dr_o}{dy} \tag{5.234}$$

Combining 5.201, 5.233, and 5.234,

$$r_2 = r_o \sqrt{1 + \cot^2 \phi} = r_o \sqrt{1 + \left(\frac{dr_o}{dy} \right)^2} \tag{5.235}$$

Substituting 5.213,

$$r_2 = r_o \sqrt{1 + \left(\frac{a^2 y}{b^2 r_o} \right)^2} \tag{5.236}$$

$$r_2 = r_o \left[1 + \frac{a^4 y^2}{b^4 r_o^2} \right]^{1/2} \tag{5.237}$$

$$r_2 = \left(r_o^2 + \frac{a^4 y^2}{b^4} \right)^{1/2} \tag{5.238}$$

Substituting 5.208,

$$r_2 = \left[\frac{a^2}{b^2}\left(b^2 + y^2\right) + \frac{a^4y^2}{b^4}\right]^{1/2} \tag{5.239}$$

$$r_2 = \left[a^2 + \frac{a^2y^2}{b^2} + \frac{a^4y^2}{b^4}\right]^{1/2} \tag{5.240}$$

$$r_2 = a\left[1 + \left(\frac{1}{b^2} + \frac{a^2}{b^4}\right)y^2\right]^{1/2} \tag{5.241}$$

or

$$r_2 = a\sqrt{1 + \left(\frac{1}{b^2} + \frac{a^2}{b^4}\right)y^2} \tag{5.242}$$

Substituting 5.238 into 5.232,

$$r_1 = -\frac{b^2}{a^4}(r_2)^3 \tag{5.243}$$

5.5.5 Membrane Stresses Due to Gravity Load

From Fig. 5.16, the gravity load is expressed as follows:

$$P_y = -q\sin\phi \tag{5.244}$$
$$P_z = -q\cos\phi \tag{5.245}$$

Similar to circular dome theory, the vertical stress may be expressed as:

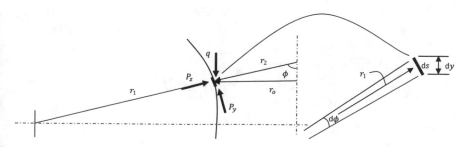

Fig. 5.16 Geometry of gravity loads

$$N'_\phi = \frac{-R}{2\pi r_o \sin\phi} \qquad (4.35)$$

Since,

$$r_o = r_2 \sin\phi \qquad (5.200)$$

Substitute 5.200 into 4.35.

$$N'_\phi = \frac{-R}{2\pi r_2 \sin^2\phi} \qquad (5.246)$$

Furthermore,

$$R = \int_0^\phi q 2\pi r_o r_1 d\phi \qquad (5.247)$$

Substituting 5.200,

$$R = \int_0^\phi q 2\pi r_2 \sin\phi r_1 d\phi \qquad (5.248)$$

Substituting 5.248 into 5.246,

$$N'_\phi = -\frac{1}{2\pi r_2 \sin^2\phi} \int_0^\phi q 2\pi r_2 r_1 \sin\phi d\phi \qquad (5.249)$$

$$N'_\phi = -\frac{1}{r_2 \sin^2\phi} \int_0^\phi q r_2 r_1 \sin\phi d\phi \qquad (5.250)$$

From Fig. 5.16,

$$r_1 d\phi = \frac{dy}{\sin\phi} \qquad (5.251)$$

$$dy = r_1 \sin\phi d\phi \qquad (5.252)$$

Therefore,

$$N'_\phi = -\frac{1}{r_2 \sin^2\phi} \int_0^y q r_2 dy \qquad (5.253)$$

$$N'_\phi = -\frac{r_2}{r_2^2 \sin^2 \phi} \int_0^y qr_2 dy \qquad (5.254)$$

Substituting 5.200,

$$N'_\phi = -\frac{r_2}{r_o^2} \int_0^y qr_2 dy \qquad (5.255)$$

From the equilibrium equation,

$$\frac{N'_\phi}{r_1} + \frac{N'_\theta}{r_2} + P_z = 0 \qquad (4.18)$$

and substituting 5.243,

$$\frac{N'_\phi}{-\frac{b^2}{a^4}(r_2)^3} + \frac{N'_\theta}{r_2} + P_z = 0 \qquad (5.256)$$

Rearranging the above equation,

$$N'_\theta = \frac{N'_\phi a^4}{b^2 r_2^2} - P_z r_2 \qquad (5.257)$$

Substitute equations 5.245 and 5.255.

$$N'_\theta = -\frac{a^4}{b^2 r_2 r_o^2} \left(\int_0^y qr_2 dy + c \right) + q \cos \phi r_2 \qquad (5.258)$$

Since,

$$r_o = r_2 \sin \phi \qquad (5.200)$$
$$r_2 \cos \phi = r_o \cot \phi \qquad (5.259)$$
$$N'_\theta = -\frac{a^4}{b^2 r_2 r_o^2} \left(\int_0^y qr_2 dy + c \right) + q r_o \cot \phi \qquad (5.260)$$

Substituting 5.234 and 5.213,

$$N'_\theta = -\frac{a^4}{b^2 r_2 r_o^2} \left(\int_0^y qr_2 dy + c \right) + \frac{qa^2 y}{b^2} \qquad (5.261)$$

Due to the complexity of the integrand, the above equation is integrated numerically.

5.6   Example Solutions and a Comparison of the Circular, Conoidal, Elliptical, and Conical Domes

To determine the efficiency of each dome, a comparison is drawn for shapes that have identical heights and lengths. The height from floor level to apex is 5 m and the length is 20 m. All of the solutions will only include the membrane vertical and hoop stresses for gravity load (self-weight). For simplicity, the loading will be assumed to be equal to 1.

Conoidal Dome

The solution requires the choice of five parameters, of which three are the same for each of the solutions. Referring to Fig. 5.13, the other two parameters (r' and α) are arbitrarily chosen—these two parameters define the shape of the dome (i.e., how pointed the shell is and the slope at the base of the shell).

Near the apex of the shell, the equations deteriorate rapidly as ϕ goes to zero. Thus, the solution near the apex should be excluded and the stresses extrapolated to zero (for both the vertical and hoop stresses).

$H = 5$ m
$L = 20$ m
$q = 1$ kN/m^2
$\alpha = 45°$
$r' = 2.5$ m

With the above parameters, the radius (a) and the difference between the height H and the peak of the circular arc (c) are determined from the following equations:

$$a^2 = (r')^2 + (a - c)^2 \tag{5.262}$$

$$a^2 = \left(\frac{L}{2} + r'\right)^2 + (a - c - H)^2 \tag{5.263}$$

Substituting known values,

$$a^2 = (2.5)^2 + (a - c)^2$$

$$a^2 = (10 + 2.5)^2 + (a - c - 5)^2$$

With two equations and two unknowns, a solution is possible.

$a = 17.678$ m
$c = 0.178$ m

The angle (ϕ_o) is therefore solvable from a and c.

$$\phi_o = \sin^{-1}\left(\frac{a-c}{a}\right) \qquad (5.264)$$

Solving for ϕ_o,

$$\phi_o = \sin^{-1}\left(\frac{17.678 - 0.178}{17.678}\right) = 8.13°$$

The vertical and hoop stresses, at various angles ϕ, may now be solved. At the base of the shell:

$$N'_\phi = \frac{qa[(\cos\phi - \cos\phi_o) + \sin\phi_o(\phi - \phi_o)]}{(\sin\phi - \sin\phi_o)\sin\phi} \qquad (5.186)$$

$$N'_\phi = \frac{(1)17.678[(\cos 45 - \cos 8.13) + \sin 8.13(45 - 8.13)]}{(\sin 45 - \sin 8.13)\sin 45} = -8.478 \text{ kN/m}$$

$$N'_\theta = -\frac{qa}{\sin^2\phi}$$

$$\times [\cos\phi \sin\phi(\sin\phi - \sin\phi_o) + \sin\phi_0(\phi - \phi_o) - (\cos\phi_o - \cos\phi)] \qquad (5.192)$$

$$N'_\theta = -\frac{(1)17.678}{\sin^2 45}$$

$$\times [\cos 45 \sin 45(\sin 45 - \sin 8.13) + \sin 8.13(45 - 8.13) - (\cos 8.13 - \cos 45)]$$
$$= -3.218 \text{ kN/m}$$

Elliptical Dome
Referring to Fig. 5.10, the major and minor axis of the elliptical shape must be defined. For this case,

$H = b = 5$ m
$L = 2a = 20$ m
$q = 1$ kN/m^2

The only restriction with an elliptical dome is that the major axis must be larger than the minor axis (i.e., $a > b$).

The membrane stresses are given in Eqs. 5.155 and 5.165.

$$N'_\phi = -\frac{qa^2}{b}\frac{CQ}{\left(1 - \frac{y^2}{b^2}\right)} \qquad (5.159)$$

$$N'_\theta = -\frac{qa^2}{b}\left[\frac{y}{b} - \frac{Cb^2}{Q(b^2 - y^2)}\right]$$ (5.170)

where

$$C = \frac{1}{2} - \frac{y}{2ab^2}\sqrt{b^4 + y^2(a^2 - b^2)}$$

$$+ \frac{b^2}{2a\sqrt{a^2 - b^2}}\log\frac{b\left(a + \sqrt{a^2 - b^2}\right)}{y\sqrt{a^2 - b^2} + \sqrt{b^4 + y^2(a^2 - b^2)}}$$ (5.146)

$$Q = \frac{\sqrt{b^4 + y^2(a^2 - b^2)}}{ab}$$ (5.157)

Solving for the stresses at the base of the shell ($y = 0$),

$$C = \frac{1}{2} - \frac{0}{2(10)5^2}\sqrt{5^4 + 0^2(10^2 - 5^2)}$$

$$+ \frac{5^2}{2(10)\sqrt{10^2 - 5^2}}\log\frac{5\left(10 + \sqrt{10^2 - 5^2}\right)}{0\sqrt{10^2 - 5^2} + \sqrt{5^4 + 0^2(10^2 - 5^2)}}$$

$$= 0.583$$

$$Q = \frac{\sqrt{5^4 + 0^2(10^2 - 5^2)}}{(10)5} = 0.5$$

$$N'_\phi = -\frac{(1)10^2}{5}\frac{(0.583)0.5}{\left(1 - \frac{0^2}{5^2}\right)} = -5.826 \text{ kN/m}$$

$$N'_\theta = -\frac{(1)10^2}{5}\left[\frac{0}{5} - \frac{(0.583)5^2}{0.5(5^2 - 0^2)}\right] = 23.302 \text{ kN/m}$$

Conical Dome

The only parameters that must be defined are the height and base radius.

$H = 5$ m

$L = 2r_o = 20$ m

$q = 1$ kN/m^2

The angle α is a constant, defined by:

$$\alpha = \tan^{-1}\left(\frac{L}{2H}\right) \tag{5.265}$$

And the length of the sloped walls of the shell,

$$\text{Slope length} = \sqrt{\left(\frac{L}{2}\right)^2 + H^2} \tag{5.266}$$

Solving for the angle and slope length,

$$\alpha = \tan^{-1}\left(\frac{20}{2(5)}\right) = 63.435^\circ$$

$$\text{Slope length} = \sqrt{\left(\frac{20}{2}\right)^2 + 5^2} = 11.180 \text{ m}$$

Solving for the stresses at the base of the shell,

$$N'_\phi = -\frac{q\,y}{2\cos\alpha} \tag{5.104}$$

$$N'_\phi = -\frac{(1)11.180}{2\cos 63.435} = -12.5 \text{ kN/m}$$

$$N'_\theta = -qy\,\sin\alpha\tan\alpha \tag{5.109}$$

$$N'_\theta = -(1)11.180\,\sin 63.435\tan 63.435 = -20 \text{ kN/m}$$

Circular Dome

Although a solution that includes boundary effect is available, only the membrane stresses are solved to enable a direct comparison with the other shell shapes. The solution of the circular dome assumes the same parameters as the other shell shapes.

$H = 5$ m
$L = 20$ m
$q = 1$ kN/m^2

With only the two geometrical dimensions, the other parameters are solved by the following equations:

$$\alpha = 180 - 2\,\tan^{-1}\left(\frac{L}{2H}\right) \tag{4.263}$$

$$a = \frac{H^2 + \left(\frac{L}{2}\right)^2}{2H} \tag{4.262}$$

Solving these two equations,

$$\alpha = 180 - 2 \ \tan^{-1}\left[\frac{20}{2(5)}\right] = 53.13^{\circ}$$

$$a = \frac{5^2 + \left(\frac{20}{2}\right)^2}{2(5)} = 12.5 \ \text{m}$$

The stresses are solved by Eqs. 4.50 and 4.53.

$$N'_{\phi} = -\frac{qa}{(1 + \cos\phi)} \qquad (4.50)$$

$$N'_{\theta} = aq\left(\frac{1}{1 + \cos\phi} - \cos\phi\right) \qquad (4.53)$$

Solving for the stresses at the base of the shell,

$$N'_{\phi} = -\frac{(1)12.5}{(1 + \cos 53.13)} = -7.81 \ \text{kN/m}$$

$$N'_{\theta} = 12.5(1)\left(\frac{1}{1 + \cos 53.13} - \cos 53.13\right) = 0.31 \ \text{kN/m}$$

Comparison of Stresses

Graphs of N'_{ϕ} and N'_{θ} are given in Figs. 5.17 and 5.18.

As seen in the figures, the errors of the conoidal dome analysis are apparent in both figures; the stresses diverge rapidly near the apex of the dome. The analysis should take this into account and extrapolate the stress to a zero point. By making this adjustment to the graphs, it is evident that the conoidal and circular domes are the most efficient in carrying the self-weight of the shell. In comparison, the conical and elliptical domes show very high peak stresses at the base of the shell.

The membrane stresses are the actual stress distribution in the majority of the shell. A design, based only on a membrane stress analysis, is an acceptable method. For example, silo designs are usually only based on a membrane solution. These simplified methods are endorsed by the majority of codes of practice. However, the designer must be mindful that the stress distribution is not entirely correct. At the base of the shell, fixity will cause boundary effects. In other words, bending moments and shears will exist at the foundation level. The magnitude of the moment is dependent on the degree of fixity and Gaussian curvature. However, moments and shears usually dissipate rapidly, depending on the stiffness and curvature of the shell. Stiffer shells, such as those constructed of reinforced concrete, may have moments and shears that travel up to a third of the height of the shell. On the other hand, flexible shells, such as those that are constructed of steel, may dissipate the moments and shears relatively close to the base. Whatever the case may be, the designer must be aware of the limitations of a purely membrane solution.

5.7 Exercises

5.7.1. Program the membrane theory for the conoidal, elliptical, and conical domes into a spreadsheet, or an alternative programing language. Reproduce Figs. 5.17 and 5.18, but in tabular form in no less than 10 height divisions. Compare with circular dome theory of Chap. 4, and assume $r' = 2.5$ for the conoidal dome.

5.7.2. Repeat exercise 5.7.1, but set $H = 1$ m. Graph the results and compare to Figs. 5.17 and 5.18.

5.7.3. Compare an inverted cone and a cylindrical water tank for water loads. Both tanks are full and measure 10 m high and 10 m in diameter. The thickness of each shell is 200 mm, constructed of reinforced concrete. Assume that the weight of concrete is 24 kN/m^3 and water 10 kN/m^3. Determine and graph the stress distribution. Draw conclusions concerning the efficiency of each shape, and why a shape is more efficient than the other.

5.7.4. Comment how to incorporate into the design the boundary effects (Fig. 5.19).

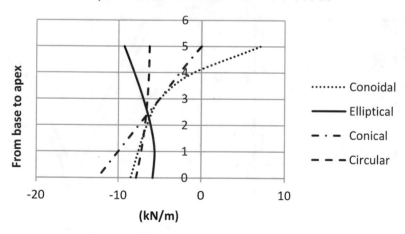

Fig. 5.17 Comparison of the membrane stresses in the vertical direction

N'θ Hoop Stresses

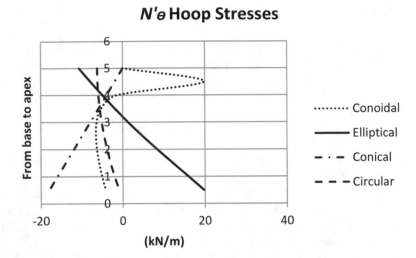

Fig. 5.18 Comparison of the membrane stresses in the hoop direction

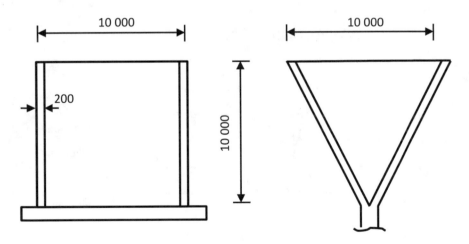

Fig. 5.19 Cylindrical and conical water tanks

References

Beles, A. A. & Soare, M. V., 1976. *Elliptic and Hyperbolic Paraboloidal Shells Used in Construction.* London: S P Christie and Partners.

Billington, D. P., 1982. *Thin Shell Concrete Structures.* 2 ed. New York: McGraw-Hill.

Chronowicz, A., 1959. *The Design of Shells: A Practical Approach.* London: William Clowes and Sons.

Flugge, W., 1960. *Stresses in Shells.* Berlin: Springer-Verag.

Timoshenko, S., 1959. *Theory of Plates and Shells.* 2 ed. New York: McGraw-Hill Book Co..

Wilson, A., 2005. *Practical Design of Concrete Shells.* Texas: Monolithic Dome Institute.

Chapter 6
The Circular Barrel Vault

Mitchell Gohnert

Abstract This chapter develops the membrane and deformation theories for barrel vaults, for various loadings such as uniform vertical and horizontal loads, distributed evenly or assuming a sin wave variation. Boundary effects theory is developed to correct membrane errors that occur along the free edge. Boundary theory is also derived for the case of edge beams; compatibility equations are formulated to determine the reactions between the edge beam and shell walls, which are used to distribute the reaction stresses in the walls of the vault. The mathematical derivation is fully developed for this shell. However, the ASCE Manuals of Engineering Practice, No. 31, on the Design of Cylindrical Concrete Shell Roofs is given as an alternative analysis procedure, which is based on the developed theory. The ASCE method is given to ease the complex nature of the analysis.

6.1 Introduction

The circular arch shell is commonly known as a barrel shell, or barrel vault. The cross-section is circular, and ranges from a semicircle to a segment of a circle. The barrel vault is a shell of single curvature, and resembles a slice of a full cylinder. The corners of the vault are supported by columns, and longitudinal edge beams often span between columns, supporting the free edge. The basic configuration is given Fig. 6.1. This type of shell is characteristically used as a roof structure. The plan dimensions of a barrel vault is rectangular; thus, the shell will fit onto a rectangular arrangement of vertical walls. It is therefore a favorable solution, and integrates shell structures into living spaces we are accustomed too.

The mathematical derivation of the barrel shell is, perhaps, the most complex of all the shell theories, and the membrane solution is the least accurate through a range of configurations. For barrel shells of long lengths, the real stresses deviate significantly from the membrane theory, but tends to be closer to beam theory (i.e., modeling a barrel vault as beam, with a curved cross section) (Billington 1982;

M. Gohnert
University of the Witwatersrand, Johannsesburg, South Africa

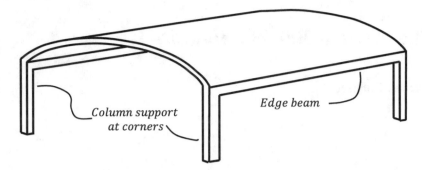

Fig. 6.1 Barrel vault with edge beams and supports at the corners

Flugge 1960). In shorter shells, the real stress distribution is closer to the membrane theory, but a deviation is still evident, especially near the boundaries. Unlike other shell shapes, where a membrane solution is a good representation of actual stress, the barrel shell membrane solution should be used with caution, coupled with an understanding of boundary effects.

If the vault is supported on a foundation, or if the edge beams are rigidly supported, arch theory is applied; arches, however, are considered in Chap. 7.

The theory is developed in three parts: The first part solves for the membrane stresses, assuming that the shell is free along the edge. The second part considers the influence of stresses, bending moments, and shears in the real system (i.e., corrections to the membrane solution). The third part determines the influence of edge beams.

6.2 Membrane Theory of the Barrel Vault

6.2.1 External Loading on the Shell

The geometry and membrane forces are defined in Fig. 6.2.

From cylindrical shell theory, the membrane equilibrium equations are defined by Eqs. 3.11, 3.13 and 3.14.

$$N'_{y\theta} = \int \left(-\frac{1}{r} \frac{\partial N'_\theta}{\partial \theta} - P_\theta \right) dy + f_1(\theta) \tag{3.11}$$

$$N'_y = \int \left(-\frac{1}{r} \frac{\partial N'_{\theta y}}{\partial \theta} - P_y \right) dy + f_2(\theta) \tag{3.13}$$

$$N'_\theta = -P_z r \tag{3.14}$$

Change the subscript θ to ϕ and y to x, to match the geometry of Fig. 6.2

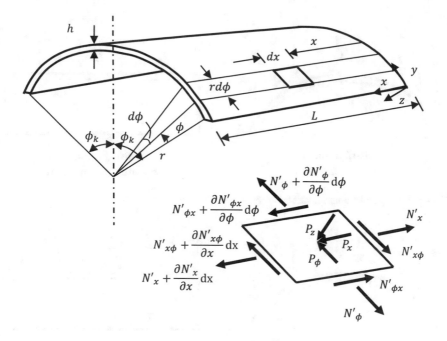

Fig. 6.2 Geometry and membrane stresses in a differential element

$$N'_{x\phi} = \int \left(-\frac{1}{r} \frac{\partial N'_\phi}{\partial \phi} - P_\phi \right) dx + f_1(\phi) \tag{6.1}$$

$$N'_x = \int \left(-\frac{1}{r} \frac{\partial N'_{\phi x}}{\partial \phi} - P_x \right) dx + f_2(\phi) \tag{6.2}$$

$$N'_\phi = -P_z r \tag{6.3}$$

The load pressures (P) are defined according to how the load is spread over the shell. The expressions for P_z, P_ϕ, and P_x are a function of the load P_d.

Uniform distributed load in the transverse and longitudinal directions (gravity load) (Fig. 6.3)

$$P_z = P_d \cos(\phi_k - \phi) \tag{6.4}$$

$$P_\phi = -P_d \sin(\phi_k - \phi) \tag{6.5}$$

$$P_x = 0 \tag{6.6}$$

where P_d represents the self-weight of the shell.

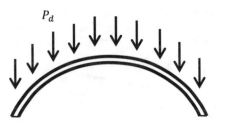

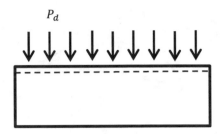

Load distribution in the
transverse direction

Load distribution in the
longitudinal direction

Fig. 6.3 Gravity loading on the shell

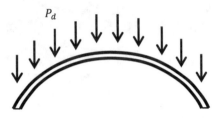

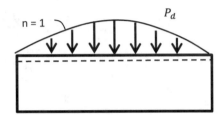

Load distribution in transverse
direction

Load distribution in longitudinal
direction

Fig. 6.4 UDL distributed according to a sin wave in the longitudinal direction

Uniform distributed load in the transverse direction, but a sin wave variation in the longitudinal direction (Fig. 6.4)

$$P_z = P_d \cos{(\phi_k - \phi)} \frac{1}{n} \sin{\left(\frac{n\pi x}{L}\right)} \qquad (6.7)$$

$$P_\phi = -P_d \sin{(\phi_k - \phi)} \frac{1}{n} \sin{\left(\frac{n\pi x}{L}\right)} \qquad (6.8)$$

$$P_x = 0 \qquad (6.9)$$

where $n = 1, 3, 5, \ldots$

If $n = 1$, the shape of the loading is a half of a sign wave. If $n = 3$, the shape of the loading has three halves of a sign wave.

Horizontal uniformly distributed load in the transverse and longitudinal directions (Fig. 6.5)

$$P_z = P_d \cos^2{(\phi_k - \phi)} \qquad (6.10)$$

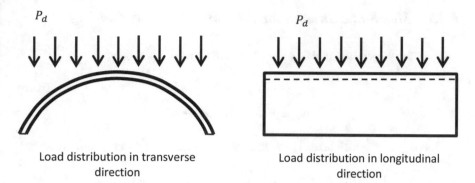

Load distribution in transverse
direction

Load distribution in longitudinal
direction

Fig. 6.5   Horizontal distribution of load in the transverse and longitudinal directions

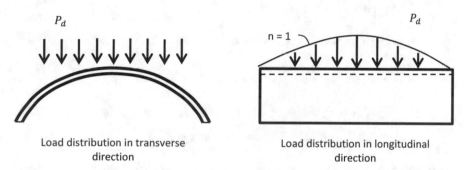

Load distribution in transverse
direction

Load distribution in longitudinal
direction

Fig. 6.6   Horizontal distribution of load, varying according to a sin wave in the longitudinal direction

$$P_\phi = -P_d \cos\left(\phi_k - \phi\right) \sin\left(\phi_k - \phi\right) \tag{6.11}$$

$$P_x = 0 \tag{6.12}$$

Horizontal uniformly distributed load in the transverse direction, but a sin wave variation in the longitudinal direction (Fig. 6.6)

$$P_z = P_d \cos^2(\phi_k - \phi)\frac{1}{n}\sin\left(\frac{n\pi x}{L}\right) \tag{6.13}$$

$$P_\phi = -P_d \cos\left(\phi_k - \phi\right) \sin\left(\phi_k - \phi\right)\frac{1}{n}\sin\left(\frac{n\pi x}{L}\right) \tag{6.14}$$

$$P_x = 0 \tag{6.15}$$

where $n = 1, 3, 5, \ldots$

The above equations solve for the pressure in the z- and ϕ-directions. The most common distribution is the first case, Eqs. 6.4, 6.5, and 6.6.

6.2.2 Membrane Stresses for the Case of Gravity Loading

Solving for the stresses in the ϕ-direction, substitute 6.4 into 6.3.

$$N'_\phi = -[P_d \ \cos{(\phi_k - \phi)}]r \tag{6.16}$$

$$N'_\phi = -P_d \ r \ \cos{(\phi_k - \phi)} \tag{6.17}$$

Solving for the membrane shear stresses, substitute 6.17 and 6.5 into 6.1.

$$N'_{x\phi} = -\frac{1}{r} \int \left[\frac{\partial(-P_d \ r \ \cos{(\phi_k - \phi)})}{\partial \phi} \right] dx - \int [-P_d \sin{(\phi_k - \phi)}] dx$$
$$+ f_1(\phi) \tag{6.18}$$

$$N'_{x\phi} = P_d \int \sin{(\phi_k - \phi)} dx + P_d \int \sin{(\phi_k - \phi)} dx + f_1(\phi) \tag{6.19}$$

$$N'_{x\phi} = P_d \sin{(\phi_k - \phi)} \int dx + P_d \sin{(\phi_k - \phi)} \int dx + f_1(\phi) \tag{6.20}$$

$$N'_{x\phi} = P_d \sin{(\phi_k - \phi)}x + P_d \sin{(\phi_k - \phi)}x + f_1(\phi) \tag{6.21}$$

$$N'_{x\phi} = 2P_d \sin{(\phi_k - \phi)}x + f_1(\phi) \tag{6.22}$$

To solve for $f_1(\phi)$, we consider the shears in the shell. We assume that the barrel shell in Fig. 6.2 is simply supported at $x = 0$ and $x = L$. At $x = \frac{L}{2}$, the shear is equal to zero. Substituting these conditions into 6.22,

$$0 = 2P_d \sin{(\phi_k - \phi)}\frac{L}{2} + f_1(\phi) \tag{6.23}$$

Solving for $f_1(\phi)$,

$$f_1(\phi) = -P_d L \sin{(\phi_k - \phi)} \tag{6.24}$$

Therefore, Eq. 6.22 becomes

$$N'_{x\phi} = 2P_d \sin{(\phi_k - \phi)}x - P_d L \sin{(\phi_k - \phi)} \tag{6.25}$$

Rearranged,

$$N'_{x\phi} = P_d \sin{(\phi_k - \phi)}(2x - L) \tag{6.26}$$

or

$$N'_{x\phi} = -P_d \sin{(\phi_k - \phi)}L\left(1 - \frac{2x}{L}\right) \tag{6.27}$$

To solve for the stresses in the x direction, substitute 6.27 and 6.6 into Eq. 6.2.

$$N'_x = \int\left[-\frac{1}{r}\frac{\partial\left(-P_d \sin{(\phi_k - \phi)}L\left(1 - \frac{2x}{L}\right)\right)}{\partial\phi} - 0\right]dx + f_2(\phi) \tag{6.28}$$

Using the product rule to take the derivative,

$$(fg)' = f'g + fg'$$

$$f = \sin{(\phi_k - \phi)}\quad f' = -\cos{(\phi_k - \phi)}$$

$$g = -P_d L\left(1 - \frac{2x}{L}\right)\quad g' = 0$$

Therefore,

$$N'_x = -\frac{1}{r}\int P_d L \cos{(\phi_k - \phi)}\left(1 - \frac{2x}{L}\right)dx + f_2(\phi) \tag{6.29}$$

$$N'_x = -\frac{P_d L \cos{(\phi_k - \phi)}}{r}\int\left(1 - \frac{2x}{L}\right)dx + f_2(\phi) \tag{6.30}$$

Since,

$$\int\left(1 - \frac{2x}{L}\right)dx = \left(x - \frac{x^2}{L}\right) = x\left(1 - \frac{x}{L}\right)$$

$$N'_x = -\frac{P_d L}{r}\cos{(\phi_k - \phi)}x\left(1 - \frac{x}{L}\right) + f_2(\phi) \tag{6.31}$$

To solve for $f_2(\phi)$, boundary conditions are again applied. At $x = 0$ and $x = L$, the stress is equal to zero for the case of simple supports.

$$0 = -\frac{P_d L}{r}\cos{(\phi_k - \phi)}(0)\left(1 - \frac{0}{L}\right) + f_2(\phi) \tag{6.32}$$

Thus,

$$f_2(\phi) = 0 \tag{6.33}$$

$$N'_x = -\frac{P_d L}{r}\cos{(\phi_k - \phi)}x\left(1 - \frac{x}{L}\right) \tag{6.34}$$

The other membrane stresses for different load cases are solved in a similar manner.

The real internal stresses may differ significantly from the membrane stresses in long shells. In shorter shells, the real internal stresses are closer to the membrane solution, but a deviation is still evident, especially near the free edge. In barrel vaults, the stresses of the membrane solution are incorrect along the free edge of the shell (the stresses should reduce to zero along the free edge). Corrections to the membrane solution are therefore required by considering boundary effects, or by applying line loads that are equivalent to the errors along the free edge. The boundary effects theory is derived, but the line load method is given as a design procedure. Both methods are based on the same theory, but expressed in a different form.

6.3   Deformation Theory

A differential element is shown in Fig. 6.7, before loading is applied to the shell, and after. As depicted, the total deformation is composed of two parts—translations and rotations. The differential element may deform in three directions, i.e., u in the x-direction, v in the ϕ-direction, and w in the z-direction.

In Fig. 6.8, only the translations and rotations in the v-directions are considered. The dashed curve represents the edge of the element, which has undergone a translation v and a rotation ϕ. The rotation is represented by the following equation:

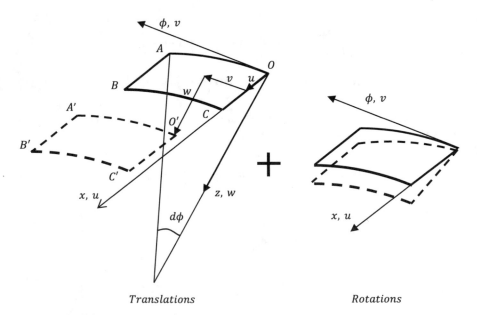

Translations   *Rotations*

Fig. 6.7 Deformations of a differential element

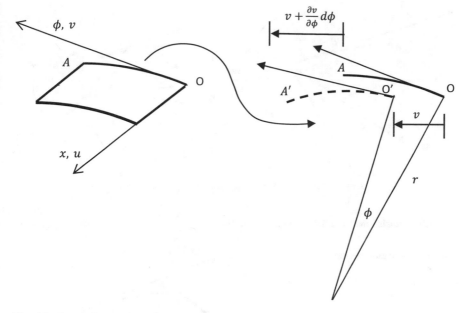

Fig. 6.8 Translations and rotations in the v-direction

$$\phi_\phi = \frac{v}{r} + \frac{\frac{\partial v}{\partial \phi} d\phi}{r} \tag{6.35}$$

Since the value of the second term of 6.35 is small compared to the first term, the rotation may be expressed in a simpler form:

$$\phi_\phi = \frac{v}{r} \tag{6.36}$$

Referring to Fig. 6.9, the deformations that are illustrated are only those in the w-direction. The rotation due to the deformation w is given by Eq. 6.37.

$$\phi_\phi = \frac{\frac{\partial w}{\partial \phi} d\phi}{r d\phi} = \frac{\partial w}{r \partial \phi} \tag{6.37}$$

The total rotational deformation of the element is the sum of Eqs. 6.36 and 6.37.

$$\phi_\phi = \frac{v}{r} + \frac{\partial w}{r \partial \phi} \tag{6.38}$$

In Fig. 6.10, the rotations are given in the x-direction, which is expressed as:

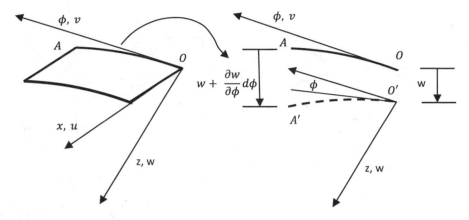

Fig. 6.9 Translations and rotations in the z-direction

Fig. 6.10 Rotations in the
x-direction

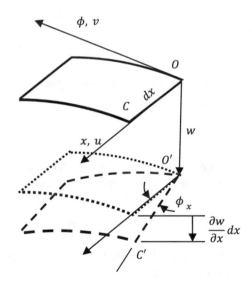

$$\phi_x = \frac{\partial w}{\partial x} \qquad (6.39)$$

6.3.1 Strains in the x-Direction

The strain in the x-direction is simply the extension divided by the original length
(see Fig. 6.11). The strain in the x-direction is therefore,

Fig. 6.11 Strains in the x-direction

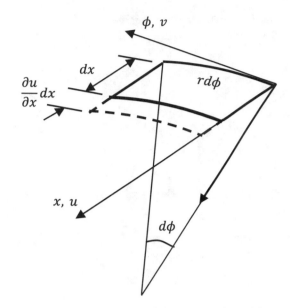

$$\varepsilon_x = \frac{\frac{\partial u}{\partial x} dx}{dx} = \frac{\partial u}{\partial x} \tag{6.40}$$

6.3.2 Strains in the ϕ-Direction, Due to an Extension v

Part of the strain in the v-direction is due to the extension of the differential element in the same direction (see Fig. 6.12).

$$\varepsilon_\phi = \frac{\frac{\partial v}{\partial \phi} d\phi}{rd\phi} = \frac{\partial v}{r\partial \phi} \tag{6.41}$$

6.3.3 Strains in the ϕ-Direction, Due to a Change in Radius

Referring to Fig. 6.13, a part of the strain may be caused by a change in the length of the radius.

The original length of the edge of the differential element:

Fig. 6.12 Strains in the
ϕ-direction

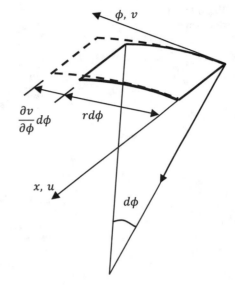

Fig. 6.13 Strains caused by
a change in the radius of the
differential element

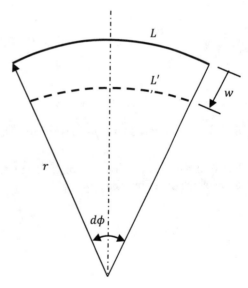

$$L = rd\phi \tag{6.42}$$

$$\frac{L}{r} = \frac{L'}{r - w} \tag{6.43}$$

Therefore,

$$L' = \frac{L(r - w)}{r} \tag{6.44}$$

where the original length is L and L' is the length after deformation. The strain is therefore

$$\varepsilon_\phi = -\frac{L - L'}{rd\phi} \tag{6.45}$$

Substitute 6.42 and 6.44 into 6.45.

$$\varepsilon_\phi = -\frac{rd\phi - \frac{rd\phi(r-w)}{r}}{rd\phi} \tag{6.46}$$

A minus sign is added, since a contraction of the differential element will cause a compression.

$$\varepsilon_\phi = -1 + \frac{(r - w)}{r} \tag{6.47}$$

$$\varepsilon_\phi = -\frac{w}{r} \tag{6.48}$$

6.3.4 Total Strain in the φ-Direction

The total strain in the φ-direction is the sum of the two strains, Eqs. 6.41 and 6.48.

$$\varepsilon_\phi = \frac{\partial v}{r\partial \phi} - \frac{w}{r} \tag{6.49}$$

6.3.5 Shearing Strain

Referring to Fig. 6.14, the shear strain is the maximum deformation along an edge and divided by the perpendicular length, added in both the x- and y-directions.

$$\gamma_{\phi x} = \frac{\frac{\partial u}{\partial x} dx}{rd\phi} + \frac{\frac{\partial v}{\partial \phi} d\phi}{dx} \tag{6.50}$$

Fig. 6.14 Shearing strains
in a differential element

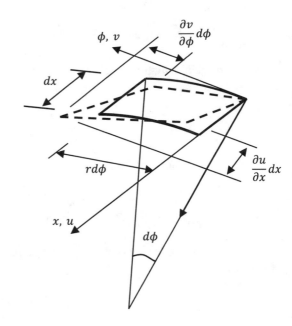

$$\gamma_{\phi x} = \frac{\partial u}{rd\phi} + \frac{\partial v}{\partial x} \qquad (6.51)$$

6.3.6 Membrane Stresses

From cylindrical shell theory,

$$\varepsilon_\theta = \frac{1}{Eh}\left[N'_\theta - \nu N'_y\right] \qquad (3.33)$$

$$\varepsilon_y = \frac{1}{Eh}\left[N'_y - \nu N'_\theta\right] \qquad (3.34)$$

$$\gamma_{\theta y} = \frac{1}{Gh}N'_{\theta y} \qquad (6.52)$$

To be consistent, θ is changed to ϕ and y is changed to x. Furthermore, the strain Eqs. 6.40, 6.49, and 6.51 are equated to 3.33, 3.34, and 6.52.

$$\varepsilon_x = \frac{\partial u}{\partial x} = \frac{1}{Eh}\left[N'_x - \nu N'_\phi\right] \qquad (6.53)$$

$$\varepsilon_\phi = \frac{\partial v}{r\partial \phi} - \frac{w}{r} = \frac{1}{Eh}\left[N'_\phi - \nu N'_x\right] \tag{6.54}$$

$$\gamma_{x\phi} = \frac{\partial u}{r\partial \phi} + \frac{\partial v}{\partial x} = \frac{1}{Gh}N'_{x\phi} \tag{6.55}$$

Since $G = \frac{E}{2(1+\nu)}$,

$$\gamma_{x\phi} = \frac{\partial u}{r\partial \phi} + \frac{\partial v}{\partial x} = \frac{2(1+\nu)}{Eh}N'_{x\phi} \tag{6.56}$$

If we integrate 6.53 and assume $\nu = 0$ to simplify the equation,

$$u = \int \frac{\partial u}{\partial x}dx = \frac{1}{Eh}\int N'_x dx + f_3(\phi) \tag{6.57}$$

Rearrange 6.56.

$$\frac{\partial v}{\partial x} = -\frac{\partial u}{r\partial \phi} + \frac{2}{Eh}N'_{x\phi} \tag{6.58}$$

Integrating 6.58 and assume $\nu = 0$,

$$v = \int \frac{\partial v}{\partial x}dx = -\frac{1}{r}\int \frac{\partial u}{\partial \phi}dx + \frac{2}{Eh}\int N'_{x\phi}dx + f_4(\phi) \tag{6.59}$$

From 6.54,

$$\frac{w}{r} = \frac{\partial v}{r\partial \phi} - \frac{N'_\phi}{Eh} \tag{6.60}$$

or

$$w = \frac{\partial v}{\partial \phi} - \frac{rN'_\phi}{Eh} \tag{6.61}$$

A Fourier series is used to represent the load distribution, with $n = 1$. A Fourier series is used to represent the distribution of the load and overcome mathematical difficulties in the derivation.

Use Eq. 6.3 to solve for the stresses in the ϕ-direction. Substitute 6.7 into 6.3.

$$N'_\phi = -Pr\cos(\phi_k - \phi)\sin\left(\frac{\pi x}{L}\right) \tag{6.62}$$

where

$$P = \frac{4P_d}{\pi}$$

The shear strains are solved from Eq. 6.1.

$$N'_{x\phi} = \int \left(-\frac{1}{r} \frac{\partial N'_{\phi}}{\partial \phi} - P_{\phi} \right) dx + f_1(\phi) \tag{6.1}$$

$$N'_{x\phi} = -\int \frac{1}{r} \frac{\partial N'_{\phi}}{\partial \phi} dx - \int P_{\phi} dx + f_1(\phi)$$

Substituting 6.62 and 6.8,

$$N'_{x\phi} = -\int \frac{1}{r} \frac{\partial \left[-Pr \cos (\phi_k - \phi) \sin \left(\frac{\pi x}{L} \right) \right]}{\partial \phi} dx$$

$$- \int \left[-P \sin (\phi_k - \phi) \sin \left(\frac{\pi x}{L} \right) \right] dx + f_1(\phi) \tag{6.63}$$

Taking the derivative of the first term,

$$N'_{x\phi} = \int P \sin (\phi_k - \phi) \sin \left(\frac{\pi x}{L} \right) dx + \int \left[P \sin (\phi_k - \phi) \sin \left(\frac{\pi x}{L} \right) \right] dx$$

$$+ f_1(\phi) \tag{6.64}$$

$$N'_{x\phi} = P \sin (\phi_k - \phi) \int \sin \left(\frac{\pi x}{L} \right) dx + P \sin (\phi_k - \phi) \int \sin \left(\frac{\pi x}{L} \right) dx$$

$$+ f_1(\phi) \tag{6.65}$$

$$N'_{x\phi} = P \sin (\phi_k - \phi) \left[\frac{-L \cos \left(\frac{\pi x}{L} \right)}{\pi} \right] + P \sin (\phi_k - \phi) \left[\frac{-L \cos \left(\frac{\pi x}{L} \right)}{\pi} \right]$$

$$+ f_1(\phi) \tag{6.66}$$

$$N'_{x\phi} = -\frac{2PL}{\pi} \sin (\phi_k - \phi) \cos \left(\frac{\pi x}{L} \right) + f_1(\phi) \tag{6.67}$$

Using Eq. 6.2, the stress is solved in the x-direction.

$$N'_x = \int \left(-\frac{1}{r} \frac{\partial N'_{\phi x}}{\partial \phi} - P_x \right) dx + f_2(\phi) \tag{6.2}$$

$$N'_x = -\frac{1}{r} \int \frac{\partial N'_{\phi x}}{\partial \phi} dx - \int P_x dx + f_2(\phi)$$

Substituting Eqs. 6.67 and 6.9 into 6.2,

$$N'_x = -\frac{1}{r} \int \frac{\partial \left[-\frac{2PL}{\pi} \sin (\phi_k - \phi) \cos \left(\frac{\pi x}{L} \right) + f_1(\phi) \right]}{\partial \phi} dx + f_2(\phi) \qquad (6.68)$$

Taking the derivative,

$$N'_x = -\frac{1}{r} \int \left[\frac{2PL}{\pi} \cos (\phi_k - \phi) \cos \left(\frac{\pi x}{L} \right) + \frac{\partial f_1(\phi)}{\partial \phi} \right] dx + f_2(\phi) \qquad (6.69)$$

Integrating the above equation,

$$N'_x = -\frac{2PL^2}{r\pi^2} \cos (\phi_k - \phi) \sin \left(\frac{\pi x}{L} \right) - \frac{x}{r} \frac{\partial f_1(\phi)}{\partial \phi} + f_2(\phi) \qquad (6.70)$$

Solve for the deformation in the u-direction using Eq. 6.57. Substituting 6.70 into 6.57,

$$u = \frac{1}{Eh} \int \left[-\frac{2PL^2}{r\pi^2} \cos (\phi_k - \phi) \sin \left(\frac{\pi x}{L} \right) - \frac{x}{r} \frac{\partial f_1(\phi)}{\partial \phi} + f_2(\phi) \right] dx$$
$$+ f_3(\phi) \qquad (6.71)$$

Integrating,

$$u = \frac{1}{Eh} \left\{ -\frac{2PL^2}{r\pi^2} \cos (\phi_k - \phi) \left[\frac{-\cos \left(\frac{\pi x}{L} \right)}{\frac{\pi}{L}} \right] - \frac{x^2}{2r} \frac{\partial f_1(\phi)}{\partial \phi} + x f_2(\phi) \right\}$$
$$+ f_3(\phi) \qquad (6.72)$$

$$u = \frac{1}{Eh} \left[\frac{2PL^3}{r\pi^3} \cos (\phi_k - \phi) \cos \left(\frac{\pi x}{L} \right) - \frac{x^2}{2r} \frac{\partial f_1(\phi)}{\partial x} + x f_2(\phi) \right] + f_3(\phi) \qquad (6.73)$$

Solve for the deformation in the v-direction using Eq. 6.59. Substitute 6.73 and 6.67 into 6.59.

$$v = -\frac{1}{r}$$

$$\times \int \frac{\partial \left(\frac{1}{Eh} \left[\frac{2PL^3}{r\pi^3} \cos (\phi_k - \phi) \cos \left(\frac{\pi x}{L} \right) - \frac{x^2}{2r} \frac{\partial f_1(\phi)}{\partial \phi} + x f_2(\phi) \right] + f_3(\phi) \right)}{\partial \phi} dx$$

$$+ \frac{2}{Eh} \int \left[-\frac{2PL}{\pi} \sin (\phi_k - \phi) \cos \left(\frac{\pi x}{L} \right) + f_1(\phi) \right] dx + f_4(\phi)$$

$$(6.74)$$

Taking the derivative of the first term,

$$v = -\frac{1}{r}$$

$$\times \int \left\{ \frac{1}{Eh} \left[\frac{2PL^3}{r\pi^3} \sin(\phi_k - \phi) \cos\left(\frac{\pi x}{L}\right) - \frac{x^2}{2r} \frac{\partial^2 f_1(\phi)}{\partial \phi^2} + x \frac{\partial f_2(\phi)}{\partial \phi} \right] + \frac{\partial f_3(\phi)}{\partial \phi} \right\} dx$$

$$+ \frac{2}{Eh} \int \left[-\frac{2PL}{\pi} \sin(\phi_k - \phi) \cos\left(\frac{\pi x}{L}\right) + f_1(\phi) \right] dx + f_4(\phi)$$

$$(6.75)$$

and integrate

$$v = -\frac{1}{r}$$

$$\times \left\{ \frac{1}{Eh} \left[\frac{2PL^3}{r\pi^3} \sin(\phi_k - \phi) \left(\frac{\sin\left(\frac{\pi x}{L}\right)}{\frac{\pi}{L}} \right) - \frac{x^3}{6r} \frac{\partial^2 f_1(\phi)}{\partial \phi^2} + \frac{x^2}{2} \frac{\partial f_2(\phi)}{\partial \phi} \right] + x \frac{\partial f_3(\phi)}{\partial \phi} \right\}$$

$$+ \frac{2}{Eh} \left[\frac{-2PL}{\pi} \sin(\phi_k - \phi) \left(\frac{\sin\left(\frac{\pi x}{L}\right)}{\frac{\pi}{L}} \right) + x f_1(\phi) \right] + f_4(\phi)$$

$$(6.76)$$

$$v = -\frac{1}{Ehr} \left[\frac{2PL^4}{r\pi^4} \sin(\phi_k - \phi) \sin\left(\frac{\pi x}{L}\right) - \frac{x^3}{6r} \frac{\partial^2 f_1(\phi)}{\partial \phi^2} + \frac{x^2}{2} \frac{\partial f_2(\phi)}{\partial \phi} \right]$$

$$- \frac{x}{r} \frac{\partial f_3(\phi)}{\partial \phi} - \frac{2}{Eh} \left[\frac{2PL^2}{\pi^2} \sin(\phi_k - \phi) \sin\left(\frac{\pi x}{L}\right) - x f_1(\phi) \right] + f_4(\phi) \quad (6.77)$$

Using Eq. 6.61, substitute equations 6.62 and 6.77.

$$w = \frac{\partial}{\partial \phi}$$

$$\times \left\{ -\frac{1}{Ehr} \left[\frac{2PL^4}{r\pi^4} \sin(\phi_k - \phi) \sin\left(\frac{\pi x}{L}\right) - \frac{x^3}{6r} \frac{\partial^2 f_1(\phi)}{\partial \phi^2} + \frac{x^2}{2} \frac{\partial f_2(\phi)}{\partial \phi} \right] \right.$$

$$\left. \times -\frac{x}{r} \frac{\partial f_3(\phi)}{\partial \phi} - \frac{2}{Eh} \left[\frac{2PL^2}{\pi^2} \sin(\phi_k - \phi) \sin\left(\frac{\pi x}{L}\right) - x f_1(\phi) \right] + f_4(\phi) \right\}$$

$$+ \frac{Pr^2}{Eh} \cos(\phi_k - \phi) \sin\left(\frac{\pi x}{L}\right)$$

$$(6.78)$$

Taking the derivative of the first term,

$$w = -\frac{1}{Ehr}\left[\frac{-2PL^4}{r\pi^4}\cos(\phi_k - \phi)\sin\left(\frac{\pi x}{L}\right) - \frac{x^3}{6r}\frac{\partial^3 f_1(\phi)}{\partial\phi^3} + \frac{x^2}{2}\frac{\partial^2 f_2(\phi)}{\partial\phi^2}\right]$$

$$-\frac{x}{r}\frac{\partial^2 f_3(\phi)}{\partial\phi^2} - \frac{2}{Eh}\left[\frac{-2PL^2}{\pi^2}\cos(\phi_k - \phi)\sin\left(\frac{\pi x}{L}\right) - x\frac{\partial f_1(\phi)}{\partial\phi}\right]$$

$$+\frac{\partial f_4(\phi)}{\partial\phi} + \frac{Pr^2}{Eh}\cos(\phi_k - \phi)\sin\left(\frac{\pi x}{L}\right) \tag{6.79}$$

$$w = \frac{1}{Ehr}\left[\frac{2PL^4}{r\pi^4}\cos(\phi_k - \phi)\sin\left(\frac{\pi x}{L}\right) + \frac{x^3}{6r}\frac{\partial^3 f_1(\phi)}{\partial\phi^3} - \frac{x^2}{2}\frac{\partial^2 f_2(\phi)}{\partial\phi^2}\right]$$

$$-\frac{x}{r}\frac{\partial^2 f_3(\phi)}{\partial\phi^2} + \frac{2}{Eh}\left[\frac{2PL^2}{\pi^2}\cos(\phi_k - \phi)\sin\left(\frac{\pi x}{L}\right) + x\frac{\partial f_1(\phi)}{\partial\phi}\right] \tag{6.80}$$

$$+\frac{\partial f_4(\phi)}{\partial\phi} + \frac{Pr^2}{Eh}\cos(\phi_k - \phi)\sin\left(\frac{\pi x}{L}\right)$$

If the barrel shell is simply supported and subjected to a uniform distributed load, the deformation equations are significantly simplified. The integration constants, which represent the boundary conditions, are assumed to be equal to zero.

$$f_1(\phi) = f_2(\phi) = f_3(\phi) = f_4(\phi) = 0 \tag{6.81}$$

Therefore, simplifying Eqs. 6.73, 6.77, and 6.80,

$$u = \frac{2PL^3}{Ehr\pi^3}\cos(\phi_k - \phi)\cos\left(\frac{\pi x}{L}\right) \tag{6.82}$$

$$v = -\frac{2PL^4}{Ehr^2\pi^4}\sin(\phi_k - \phi)\sin\left(\frac{\pi x}{L}\right) - \frac{4PL^2}{Eh\pi^2}\sin(\phi_k - \phi)\sin\left(\frac{\pi x}{L}\right)$$

$$v = -\frac{2P}{Eh}\left(\frac{L^4}{r^2\pi^4} + \frac{2L^2}{\pi^2}\right)\sin(\phi_k - \phi)\sin\left(\frac{\pi x}{L}\right) \tag{6.83}$$

$$w = \frac{1}{Ehr}\left[\frac{2PL^4}{r\pi^4}\cos(\phi_k - \phi)\sin\left(\frac{\pi x}{L}\right)\right] + \frac{2}{Eh}$$

$$\times\left[\frac{2PL^2}{\pi^2}\cos(\phi_k - \phi)\sin\left(\frac{\pi x}{L}\right)\right] + \frac{Pr^2}{Eh}\cos(\phi_k - \phi)\sin\left(\frac{\pi x}{L}\right) \tag{6.84}$$

$$w = \frac{2PL^4}{Ehr^2\pi^4}\cos(\phi_k - \phi)\sin\left(\frac{\pi x}{L}\right) + \frac{4PL^2}{Eh\pi^2}\cos(\phi_k - \phi)\sin\left(\frac{\pi x}{L}\right)$$

$$+\frac{Pr^2}{Eh}\cos(\phi_k - \phi)\sin\left(\frac{\pi x}{L}\right) \tag{6.85}$$

$$w = \frac{2PL^4}{Ehr^2\pi^4}\left(1 + \frac{2r^2\pi^2}{L^2} + \frac{r^4\pi^4}{2L^4}\right)\cos(\phi_k - \phi)\sin\left(\frac{\pi x}{L}\right) \tag{6.86}$$

6.3.7 Horizontal and Vertical Deformations

From Fig. 6.15, the horizontal and vertical deformations are components of the deformations in the v- and w-directions.

$$\Delta'_v = -v \sin(\phi_k - \phi) + w \cos(\phi_k - \phi) \tag{6.87}$$

Substitute 6.83 and 6.86 into 6.87.

$$\Delta'_v = \frac{2P}{Eh}\left(\frac{L^4}{r^2\pi^4} + \frac{2L^2}{\pi^2}\right)\sin(\phi_k - \phi)\sin\left(\frac{\pi x}{L}\right)\sin(\phi_k - \phi) + \frac{2PL^4}{Ehr^2\pi^4}$$
$$\times \left(1 + \frac{2r^2\pi^2}{L^2} + \frac{\pi^4 r^4}{2L^4}\right)\cos(\phi_k - \phi)\sin\left(\frac{\pi x}{L}\right)\cos(\phi_k - \phi) \tag{6.88}$$

$$\Delta'_v = \frac{2P}{Eh}\left(\frac{L^4}{r^2\pi^4} + \frac{2L^2}{\pi^2}\right)\sin^2(\phi_k - \phi)\sin\left(\frac{\pi x}{L}\right) + \frac{2PL^4}{Ehr^2\pi^4}$$
$$\times \left(1 + \frac{2r^2\pi^2}{L^2} + \frac{\pi^4 r^4}{2L^4}\right)\cos^2(\phi_k - \phi)\sin\left(\frac{\pi x}{L}\right) \tag{6.89}$$

Rearrange 6.89.

Fig. 6.15 Orientation of the horizontal and vertical deformations

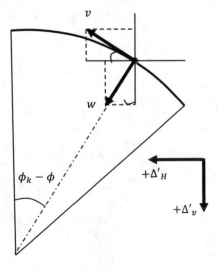

$$\Delta'_v = \frac{2PL^4}{Ehr^2\pi^4} \sin\left(\frac{\pi x}{L}\right)$$

$$\times \left[\left(1 + \frac{2\pi^2 r^2}{L^2}\right) \sin^2(\phi_k - \phi) + \left(1 + \frac{2r^2\pi^2}{L^2} + \frac{\pi^4 r^4}{2L^4}\right) \cos^2(\phi_k - \phi)\right]$$

$$(6.90)$$

Splitting the second term,

$$\Delta'_v = \frac{2PL^4}{Ehr^2\pi^4} \sin\left(\frac{\pi x}{L}\right)\left[\left(1 + \frac{2\pi^2 r^2}{L^2}\right) \sin^2(\phi_k - \phi)\right.$$

$$\left. \times + \left(1 + \frac{2r^2\pi^2}{L^2}\right) \cos^2(\phi_k - \phi) + \left(\frac{\pi^4 r^4}{2L^4}\right) \cos^2(\phi_k - \phi)\right]$$

$$(6.91)$$

Since $\sin^2 + \cos^2 = 1$,

$$\Delta'_v = \frac{2PL^4}{Ehr^2\pi^4} \sin\left(\frac{\pi x}{L}\right)\left[1 + \frac{2\pi^2 r^2}{L^2} + \left(\frac{\pi^4 r^4}{2L^4}\right) \cos^2(\phi_k - \phi)\right]$$

$$(6.92)$$

Similarly in the horizontal direction,

$$\Delta'_H = v\cos(\phi_k - \phi) + w\sin(\phi_k - \phi)$$

$$(6.93)$$

Substitute 6.83 and 6.86 into 6.92.

$$\Delta'_H = \left[-\frac{2P}{Eh}\left(\frac{L^4}{r^2\pi^4} + \frac{2L^2}{\pi^2}\right) \sin(\phi_k - \phi) \sin\left(\frac{\pi x}{L}\right)\right] \cos(\phi_k - \phi)$$

$$+ \left[\frac{2PL^4}{Ehr^2\pi^4}\left(1 + \frac{2r^2\pi^2}{L^2} + \frac{r^4\pi^4}{2L^4}\right) \cos(\phi_k - \phi) \sin\left(\frac{\pi x}{L}\right)\right] \sin(\phi_k - \phi)$$

$$(6.94)$$

$$\Delta'_H = \frac{2P}{Ehr^2}\left(-\frac{L^4}{\pi^4} - \frac{2r^2 L^2}{\pi^2}\right) \sin(\phi_k - \phi) \sin\left(\frac{\pi x}{L}\right) \cos(\phi_k - \phi) + \frac{2P}{Ehr^2}$$

$$\times \left(\frac{L^4}{\pi^4} + \frac{2r^2 L^2}{\pi^2} + \frac{r^4}{2}\right) \cos(\phi_k - \phi) \sin\left(\frac{\pi x}{L}\right) \sin(\phi_k - \phi)$$

$$(6.95)$$

Combine the two parts of the equation.

$$\Delta'_H = \frac{2P}{Ehr^2}\left(-\frac{L^4}{\pi^4} - \frac{2r^2 L^2}{\pi^2} + \frac{L^4}{\pi^4} + \frac{2r^2 L^2}{\pi^2} + \frac{r^4}{2}\right) \sin(\phi_k - \phi) \sin\left(\frac{\pi x}{L}\right) \cos(\phi_k - \phi)$$

$$(6.96)$$

Simplify.

$$\Delta'_H = \frac{2P}{Ehr^2}\left(\frac{r^4}{2}\right)\sin\left(\phi_k - \phi\right)\sin\left(\frac{\pi x}{L}\right)\cos\left(\phi_k - \phi\right)$$

or

$$\Delta'_H = \frac{Pr^2}{Eh}\sin\left(\phi_k - \phi\right)\sin\left(\frac{\pi x}{L}\right)\cos\left(\phi_k - \phi\right) \tag{6.97}$$

6.4 Shallow Shell Theory to Solve for the Boundary Effects

As depicted in Fig. 6.16, the bending moments, twisting moments, and out-of-plane shears are given. These forces were neglected in the membrane solution but are now included to correct the boundary errors or boundary conditions.

6.4.1 Stresses from Boundary Effects

From cylindrical shell theory, the following equilibrium equations were derived:

Fig. 6.16 Moments and shears on the differential element

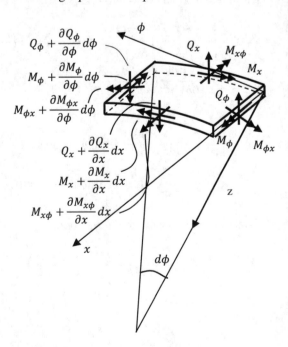

$$\frac{\partial N_y}{\partial y} r + \frac{\partial N_{\theta y}}{\partial \theta} + P_y r = 0 \tag{3.47}$$

$$\frac{\partial N_\theta}{\partial \theta} + \frac{\partial N_{y\theta}}{\partial y} r + P_\theta r - Q_\theta = 0 \tag{3.46}$$

$$N_\theta + \frac{\partial Q_\theta}{\partial \theta} + \frac{\partial Q_y}{\partial y} r + P_z r = 0 \tag{3.51}$$

$$\frac{\partial M_\theta}{\partial \theta} + \frac{\partial M_{y\theta}}{dy} r - Q_\theta r = 0 \tag{3.57}$$

$$-\frac{\partial M_y}{\partial y} r + \frac{\partial M_{\theta y}}{\partial \theta} + Q_y r = 0 \tag{3.54}$$

These equations are adapted to barrel vaults by changing the subscripts (i.e., θ to ϕ and y to x).

$$\frac{\partial N_x}{\partial x} r + \frac{\partial N_{\phi x}}{\partial \phi} + P_x r = 0 \tag{6.98}$$

$$\frac{\partial N_\phi}{\partial \phi} + \frac{\partial N_{x\phi}}{\partial x} r - Q_\phi + P_\phi r = 0 \tag{6.99}$$

$$N_\phi + \frac{\partial Q_x}{\partial x} r + \frac{\partial Q_\phi}{\partial \phi} + P_z r = 0 \tag{6.100}$$

$$-\frac{\partial M_\phi}{\partial \phi} + \frac{\partial M_{x\phi}}{\partial x} r + Q_\phi r = 0 \tag{6.101}$$

$$\frac{\partial M_x}{\partial x} r + \frac{\partial M_{\phi x}}{d\phi} - Q_x r = 0 \tag{6.102}$$

In the last two equations, the sign was changed (except for the twisting moment) to reflect the change in the coordinate system.

From plate theory (Szilard 1974)),

$$N_x = \frac{Eh}{(1 - \nu^2)} \left[\varepsilon_x + \nu \varepsilon_\phi \right] \tag{6.103}$$

$$N_\phi = \frac{Eh}{(1 - \nu^2)} \left[\varepsilon_\phi + \nu \varepsilon_x \right] \tag{6.104}$$

$$N_{x\phi} = N_{\phi x} = Gh\gamma_{x\phi} \tag{6.105}$$

where,

$$G = \frac{E}{2(1+\nu)} \tag{6.106}$$

As previously defined,

$$\varepsilon_x = \frac{\partial u}{\partial x} \tag{6.40}$$

$$\varepsilon_\phi = \frac{\partial v}{r\partial \phi} - \frac{w}{r} \tag{6.49}$$

$$\gamma_{x\phi} = \frac{\partial u}{r\partial \phi} + \frac{\partial v}{\partial x} \tag{6.51}$$

Substituting 6.40, 6.49, and 6.51 into 6.103, 6.104, and 6.105,

$$N_x = \frac{Eh}{(1-\nu^2)} \left[\frac{\partial u}{\partial x} + \nu \left(\frac{\partial v}{r\partial \phi} - \frac{w}{r} \right) \right] \tag{6.107}$$

$$N_\phi = \frac{Eh}{(1-\nu^2)} \left[\frac{\partial v}{r\partial \phi} - \frac{w}{r} + \nu \frac{\partial u}{\partial x} \right] \tag{6.108}$$

$$N_{x\phi} = N_{\phi x} = \frac{Eh}{2(1+\nu)} \left(\frac{\partial u}{r\partial \phi} + \frac{\partial v}{\partial x} \right) \tag{6.109}$$

Similarly, the slope of the differential element was also previously defined.

$$\phi_x = \frac{\partial w}{\partial x} \tag{6.39}$$

$$\phi_\phi = \frac{v}{r} + \frac{\partial w}{r\partial \phi} \tag{6.38}$$

The curvature is the change in slope, or rather the derivative of the slope. Therefore,

$$\chi_x = \frac{\partial \phi_x}{\partial x} = \frac{\partial^2 w}{\partial x^2} \tag{6.110}$$

In the ϕ-direction,

$$\chi_\phi = \frac{\partial \phi_\phi}{r \partial \phi} = \frac{\partial v}{r^2 \partial \phi} + \frac{\partial^2 w}{r^2 \partial \phi^2} \tag{6.111}$$

The twisting curvature, however, measures the change in slope at right angles to the plane of curvature (Szilard 1974; Timoshenko 1959). The twisting curvature is defined on each edge of the differential element.

$$\chi_{\phi x} = \frac{\partial \phi_\phi}{\partial x} \tag{6.112}$$

$$\chi_{x\phi} = \frac{\partial \phi_x}{r \partial \phi} \tag{6.113}$$

The twisting curvature of the differential element is the average of the two curvatures.

$$\chi_{x\phi} = \frac{1}{2}\left(\frac{\partial \phi_x}{r \partial \phi} + \frac{\partial \phi_\phi}{\partial x}\right) \tag{6.114}$$

Substituting 6.39 and 6.38,

$$\chi_{x\phi} = \frac{\partial \left(\frac{\partial w}{\partial x}\right)}{2r \partial \phi} + \frac{\partial \left(\frac{v}{r} + \frac{\partial w}{r \partial \phi}\right)}{2 \partial x} \tag{6.115}$$

Expanding 6.115,

$$\chi_{x\phi} = \frac{\partial^2 w}{2r\, \partial x\, \partial \phi} + \frac{\partial v}{2r\, \partial x} + \frac{\partial^2 w}{2r\, \partial x\, \partial \phi} \tag{6.116}$$

or

$$\chi_{x\phi} = \frac{\partial v}{2r\, \partial x} + \frac{\partial^2 w}{r\, \partial x\, \partial \phi} \tag{6.117}$$

From plate theory (Jaeger 1969),

$$M_x = -D\left(\chi_x + \nu\chi_\phi\right) \tag{6.118}$$

$$M_\phi = -D\left(\chi_\phi + \nu\chi_x\right) \tag{6.119}$$

$$M_{x\phi} = -M_{\phi x} = D(1 - \nu)\chi_{x\phi} \tag{6.120}$$

where

$$D = \frac{Eh^3}{12(1 - \nu^2)} \tag{6.121}$$

Substituting 6.110, 6.111, and 6.117 into 6.118, 6.119, and 6.120.

$$M_x = -D\left[\frac{\partial^2 w}{\partial x^2} + \nu\left(\frac{\partial v}{r^2 \partial \phi} + \frac{\partial^2 w}{r^2 \partial \phi^2}\right)\right] \tag{6.122}$$

or

$$M_x = -D\left[\frac{\partial^2 w}{\partial x^2} + \frac{\nu}{r^2}\left(\frac{\partial v}{\partial \phi} + \frac{\partial^2 w}{\partial \phi^2}\right)\right] \tag{6.123}$$

$$M_\phi = -D\left[\frac{\partial v}{r^2 \partial \phi} + \frac{\partial^2 w}{r^2 \partial \phi^2} + \nu\left(\frac{\partial^2 w}{\partial x^2}\right)\right] \tag{6.124}$$

$$M_{x\phi} = -M_{\phi x} = D(1 - \nu)\left[\frac{\partial v}{2r \partial x} + \frac{\partial^2 w}{r \partial x \partial \phi}\right] \tag{6.125}$$

Taking the derivative of Eq. 6.101 with respect to ϕ,

$$-\frac{\partial^2 M_\phi}{\partial \phi^2} + \frac{\partial^2 M_{x\phi}}{\partial x \partial \phi}r + \frac{\partial Q_\phi}{\partial \phi}r = 0 \tag{6.126}$$

Multiply by $\frac{1}{r^2}$ and rearrange.

$$\frac{\partial Q_\phi}{r \partial \phi} = \frac{\partial^2 M_\phi}{r^2 \partial \phi^2} - \frac{\partial^2 M_{x\phi}}{r \partial x \partial \phi} \tag{6.127}$$

Similarly, taking the derivative of 6.102 with respect to x,

$$\frac{\partial^2 M_x}{\partial x^2}r + \frac{\partial^2 M_{\phi x}}{\partial x \partial \phi} - \frac{\partial Q_x}{\partial x}r = 0 \tag{6.128}$$

Multiply the above equation by $\frac{1}{r}$ and rearrange.

$$\frac{\partial Q_x}{\partial x} = \frac{\partial^2 M_x}{\partial x^2} + \frac{\partial^2 M_{\phi x}}{r \partial x \partial \phi} \tag{6.129}$$

Since the in-plane deformations are significantly less than the out-of-plane deformations, the in-plane deformations are neglected from Eqs. 6.123, 6.124, and 6.125.

$$M_x = -D\left[\frac{\partial^2 w}{\partial x^2} + \frac{\nu}{r^2}\left(\frac{\partial^2 w}{\partial \phi^2}\right)\right] \tag{6.130}$$

$$M_\phi = -D\left[\frac{\partial^2 w}{r^2 \partial \phi^2} + \nu\left(\frac{\partial^2 w}{\partial x^2}\right)\right] \tag{6.131}$$

$$M_{x\phi} = -M_{\phi x} = D(1-\nu)\left[\frac{\partial^2 w}{r\,\partial x\,\partial \phi}\right] \tag{6.132}$$

Equations 6.131 and 6.132 are substituted into 6.127.

$$\frac{\partial Q_\phi}{r\partial\phi} = \frac{\partial^2\left\{-D\left[\frac{\partial^2 w}{r^2\partial\phi^2}+\nu\left(\frac{\partial^2 w}{\partial x^2}\right)\right]\right\}}{r^2\partial\phi^2} - \frac{\partial^2\left\{D(1-\nu)\left[\frac{\partial^2 w}{r\,\partial x\,\partial\phi}\right]\right\}}{r\,\partial x\,\partial\phi} \tag{6.133}$$

$$\frac{\partial Q_\phi}{r\partial\phi} = -D\left[\frac{\partial^4 w}{r^4\partial\phi^4}+\frac{\nu}{r^2}\left(\frac{\partial^4 w}{\partial\phi^2\partial x^2}\right)\right] - D(1-\nu)\left[\frac{\partial^4 w}{r^2\,\partial x^2\,\partial\phi^2}\right] \tag{6.134}$$

$$\frac{\partial Q_\phi}{r\partial\phi} = -D\left[\frac{\partial^4 w}{r^4\partial\phi^4}+\frac{\nu}{r^2}\left(\frac{\partial^4 w}{\partial\phi^2\partial x^2}\right)+(1-\nu)\left(\frac{\partial^4 w}{r^2\,\partial x^2\,\partial\phi^2}\right)\right] \tag{6.135}$$

Similarly, substituting 6.130 and 6.132 into 6.129,

$$\frac{\partial Q_x}{\partial x} = \frac{\partial^2\left(-D\left[\frac{\partial^2 w}{\partial x^2}+\frac{\nu}{r^2}\left(\frac{\partial^2 w}{\partial\phi^2}\right)\right]\right)}{\partial x^2} + \frac{\partial^2\left(-D(1-\nu)\left[\frac{\partial^2 w}{r\,\partial x\,\partial\phi}\right]\right)}{r\,\partial x\,\partial\phi} \tag{6.136}$$

$$\frac{\partial Q_x}{\partial x} = -D\left[\frac{\partial^4 w}{\partial x^4}+\frac{\nu}{r^2}\left(\frac{\partial^4 w}{\partial\phi^2\partial x^2}\right)\right] - D(1-\nu)\left[\frac{\partial^4 w}{r^2\,\partial x^2\,\partial\phi^2}\right] \tag{6.137}$$

$$\frac{\partial Q_x}{\partial x} = -D\left[\frac{\partial^4 w}{\partial x^4}+\frac{\nu}{r^2}\left(\frac{\partial^4 w}{\partial\phi^2\partial x^2}\right)+(1-\nu)\left(\frac{\partial^4 w}{r^2\,\partial x^2\,\partial\phi^2}\right)\right] \tag{6.138}$$

If we add Eqs. 6.135 and 6.138,

$$\frac{\partial Q_\phi}{r\partial\phi}+\frac{\partial Q_x}{\partial x} = -D\left[\frac{\partial^4 w}{r^4\partial\phi^4}+\frac{\nu}{r^2}\left(\frac{\partial^4 w}{\partial\phi^2\partial x^2}\right)+(1-\nu)\left(\frac{\partial^4 w}{r^2\,\partial x^2\,\partial\phi^2}\right)\right]$$
$$- D\left[\frac{\partial^4 w}{\partial x^4}+\frac{\nu}{r^2}\left(\frac{\partial^4 w}{\partial\phi^2\partial x^2}\right)+(1-\nu)\left(\frac{\partial^4 w}{r^2\,\partial x^2\,\partial\phi^2}\right)\right] \tag{6.139}$$

Simplify by combining like terms.

$$\frac{\partial Q_\phi}{r\partial\phi} + \frac{\partial Q_x}{\partial x} = -D\left[\frac{\partial^4 w}{r^4\partial\phi^4} + \frac{\partial^4 w}{\partial x^4} + \frac{2\nu\partial^4 w}{r^2\partial\phi^2\partial x^2} + \frac{2(1-\nu)\partial^4 w}{r^2\,\partial x^2\,\partial\phi^2}\right] \quad (6.140)$$

Expand the last term.

$$\frac{\partial Q_\phi}{r\partial\phi}$$

$$+\frac{\partial Q_x}{\partial x}$$

$$= -D\left[\frac{\partial^4 w}{r^4\partial\phi^4} + \frac{\partial^4 w}{\partial x^4} + \frac{2\nu\partial^4 w}{r^2\partial\phi^2\partial x^2} + \frac{2\partial^4 w}{r^2\,\partial x^2\,\partial\phi^2} - \frac{2\nu\partial^4 w}{r^2\,\partial x^2\,\partial\phi^2}\right]$$

$$(6.141)$$

Eliminate the third and the fifth terms, due to a lack of significance.

$$\frac{\partial Q_\phi}{r\partial\phi} + \frac{\partial Q_x}{\partial x} = -D\left[\frac{\partial^4 w}{r^4\partial\phi^4} + \frac{2\partial^4 w}{r^2\,\partial x^2\,\partial\phi^2} + \frac{\partial^4 w}{\partial x^4}\right] \quad (6.142)$$

Equation 6.142 may be expressed in the following form:

$$\frac{\partial Q_\phi}{r\partial\phi} + \frac{\partial Q_x}{\partial x} = -D\nabla^4 w \quad (6.143)$$

where ∇^4 is called the biharmonic operator, which is a fourth-order partial differential equation.

Dividing 6.100 by r,

$$\frac{\partial Q_x}{\partial x} + \frac{\partial Q_\phi}{r\partial\phi} + \frac{N_\phi}{r} + P_z = 0 \quad (6.144)$$

The first two terms are equivalent to Eq. 6.143

$$-D\nabla^4 w + \frac{N_\phi}{r} + P_z = 0 \quad (6.145)$$

or rearranged

$$P_z = D\nabla^4 w - \frac{N_\phi}{r} \quad (6.146)$$

The above equation is identical to the plate bending equation (Szilard 1974; Timoshenko 1959), except for the additional term $\frac{N_\phi}{r}$. Using the biharmonic deep-beam equations (Billington 1982),

$$N_x = \frac{\partial^2 F}{r^2 \partial \phi^2} - \int P_x dx \tag{6.147}$$

$$N_\phi = \frac{\partial^2 F}{\partial x^2} - \int P_\phi r d\phi \tag{6.148}$$

$$N_{x\phi} = \frac{-\partial^2 F}{r \partial x \, \partial \phi} \tag{6.149}$$

where F is called the potential function, and represents all of the in-plane stress resultants. These functions are used to solve Eqs. 6.98 and 6.99. However, the term Q_ϕ is dropped from 6.99, assuming the effect is negligible in carrying the loads.

$$\frac{\partial N_x}{\partial x} r + \frac{\partial N_{\phi x}}{\partial \phi} + P_x r = 0 \tag{6.98}$$

$$\frac{\partial N_\phi}{\partial \phi} + \frac{\partial N_{x\phi}}{\partial x} r + P_\phi r = 0 \tag{6.99}$$

Referring to Eqs. 6.53, 6.54 and 6.55,

$$\varepsilon_x = \frac{\partial u}{\partial x} = \frac{1}{Eh} \left[N'_x - \nu N'_\phi \right] \tag{6.53}$$

$$\varepsilon_\phi = \frac{\partial v}{r \partial \phi} - \frac{w}{r} = \frac{1}{Eh} \left[N'_\phi - \nu N'_x \right] \tag{6.54}$$

$$\gamma_{x\phi} = \frac{\partial u}{r \partial \phi} + \frac{\partial v}{\partial x} = \frac{1}{Gh} N'_{x\phi} \tag{6.55}$$

Multiplying each of the above equations by Eh,

$$Eh\varepsilon_x = Eh\frac{\partial u}{\partial x} = \left[N'_x - \nu N'_\phi \right] \tag{6.150}$$

$$Eh\varepsilon_\phi = Eh\left(\frac{\partial v}{r \partial \phi} - \frac{w}{r} \right) = \left[N'_\phi - \nu N'_x \right] \tag{6.151}$$

$$Eh\gamma_{x\phi} = Eh\left(\frac{\partial u}{r \partial \phi} + \frac{\partial v}{\partial x} \right) = 2(1 + \nu)N'_{x\phi} \tag{6.152}$$

where $G = \frac{E}{2(1+\nu)}$

Take two derivatives of Eq. 6.150 with respect to ϕ, and divide by r^2.

$$\frac{\partial^2 (N_x - \nu N_\phi)}{r^2 \partial \phi^2} = \frac{Eh}{r^2} \frac{\partial^3 u}{\partial x \partial \phi^2} \tag{6.153}$$

Similarly, take two derivatives of Eq. 6.151 with respect to x.

$$\frac{\partial^2 (N_\phi - \nu N_x)}{\partial x^2} = Eh \left(\frac{\partial^3 v}{r \partial x^2 \partial \phi} - \frac{\partial^2 w}{r \partial x^2} \right) \tag{6.154}$$

Finally, take derivatives of 6.152 with respect to x and ϕ, and divide by r.

$$2(1 + \nu) \frac{\partial^2 N_{x\phi}}{r \partial x \partial \phi} = Eh \left(\frac{\partial^3 u}{r^2 \partial x \partial \phi^2} + \frac{\partial^3 v}{r \partial \phi \partial x^2} \right) \tag{6.155}$$

Add Eqs. 6.153 and 6.154, and subtract 6.155.

$$\frac{\partial^2 (N_x - \nu N_\phi)}{r^2 \partial \phi^2} + \frac{\partial^2 (N_\phi - \nu N_x)}{\partial x^2} - 2(1 + \nu) \frac{\partial^2 N_{x\phi}}{r \partial x \partial \phi}$$
$$= Eh \left(\frac{\partial^3 u}{r^2 \partial x \partial \phi^2} + \frac{\partial^3 v}{r \partial x^2 \partial \phi} - \frac{\partial^2 w}{r \partial x^2} - \frac{\partial^3 u}{r^2 \partial x \partial \phi^2} - \frac{\partial^3 v}{r \partial \phi \partial x^2} \right) \tag{6.156}$$

Simplifying the right-hand side of 6.156,

$$\frac{\partial^2 (N_x - \nu N_\phi)}{r^2 \partial \phi^2} + \frac{\partial^2 (N_\phi - \nu N_x)}{\partial x^2} - 2(1 + \nu) \frac{\partial^2 N_{x\phi}}{r \partial x \partial \phi} = Eh \left(-\frac{\partial^2 w}{r \partial x^2} \right) \tag{6.157}$$

Expanding the left-hand side of 6.157,

$$\frac{\partial^2 N_x}{r^2 \partial \phi^2} - \frac{\nu \partial^2 N_\phi}{r^2 \partial \phi^2} + \frac{\partial^2 N_\phi}{\partial x^2} - \frac{\nu \partial^2 N_x}{\partial x^2} - \frac{2 \partial^2 N_{x\phi}}{r \partial x \partial \phi} - \frac{2\nu \partial^2 N_{x\phi}}{r \partial x \partial \phi} = -\frac{Eh \partial^2 w}{r \partial x^2} \tag{6.158}$$

Substitute Eqs. 6.147, 6.148, and 6.149 into 6.158.

$$\frac{\partial^2 \left(\frac{\partial^2 F}{r^2 \partial \phi^2} - \int P_x dx \right)}{r^2 \partial \phi^2} - \frac{\nu \partial^2 \left(\frac{\partial^2 F}{\partial x^2} - \int P_\phi r d\phi \right)}{r^2 \partial \phi^2} + \frac{\partial^2 \left(\frac{\partial^2 F}{\partial x^2} - \int P_\phi r d\phi \right)}{\partial x^2}$$
$$- \frac{\nu \partial^2 \left(\frac{\partial^2 F}{r^2 \partial \phi^2} - \int P_x dx \right)}{\partial x^2} - \frac{2 \partial^2 \left(\frac{-\partial^2 F}{r \partial x \partial \phi} \right)}{r \partial x \partial \phi} - \frac{2\nu \partial^2 \left(\frac{-\partial^2 F}{r \partial x \partial \phi} \right)}{r \partial x \partial \phi}$$
$$= -\frac{Eh \partial^2 w}{r \partial x^2} \tag{6.159}$$

Expanding the above equation and rearranging,

$$\frac{\partial^4 F}{r^4 \partial \phi^4} - \frac{\nu \partial^4 F}{r^2 \partial \phi^2 \partial x^2} + \frac{\partial^4 F}{\partial x^4} - \frac{\nu \partial^4 F}{r^2 \partial \phi^2 \partial x^2} + \frac{2 \partial^4 F}{r^2 \partial x^2 \partial \phi^2} + \frac{2\nu \partial^4 F}{r^2 \partial x^2 \partial \phi^2}$$

$$- \frac{1}{r^2} \int \frac{\partial^2 P_x}{\partial \phi^2} dx - \int \frac{\partial^2 P_\phi}{\partial x^2} r \, d\phi + \frac{\nu}{r^2} \int \frac{\partial^2 P_\phi}{\partial \phi^2} r \, d\phi + \nu \int \frac{\partial^2 P_x}{\partial x^2} \, dx$$

$$= -\frac{Eh \partial^2 w}{r \, \partial x^2} \tag{6.160}$$

Simplify by eliminating the second, fourth, and sixth equations, since these terms are insignificant compared to the other terms. The external pressures are brought to the right-hand side of the equation, and two of the terms are integrated.

$$\frac{\partial^4 F}{r^4 \partial \phi^4} + \frac{2 \partial^4 F}{r^2 \partial x^2 \partial \phi^2} + \frac{\partial^4 F}{\partial x^4} + \frac{Eh \partial^2 w}{r \, \partial x^2}$$

$$= \frac{1}{r^2} \int \frac{\partial^2 P_x}{\partial \phi^2} dx + \int \frac{\partial^2 P_\phi}{\partial x^2} r \, d\phi - \frac{\nu}{r} \frac{\partial P_\phi}{\partial \phi} - \nu \frac{\partial P_x}{\partial x} \tag{6.161}$$

By inspection, the first part of Eq. 6.161 is equal to the biharmonic operator (Billington 1982)(see Eqs. 6.142 and 6.143). Therefore, Eq. 6.161 may be expressed as:

$$\nabla^4 F + \frac{Eh \partial^2 w}{r \, \partial x^2} = f(p) \tag{6.162}$$

where

$$\nabla^4 F = \frac{\partial^4 F}{r^4 \partial \phi^4} + \frac{2 \partial^4 F}{r^2 \partial x^2 \partial \phi^2} + \frac{\partial^4 F}{\partial x^4} \tag{6.163}$$

$$f(p) = \frac{1}{r^2} \int \frac{\partial^2 P_x}{\partial \phi^2} dx + \int \frac{\partial^2 P_\phi}{\partial x^2} r \, d\phi - \frac{\nu}{r} \frac{\partial P_\phi}{\partial \phi} - \nu \frac{\partial P_x}{\partial x} \tag{6.164}$$

The external pressure P_z was previously defined as Eq. 6.146.

$$P_z = D\nabla^4 w - \frac{N_\phi}{r} \tag{6.146}$$

Rearranged,

$$\nabla^4 w - \frac{N_\phi}{Dr} = \frac{P_z}{D} \tag{6.165}$$

Substitute Eq. 6.148 into 6.165.

$$\nabla^4 w - \frac{\left(\frac{\partial^2 F}{\partial x^2} - \int P_\phi r d\phi\right)}{Dr} = \frac{P_z}{D} \tag{6.166}$$

$$\nabla^4 w - \frac{\partial^2 F}{Dr \, \partial x^2} + \frac{1}{Dr} \int P_\phi r d\phi = \frac{P_z}{D} \tag{6.167}$$

$$\nabla^4 w - \frac{\partial^2 F}{Dr \, \partial x^2} = \frac{P_z}{D} - \frac{1}{Dr} \int P_\phi r d\phi \tag{6.168}$$

$$\nabla^4 w - \frac{\partial^2 F}{Dr \, \partial x^2} = f'(p) \tag{6.169}$$

where

$$f'(p) = \frac{P_z}{D} - \frac{1}{Dr} \int P_\phi r d\phi \tag{6.170}$$

Multiplying Eq. 6.169 by the biharmonic operator ∇^4,

$$\nabla^4 \nabla^4 w - \frac{1}{Dr} \nabla^4 \frac{\partial^2 F}{\partial x^2} = \nabla^4 f'(p) \tag{6.171}$$

Operating on 6.162 by $\frac{1}{Dr} \frac{\partial^2}{\partial x^2}$,

$$\frac{1}{Dr} \frac{\partial^2}{\partial x^2} \nabla^4 F + \frac{Eh}{Dr^2} \frac{\partial^2}{\partial x^2} \frac{\partial^2 w}{\partial x^2} = \frac{1}{Dr} \frac{\partial^2 f(p)}{\partial x^2} \tag{6.172}$$

or

$$\frac{1}{Dr} \nabla^4 \frac{\partial^2 F}{\partial x^2} + \frac{Eh}{Dr^2} \frac{\partial^4 w}{\partial x^4} = \frac{1}{Dr} \frac{\partial^2 f(p)}{\partial x^2} \tag{6.173}$$

Adding Eqs. 6.171 and 6.173,

$$\frac{1}{Dr} \nabla^4 \frac{\partial^2 F}{\partial x^2} + \frac{Eh}{Dr^2} \frac{\partial^4 w}{\partial x^4} + \nabla^4 \nabla^4 w - \frac{1}{Dr} \nabla^4 \frac{\partial^2 F}{\partial x^2}$$

$$= \nabla^4 f'(p) + \frac{1}{Dr} \frac{\partial^2 f(p)}{\partial x^2} \tag{6.174}$$

By canceling the first and fourth terms of the left hand side, the terms with the potential function are eliminated.

$$\nabla^4 \nabla^4 w + \frac{Eh}{Dr^2} \frac{\partial^4 w}{\partial x^4} = \nabla^4 f'(p) + \frac{1}{Dr} \frac{\partial^2 f(p)}{\partial x^2} \qquad (6.175)$$

or

$$\nabla^8 w + \frac{Eh}{Dr^2} \frac{\partial^4 w}{\partial x^4} = f''(p) \qquad (6.176)$$

where

$$f''(p) = \nabla^4 f'(p) + \frac{1}{Dr} \frac{\partial^2 f(p)}{\partial x^2} \qquad (6.177)$$

Equation 6.176 is the equilibrium equation of the barrel arch, represented as a partial differential equation. The structure of Eq. 6.176 is a nonhomogenous equation. The homogenous form is therefore,

$$\nabla^8 w + \frac{Eh}{Dr^2} \frac{\partial^4 w}{\partial x^4} = 0 \qquad (6.178)$$

The solution to the differential Eq. (6.176) is in two parts.

$$w = w_c + w_p \qquad (6.179)$$

where w_c is the complimentary part of the homogenous solution and w_p is the particular part of the non-homogenous solution.

6.4.2 The Complementary Solution

The complementary describes the effects of edge conditions (i.e., edge forces), or boundary effects.

As previously defined,

$$f''(p) = \nabla^4 f'(p) + \frac{1}{Dr} \frac{\partial^2 f(p)}{\partial x^2} \qquad (6.177)$$

where

$$f'(p) = \frac{P_z}{D} - \frac{1}{Dr} \int P_\phi r d\phi = \frac{P_z}{D} - \frac{1}{D} \int P_\phi d\phi \qquad (6.170)$$

$$f(p) = \frac{1}{r^2} \int \frac{\partial^2 P_x}{\partial \phi^2} dx + \int \frac{\partial^2 P_\phi}{\partial x^2} r\, d\phi - \frac{\nu}{r} \frac{\partial P_\phi}{\partial \phi} - \nu \frac{\partial P_x}{\partial x} \tag{6.164}$$

Substitute 6.170 and 6.164 into 6.177.

$$f''(p) = \nabla^4 \left(\frac{P_z}{D} - \frac{1}{D} \int P_\phi d\phi \right) + \frac{1}{Dr} \frac{\partial^2 \left(\frac{1}{r^2} \int \frac{\partial^2 P_x}{\partial \phi^2} dx + \int \frac{\partial^2 P_\phi}{\partial x^2} r\, d\phi - \frac{\nu}{r} \frac{\partial P_\phi}{\partial \phi} - \nu \frac{\partial P_x}{\partial x} \right)}{\partial x^2} \tag{6.180}$$

Expand, apply the biharmonic operator, and integrate the terms with dx.

$$f''(p) = \frac{1}{D} \left(\nabla^4 P_z - \frac{1}{r^4} \frac{\partial^3 P_\phi}{\partial \phi^3} - \frac{2}{r^2} \frac{\partial^3 P_\phi}{\partial \phi \, \partial x^2} - \frac{1}{r} \int \frac{\partial^4 P_\phi}{\partial x^4} r\, d\phi + \frac{1}{r^3} \frac{\partial^3 P_x}{\partial \phi^2 \partial x} \right.$$
$$\left. + \frac{1}{r} \int \frac{\partial^4 P_\phi}{\partial x^4} r\, d\phi - \frac{\nu}{r^2} \frac{\partial^3 P_\phi}{\partial x^2 \partial \phi} - \frac{\nu}{r} \frac{\partial^3 P_x}{\partial x^3} \right) \tag{6.181}$$

where

$$-\nabla^4 \int P_\phi d\phi = -\left(\frac{\partial^4}{r^4 \partial \phi^4} + \frac{2\partial^4}{r^2 \partial x^2 \, \partial \phi^2} + \frac{\partial^4}{\partial x^4} \right) \left(\frac{1}{r} \int P_\phi r\, d\phi \right)$$
$$= -\frac{1}{r^4} \frac{\partial^3 P_\phi}{\partial \phi^3} - \frac{2}{r^2} \frac{\partial^3 P_\phi}{\partial \phi \, \partial x^2} - \frac{1}{r} \int \frac{\partial^4 P_\phi}{\partial x^4} r\, d\phi \tag{6.182}$$

Eliminate the fourth and sixth terms.

$$f''(p) = \frac{1}{D} \left(\nabla^4 P_z - \frac{1}{r^4} \frac{\partial^3 P_\phi}{\partial \phi^3} - \frac{2}{r^2} \frac{\partial^3 P_\phi}{\partial \phi \, \partial x^2} + \frac{1}{r^3} \frac{\partial^3 P_x}{\partial \phi^2 \partial x} - \frac{\nu}{r^2} \frac{\partial^3 P_\phi}{\partial x^2 \partial \phi} - \frac{\nu}{r} \frac{\partial^3 P_x}{\partial x^3} \right) \tag{6.183}$$

$$f''(p)$$
is the particular part of the solution.

Referring back to the homogenous equation,

$$\nabla^8 w + \frac{Eh}{Dr^2} \frac{\partial^4 w}{\partial x^4} = 0 \tag{6.178}$$

This equation is simplified by assuming $\nu = 0$. Poisson's strains range from 0.1 to 0.3 for most construction materials. Thus, this assumption will have less impact for brittle materials, such as masonry or concrete, which have low Poisson ratios.

$$D = \frac{Eh^3}{12(1 - \nu^2)} \approx \frac{Eh^3}{12} \tag{6.184}$$

The eigth order of the biharmonic operator is also expressed in a different, but equal form.

$$\nabla^8 w = \left(\frac{\partial^2}{\partial x^2} + \frac{\partial^2}{r^2 \partial \phi^2} \right)^4 w \tag{6.185}$$

The homogenous equation may therefore be expressed as:

$$\left(\frac{\partial^2}{\partial x^2} + \frac{\partial^2}{r^2 \partial \phi^2} \right)^4 w + \frac{12}{r^2 h^2} \frac{\partial^4 w}{\partial x^4} = 0 \tag{6.186}$$

Multiplying each side by r^2,

$$r^2 \left(\frac{\partial^2}{\partial x^2} + \frac{\partial^2}{r^2 \partial \phi^2} \right)^4 w + \frac{12}{h^2} \frac{\partial^4 w}{\partial x^4} = 0 \tag{6.187}$$

The complimentary solution to the homogenous equation is a trigonometric series.

$$w_c = \sum_{n=1, 3, \dots}^{\infty} A_m e^{m\phi} \sin kx \tag{6.188}$$

where

$$k = \frac{n\pi}{L} \tag{6.189}$$

The constant A_m is a function of the boundary conditions. Since the shell will have eight stress resultants at the boundary (i.e., N_x, N_ϕ, $N_{x\phi}$, Q_x, Q_ϕ, M_x, M_ϕ, $M_{x\phi}$), the solution will have potentially eight constants.

The m value in $e^{m\phi}$ represents the eight roots of the solution.

Substituting the solution (Eq. 6.188) into the homogenous equation (Eq. 6.187),

$$r^2 \left[\frac{\partial^2 \left(A_m e^{m\phi} \sin kx \right)}{\partial x^2} + \frac{\partial^2 \left(A_m e^{m\phi} \sin kx \right)}{r^2 \partial \phi^2} \right]^4 + \frac{12}{h^2} \frac{\partial^4 \left(A_m e^{m\phi} \sin kx \right)}{\partial x^4} = 0 \tag{6.190}$$

$$r^2 \left[-A_m e^{m\phi} k^2 \sin kx + \frac{A_m e^{m\phi} m^2}{r^2} \sin kx \right]^4 + \frac{12}{h^2} A_m e^{m\phi} k^4 \sin kx = 0 \tag{6.191}$$

$$r^2 \left[A_m e^{m\phi} \sin kx \left(-k^2 + \frac{m^2}{r^2} \right) \right]^4 + \frac{12}{h^2} A_m e^{m\phi} k^4 \sin kx = 0 \tag{6.192}$$

$$r^2 \left(-k^2 + \frac{m^2}{r^2} \right)^4 \left(A_m e^{m\phi} \sin kx \right) + \frac{12k^4}{h^2} \left(A_m e^{m\phi} \sin kx \right) = 0 \tag{6.193}$$

Divide each side of the equation by $A_m e^{m\phi} \sin kx$.

$$r^2 \left(\frac{m^2}{r^2} - k^2 \right)^4 + \frac{12k^4}{h^2} = 0 \tag{6.194}$$

Multiply by each term r^6.

$$r^8 \left(\frac{m^2}{r^2} - k^2 \right)^4 + \frac{12k^4 r^6}{h^2} = 0 \tag{6.195}$$

$$\left(m^2 - k^2 r^2 \right)^4 + \frac{12k^4 r^6}{h^2} = 0 \tag{6.196}$$

Divide each term by $\frac{12k^4 r^6}{h^2}$.

$$\frac{h^2}{12k^4 r^6} \left(m^2 - k^2 r^2 \right)^4 + 1 = 0 \tag{6.197}$$

If we say,

$$Q^8 = \frac{3 k^4 r^6}{h^2} \tag{6.198}$$

$$Q^8 4 = \frac{12 k^4 r^6}{h^2} \tag{6.199}$$

$$\frac{1}{Q^8 4} = \frac{h^2}{12 k^4 r^6} = \left(\frac{1}{Q^2 \sqrt{2}} \right)^4 \tag{6.200}$$

Substituting 6.200 into 6.197,

$$\left(\frac{1}{Q^2 \sqrt{2}} \right)^4 \left(m^2 - k^2 r^2 \right)^4 + 1 = 0 \tag{6.201}$$

$$\left(\frac{m^2 - k^2 r^2}{Q^2 \sqrt{2}} \right)^4 + 1 = 0 \tag{6.202}$$

$$\frac{1}{4}\left[\frac{m^2 - (kr)^2}{Q^2}\right]^4 + 1 = 0 \tag{6.203}$$

$$\left[\frac{m^2 - (kr)^2}{Q^2}\right]^4 + 4 = 0 \tag{6.204}$$

Furthermore, if we make the following replacement:

$$\gamma = \left(\frac{kr}{Q}\right)^2 \tag{6.205}$$

$$\left(\frac{m^2}{Q^2} - \gamma\right)^4 + 4 = 0 \tag{6.206}$$

Based on the above expression, the eight roots of the solution are complex and conjugate.

$$m_1 = +\left(\alpha_1 + \sqrt{-1}\beta_1\right) \tag{6.207}$$

$$m_2 = +\left(\alpha_1 - \sqrt{-1}\beta_1\right) \tag{6.208}$$

$$m_3 = -\left(\alpha_1 + \sqrt{-1}\beta_1\right) \tag{6.209}$$

$$m_4 = -\left(\alpha_1 - \sqrt{-1}\beta_1\right) \tag{6.210}$$

$$m_5 = +\left(\alpha_2 + \sqrt{-1}\beta_2\right) \tag{6.211}$$

$$m_6 = +\left(\alpha_2 - \sqrt{-1}\beta_2\right) \tag{6.212}$$

$$m_7 = -\left(\alpha_2 + \sqrt{-1}\beta_2\right) \tag{6.213}$$

$$m_8 = -\left(\alpha_2 - \sqrt{-1}\beta_2\right) \tag{6.214}$$

where

$$\alpha_1 = Q\sqrt{\frac{\sqrt{(1+\gamma)^2 + 1} + (1+\gamma)}{2}} \tag{6.215}$$

$$\beta_1 = Q\sqrt{\frac{\sqrt{(1+\gamma)^2 + 1} - (1+\gamma)}{2}} \qquad (6.216)$$

$$\alpha_2 = Q\sqrt{\frac{\sqrt{(1-\gamma)^2 + 1} - (1-\gamma)}{2}} \qquad (6.217)$$

$$\beta_2 = Q\sqrt{\frac{\sqrt{(1-\gamma)^2 + 1} + (1-\gamma)}{2}} \qquad (6.218)$$

The next step in the solution is to solve for the constants A_1 to A_8. As mentioned previously, these constants are a function of the boundary effects along the edge of the shell (four constants for each side of the shell). If we assume symmetry of loading, the eight roots reduce to four.

$$m_1 = -m_5$$

$$m_2 = -m_6$$

$$m_3 = -m_7$$

$$m_4 = -m_8$$

By making this assumption, the complementary solution becomes:

$$w_c = \left(A_1 e^{m_1\phi} + A_2 e^{m_2\phi} + A_3 e^{m_3\phi} + A_4 e^{m_4\phi} + A_5 e^{-m_1\phi} + A_6 e^{-m_2\phi}\right.$$
$$\left. + A_7 e^{-m_3\phi} + A_8 e^{-m_4\phi}\right) \sin kx \qquad (6.219)$$

If we consider the first two terms, and insert values of m_1 and m_2 (6.207 and 6.208),

$$A_1 e^{m_1\phi} + A_2 e^{m_2\phi} = A_1 e^{(\alpha_1 + \sqrt{-1}\beta_1)\phi} + A_2 e^{(\alpha_1 - \sqrt{-1}\beta_1)\phi} \qquad (6.220)$$

Replace $\sqrt{-1}$ with i and expand.

$$A_1 e^{(\alpha_1 + i\beta_1)\phi} + A_2 e^{(\alpha_1 - i\beta_1)\phi} = A_1 e^{\alpha_1\phi} e^{i\beta_1\phi} + A_2 e^{\alpha_1\phi} e^{-i\beta_1\phi} \qquad (6.221)$$

Since,

$$e^{i\beta_1\phi} = \cos\beta_1\phi + i\sin\beta_1\phi$$

$$e^{-i\beta_1\phi} = \cos\beta_1\phi - i\sin\beta_1\phi$$

$$A_1 e^{\alpha_1\phi} e^{i\beta_1\phi} + A_2 e^{\alpha_1\phi} e^{-i\beta_1\phi} = A_1 e^{\alpha_1\phi}(\cos\beta_1\phi + i\sin\beta_1\phi)$$
$$+ A_2 e^{\alpha_1\phi}(\cos\beta_1\phi - i\sin\beta_1\phi) \qquad (6.222)$$

For this expression to be real,

$$A_1 + A_2 = a \qquad (6.223)$$
$$A_1 - A_2 = ib \qquad (6.224)$$

where a and b are real numbers. Therefore,

$$A_1 = a - A_2 \qquad (6.225)$$
$$A_2 = A_1 - ib \qquad (6.226)$$

Substitute 6.225 into 6.226.

$$A_1 = a - (A_1 - ib) \qquad (6.227)$$
$$A_1 = \frac{1}{2}(a + ib) \qquad (6.228)$$

Similarly,

$$A_2 = \frac{1}{2}(a - ib) \qquad (6.229)$$

A_2 is therefore the conjugate of A_1 (complex conjugate). Substituting 6.228 and 6.229 into 6.222,

$$A_1 e^{\alpha_1\phi}(\cos\beta_1\phi + i\sin\beta_1\phi) + A_2 e^{\alpha_1\phi}(\cos\beta_1\phi - i\sin\beta_1\phi)$$
$$= \left[\frac{1}{2}(a + ib)(\cos\beta_1\phi + i\sin\beta_1\phi) + \frac{1}{2}(a - ib)(\cos\beta_1\phi - i\sin\beta_1\phi)\right] e^{\alpha_1\phi}$$
$$(6.230)$$

$$\left[\frac{1}{2}(a + ib)(\cos\beta_1\phi + i\sin\beta_1\phi) + \frac{1}{2}(a - ib)(\cos\beta_1\phi - i\sin\beta_1\phi)\right] e^{\alpha_1\phi}$$
$$= \frac{1}{2}\left[a\cos\beta_1\phi + ai\sin\beta_1\phi + bi\cos\beta_1\phi + bi^2\sin\beta_1\phi + a\cos\beta_1\phi - ai\sin\beta_1\phi\right.$$
$$\left. - bi\cos\beta_1\phi + bi^2\sin\beta_1\phi + \right] e^{\alpha_1\phi}$$
$$(6.231)$$

Since $i^2 = -1$, simplify the above expression.

$$\frac{1}{2}\left[a\cos\beta_1\phi + ai\sin\beta_1\phi + bi\cos\beta_1\phi + bi^2\sin\beta_1\phi + a\cos\beta_1\phi - ai\sin\beta_1\phi\right.$$
$$\left.?-bi\cos\beta_1\phi + bi^2\sin\beta_1\phi+\right]e^{\alpha_1\phi} = (a\cos\beta_1\phi - b\sin\beta_1\phi)\,e^{\alpha_1\phi}$$

$$(6.232)$$

Equation 6.232 is only for the first two terms. The remaining terms are solved in similar manner:

$$A_3 e^{m_3\phi} + A_4 e^{m_4\phi} = (c\cos\beta_2\phi - d\sin\beta_2\phi)e^{\alpha_2\phi} \tag{6.233}$$

$$A_5 e^{m_5\phi} + A_6 e^{m_6\phi} = (e\cos\beta_1\phi + f\sin\beta_1\phi)e^{-\alpha_1\phi} \tag{6.234}$$

$$A_7 e^{m_7\phi} + A_8 e^{m_8\phi} = (g\cos\beta_2\phi + h\sin\beta_2\phi)e^{-\alpha_2\phi} \tag{6.235}$$

Collecting the terms and setting them into Eq. 6.219,

$$w_c = \left[(a\cos\beta_1\phi - b\sin\beta_1\phi)\,e^{\alpha_1\phi} + (c\cos\beta_2\phi - d\sin\beta_2\phi)e^{\alpha_2\phi}\right.$$
$$\left.+(e\cos\beta_1\phi + f\sin\beta_1\phi)e^{-\alpha_1\phi} + (g\cos\beta_2\phi + h\sin\beta_2\phi)e^{-\alpha_2\phi}\right]\sin kx$$

$$(6.236)$$

If the shell is loaded symmetrically,

$$a = e, b = f, c = g, d = h$$

Introducing this simplification,

$$w_c = \left[(a\cos\beta_1\phi - b\sin\beta_1\phi)\,e^{\alpha_1\phi} + (c\cos\beta_2\phi - d\sin\beta_2\phi)e^{\alpha_2\phi}\right]$$
$$+(a\cos\beta_1\phi + b\sin\beta_1\phi)e^{-\alpha_1\phi} + (c\cos\beta_2\phi + d\sin\beta_2\phi)e^{-\alpha_2\phi}\right]\sin kx$$

$$(6.237)$$

Rearranging the above equation,

$$w_c = \left[a\left(e^{\alpha_1\phi} + e^{-\alpha_1\phi}\right)\cos\beta_1\phi - b\left(e^{\alpha_1\phi} - e^{-\alpha_1\phi}\right)\sin\beta_1\phi\right.$$
$$\left.+c\left(e^{\alpha_2\phi} + e^{-\alpha_2\phi}\right)\cos\beta_2\phi - d\left(e^{\alpha_2\phi} - e^{-\alpha_2\phi}\right)\sin\beta_2\phi\right]\sin kx$$

$$(6.238)$$

A further simplification,

$$\cosh x = \frac{e^x + e^{-x}}{2}, \quad 2\cosh x = e^x + e^{-x}$$

$$\sinh x = \frac{e^x - e^{-x}}{2}, \quad 2\sinh x = e^x - e^{-x}$$

Applying this simplification,

$$w_c = 2[a \cosh (\alpha_1\phi) \cos \beta_1\phi - b \sinh (\alpha_1\phi) \sin \beta_1\phi + c \cosh (\alpha_2\phi) \cos \beta_2\phi$$
$$-d \sinh (\alpha_2\phi) \sin \beta_2\phi] \sin kx$$

$$(6.239)$$

Setting,

$$J = a \cosh (\alpha_1\phi) \cos \beta_1\phi - b \sinh (\alpha_1\phi) \sin \beta_1\phi + c \cosh (\alpha_2\phi) \cos \beta_2\phi$$
$$- d \sinh (\alpha_2\phi) \sin \beta_2\phi \qquad (6.240)$$

A simplified form,

$$w_c = 2 J \sin kx \qquad (6.241)$$

6.4.3 The Particular Solution

The particular solution defines the response of the shell to the loading, regardless of the boundary conditions. Both parts, the particular and complementary solutions, are necessary to describe the stress distribution and deformations in the shell.

Components of the external loading were previously derived as Eqs. 6.4, 6.5, and 6.6, for gravity loads.

$$P_z = P_d \, \cos (\phi_k - \phi) \qquad (6.4)$$
$$P_\phi = -P_d \, \sin (\phi_k - \phi) \qquad (6.5)$$
$$P_x = 0 \qquad (6.6)$$

However, a Fourier series is used to represent the loads on the shell, as a convenience to overcome mathematical difficulties in the derivation.

$$P_d(x) = P \sum_{n=1,3,5...}^{\infty} \frac{1}{n} \sin \frac{n\pi x}{L} \qquad (6.242)$$

If $n = 1$, the load takes the shape of a single *sin* curve. As n increases, the loading resembles a uniform distributed load, which is the result of overlapping *sin* curves.
If $n = 1$,

$$P_z = \frac{4P_d}{\pi} \cos (\phi_k - \phi) \sin \frac{\pi x}{L} \qquad (6.243)$$

$$P_\phi = -\frac{4P_d}{\pi} \sin (\phi_k - \phi) \sin \frac{\pi x}{L} \tag{6.244}$$

$$P_x = 0$$

where

$$P = \frac{4P_d}{\pi} \tag{6.245}$$

See Eqs. 6.7, 6.8, and 6.9.

The particular solution is determined by solving for the nonhomogenous differential equation.

$$\nabla^8 w + \frac{Eh}{Dr^2} \frac{\partial^4 w}{\partial x^4} = f''(p) \tag{6.176}$$

Simplify $f''(p)$ (Eq. 6.183) by assuming $\nu = 0$, and knowing that $P_x = 0$,

$$f''(p) = \frac{1}{EI} \left(\nabla^4 P_z - \frac{1}{r^4} \frac{\partial^3 P_\phi}{\partial \phi^3} - \frac{2}{r^2} \frac{\partial^3 P_\phi}{\partial \phi \, \partial x^2} \right) \tag{6.246}$$

Expanding the above equation and rearranging,

$$f''(p) = \frac{1}{EI} \left[\left(\frac{\partial^4}{r^4 \partial \phi^4} + \frac{2 \partial^4}{r^2 \, \partial x^2 \, \partial \phi^2} + \frac{\partial^4}{\partial x^4} \right) P_z - \left(\frac{1}{r^4} \frac{\partial^3}{\partial \phi^3} + \frac{2}{r^2} \frac{\partial^3}{\partial \phi \, \partial x^2} \right) P_\phi \right] \tag{6.247}$$

where the biharmonic operator is equal to:

$$\nabla^4 = \left(\frac{\partial^4}{r^4 \partial \phi} + \frac{2 \partial^4}{r^2 \partial x^2 \partial \phi^2} + \frac{\partial^4}{\partial x^4} \right) \tag{6.248}$$

Substituting the external load Eqs. (6.243 and 6.244) into 6.247,

$$f''(p) = \frac{1}{EI} \left[\left(\frac{\partial^4}{r^4 \partial \phi^4} + \frac{2 \partial^4}{r^2 \, \partial x^2 \, \partial \phi^2} + \frac{\partial^4}{\partial x^4} \right) \frac{4P_d}{\pi} \cos (\phi_k - \phi) \sin kx \right.$$
$$\left. + \left(\frac{1}{r^4} \frac{\partial^3}{\partial \phi^3} + \frac{2}{r^2} \frac{\partial^3}{\partial \phi \, \partial x^2} \right) \frac{4P_d}{\pi} \sin (\phi_k - \phi) \sin kx \right] \tag{6.249}$$

Operate on the load equation and temporarily replace $P = \frac{4P_d}{\pi}$.

$$f''(P) = \frac{1}{EI}\left[\frac{P}{r^4}\cos{(\phi_k - \phi)}\,\sin kx + \frac{2Pk^2}{r^2}\cos{(\phi_k - \phi)}\sin kx + Pk^4\cos{(\phi_k - \phi)}\sin kx\right.$$
$$\left. + \frac{P}{r^4}\cos{(\phi_k - \phi)}\sin kx + \frac{2Pk^2}{r^2}\cos{(\phi_k - \phi)}\,\sin kx\right]$$

$$(6.250)$$

or

$$f''(P) = \frac{4P_d}{EI\pi}\left(\frac{1}{r^4} + \frac{2k^2}{r^2} + k^4 + \frac{1}{r^4} + \frac{2k^2}{r^2}\right)\cos{(\phi_k - \phi)}\sin kx \qquad (6.251)$$

$$f''(P) = \frac{4P_d}{EI\pi}\left(\frac{2}{r^4} + k^4 + \frac{4k^2}{r^2}\right)\cos{(\phi_k - \phi)}\sin kx \qquad (6.252)$$

We state that the particular solutions is of the form:

$$w_p = \overline{K}P\cos{(\phi_k - \phi)}\sin kx \qquad (6.253)$$

The next step is to solve for the first term of the left hand side of 6.176. Knowing that

$$\nabla^8 w = \left(\frac{\partial^2}{\partial x^2} + \frac{\partial^2}{r^2\partial\phi^2}\right)^4 w$$
$$= \left(\frac{\partial^8}{\partial x^8} + \frac{4\partial^8}{r^2\partial x^6\partial\phi^2} + \frac{2\partial^8}{r^4\partial x^6\partial\phi^2} + \frac{4\partial^8}{r^6\partial x^4\partial\phi^4} + \frac{4\partial^8}{r^4\partial x^4\partial\phi^4} + \frac{\partial^8}{r^8\partial x^4\partial\phi^4}\right)w$$

$$(6.254)$$

$$\nabla^8 w_p + \frac{Eh}{Dr^2}\frac{\partial^4 w_p}{\partial x^4} = \overline{K}Pk^8\cos{(\phi_k - \phi)}\sin kx$$
$$+ \frac{4\overline{K}Pk^6}{r^2}\cos{(\phi_k - \phi)}\sin kx$$
$$+ \frac{2\overline{K}Pk^6}{r^4}\cos{(\phi_k - \phi)}\sin kx$$
$$+ \frac{4\overline{K}Pk^4}{r^6}\cos{(\phi_k - \phi)}\sin kx$$
$$+ \frac{4\overline{K}Pk^4}{r^4}\cos{(\phi_k - \phi)}\sin kx$$
$$+ \frac{\overline{K}Pk^4}{r^8}\cos{(\phi_k - \phi)}\sin kx$$
$$+ \frac{12\overline{K}Pk^4}{h^2r^2}\cos{(\phi_k - \phi)}\sin kx \qquad (6.255)$$

where

$$D = EI \; (v = 0) \tag{6.256}$$

Rearranged,

$$\nabla^8 w_p + \frac{Eh}{Dr^2} \frac{\partial^4 w_p}{\partial x^4} = \left(k^8 + \frac{4k^6}{r^2} + \frac{2k^6}{r^4} + \frac{4k^4}{r^6} + \frac{4k^4}{r^4} + \frac{k^4}{r^8} + \frac{12k^4}{h^2 r^2} \right) \overline{K} P \cos$$
$$(\phi_k - \phi) \sin kx$$

$$\nabla^8 w_p + \frac{Eh}{Dr^2} \frac{\partial^4 w_p}{\partial x^4} = \left[k^8 + k^6 \left(\frac{4}{r^2} + \frac{2}{r^4} \right) + k^4 \left(\frac{4}{r^6} + \frac{4}{r^4} + \frac{1}{r^8} + \frac{12}{h^2 r^2} \right) \right] \overline{K} P \cos$$
$$(\phi_k - \phi) \sin kx$$

$$\tag{6.257}$$

Since,

$$\nabla^8 w + \frac{Eh}{Dr^2} \frac{\partial^4 w}{\partial x^4} = f''(p) \tag{6.176}$$

Equating 6.257 and 6.252, the nonhomogenous equation is therefore,

$$\left[k^8 + k^6 \left(\frac{4}{r^2} + \frac{2}{r^4} \right) + k^4 \left(\frac{4}{r^6} + \frac{4}{r^4} + \frac{1}{r^8} + \frac{12}{h^2 r^2} \right) \right] \overline{K} P \cos (\phi_k - \phi) \sin kx$$

$$= \frac{4P_d}{EI\pi} \left(\frac{2}{r^4} + k^4 + \frac{4k^2}{r^2} \right) \cos (\phi_k - \phi) \sin kx \tag{6.258}$$

Replacing $P = \frac{4P_d}{\pi}$ and simplify the above equation,

$$\left[k^8 + k^6 \left(\frac{4}{r^2} + \frac{2}{r^4} \right) + k^4 \left(\frac{4}{r^6} + \frac{4}{r^4} + \frac{1}{r^8} + \frac{12}{h^2 r^2} \right) \right] \overline{K}$$

$$= \frac{1}{EI} \left(\frac{2}{r^4} + k^4 + \frac{4k^2}{r^2} \right) \tag{6.259}$$

Solving for $\overline{K}$,

$$\overline{K} = \frac{\left(\frac{2}{r^4} + k^4 + \frac{4k^2}{r^2} \right)}{EI \left[k^8 + k^6 \left(\frac{4}{r^2} + \frac{2}{r^4} \right) + k^4 \left(\frac{4}{r^6} + \frac{4}{r^4} + \frac{1}{r^8} + \frac{12}{h^2 r^2} \right) \right]} \tag{6.260}$$

where

$$k = \frac{n\pi}{L}$$

$$I = \frac{h^3}{12}$$

It should be noted that the particular and membrane solutions are similar, and are sometimes used interchangeably.

6.4.4   The General Solution for Deformations

Adding Eqs. 6.239 and 6.253, the general solution is solved.

$$w = w_c + w_p \tag{6.179}$$

$$w = 2[a \cosh(\alpha_1\phi)\cos\beta_1\phi - b \sinh(\alpha_1\phi)\sin\beta_1\phi + c \cosh(\alpha_2\phi)\cos\beta_2\phi$$
$$-d \sinh(\alpha_2\phi)\sin\beta_2\phi]\sin kx + \overline{K}P\cos(\phi_k - \phi)\,\sin\,kx$$

$$\tag{6.261}$$

The remaining unknowns are the arbitrary constants a, b, c, and d. These coefficients are determined from the boundary conditions (i.e., two coefficients for each edge of the shell). Along the free edge, the following conditions occur (Chatterjee 1971):
If the edge is unsupported,

$$M_\phi = 0, N_\phi = 0, N_{\phi x} = 0, Q_\phi = 0$$

If the edge is supported on an unyielding wall (pinned support),

$$M_\phi = 0, N_{\phi x} = 0, v = 0, w = 0$$

If the edge is rigidly held by the edge support (fixed support),

$$u = 0, v = 0, w = 0, \Delta_\phi = 0$$

For each of the above boundary conditions, four equations are solved in terms of the unknown coefficients. Thus, using the method of simultaneous equations, a solution is possible. The equations are simplified by assuming Poisson's ratio is zero ($\nu = 0$); this assumption will have a greater or lesser impact, depending on the shell material. Furthermore, the in-plane deformation (v) is assumed to be small compared to the out-of-plane deformations (w), and therefore the terms with v are deleted from the equation. The complexity of the equations requires these rationalizations, to enable a solution and to make the analysis more manageable.

$$D = \frac{Eh^3}{12(1-\nu^2)} = \frac{Eh^3}{12} = EI \qquad (6.121)$$

$$D' = \frac{Eh}{(1-\nu^2)} = EA \qquad (6.262)$$

6.4.5 Moments, Shears, and Stresses Due to Boundary Effects

The moments, shears, and stresses are simplified by setting $\nu = 0$.
  Stress N_x

$$N_x = \frac{Eh}{(1-\nu^2)}\left[\frac{\partial u}{\partial x} + \nu\left(\frac{\partial v}{r\partial\phi} - \frac{w}{r}\right)\right] \qquad (6.107)$$

$$N_x = EA\left(\frac{\partial u}{\partial x}\right) \qquad (6.263)$$

Stress N_ϕ

$$N_\phi = \frac{Eh}{(1-\nu^2)}\left[\frac{\partial v}{r\partial\phi} - \frac{w}{r} + \nu\frac{\partial u}{\partial x}\right] \qquad (6.108)$$

$$N_\phi = \frac{EA}{r}\left[\frac{\partial v}{\partial\phi} - w\right] \qquad (6.264)$$

In-plane shear stress $N_{x\phi}$

$$N_{x\phi} = \frac{Eh}{2(1+\nu)}\left(\frac{\partial u}{r\partial\phi} + \frac{\partial v}{\partial x}\right) \qquad (6.109)$$

$$N_{x\phi} = \frac{EA}{2}\left(\frac{\partial u}{r\partial\phi} + \frac{\partial v}{\partial x}\right) \qquad (6.265)$$

Bending moment M_x

$$M_x = -\frac{Eh^3}{12(1-\nu^2)}\left[\frac{\partial^2 w}{\partial x^2} + \nu\left(\frac{\partial v}{r^2\partial\phi} + \frac{\partial^2 w}{r^2\partial\phi^2}\right)\right] \qquad (6.122)$$

$$M_x = -EI\left(\frac{\partial^2 w}{\partial x^2}\right) \qquad (6.266)$$

Bending moment M_ϕ

$$M_\phi = -\frac{Eh^3}{12(1-\nu^2)}\left[\frac{\partial v}{r^2\partial\phi} + \frac{\partial^2 w}{r^2\partial\phi^2} + \nu\left(\frac{\partial^2 w}{\partial x^2}\right)\right] \tag{6.124}$$

$$M_\phi = -\frac{EI}{r^2}\left(\frac{\partial^2 w}{\partial\phi^2}\right) \tag{6.267}$$

Twisting moment $M_{x\phi}$

$$M_{x\phi} = D(1-\nu)\left[\frac{\partial v}{2r\,\partial x} + \frac{\partial^2 w}{r\,\partial x\,\partial\phi}\right] \tag{6.125}$$

$$M_{x\phi} = \frac{EI}{r}\left(\frac{\partial^2 w}{\partial x\,\partial\phi}\right) \tag{6.268}$$

where $\frac{\partial u}{\partial\phi}$ and $\frac{\partial v}{\partial x}$ makes a small contribution compared to the deformation w.

Shear stress Q_ϕ

Since,

$$\frac{\partial Q_\phi}{r\,\partial\phi} = \frac{\partial^2 M_\phi}{r^2\,\partial\phi^2} - \frac{\partial^2 M_{x\phi}}{r\,\partial x\,\partial\phi} \tag{6.127}$$

Multiplying each term by r and integrating each side of the above equation,

$$\int\frac{\partial Q_\phi}{\partial\phi}\,d\phi = \int\left(\frac{\partial^2 M_\phi}{r\,\partial\phi^2} - \frac{\partial^2 M_{x\phi}}{\partial x\,\partial\phi}\right)d\phi \tag{6.269}$$

$$Q_\phi = \frac{\partial M_\phi}{r\,\partial\phi} - \frac{\partial M_{x\phi}}{\partial x} \tag{6.270}$$

Take the derivatives of 6.267 with respect to ϕ and 6.268 with respect to x. Substitute the results into 6.270.

$$Q_\phi = -\frac{EI}{r^3}\left(\frac{\partial^3 w}{\partial\phi^3}\right) - \frac{EI}{r}\left(\frac{\partial^3 w}{\partial x^2\,\partial\phi}\right) \tag{6.271}$$

Furthermore,

$$\frac{\partial Q_x}{\partial x} = \frac{\partial^2 M_x}{\partial x^2} + \frac{\partial^2 M_{\phi x}}{r\,\partial x\,\partial\phi} \tag{6.129}$$

Integrating each side of the above equation,

$$\int \frac{\partial Q_x}{\partial x} dx = \int \left(\frac{\partial^2 M_x}{\partial x^2} + \frac{\partial^2 M_{\phi x}}{r \, \partial x \, \partial \phi} \right) dx \qquad (6.272)$$

$$Q_x = \frac{\partial M_x}{\partial x} + \frac{\partial M_{\phi x}}{r \, \partial \phi} \qquad (6.273)$$

Similarly, take the derivative of 6.266 with respect to x and 6.268 with respect to ϕ. Substitute the results into 6.273

$$Q_x = -EI \left(\frac{\partial^3 w}{\partial x^3} \right) + \frac{EI}{r^2} \left(\frac{\partial^3 w}{\partial x \, \partial \phi^2} \right) \qquad (6.274)$$

The deformations were previously defined by Eqs. 6.82, 6.83, and 6.261.

$$u = \frac{2PL^3}{Ehr\pi^3} \cos(\phi_k - \phi) \cos(kx) \qquad (6.82)$$

$$v = -\frac{2P}{Eh} \left(\frac{L^4}{r^2 \pi^4} + \frac{2L^2}{\pi^2} \right) \sin(\phi_k - \phi) \sin(kx) \qquad (6.83)$$

$$w = 2[a \cosh(\alpha_1 \phi) \cos \beta_1 \phi - b \sinh(\alpha_1 \phi) \sin \beta_1 \phi + c \cosh(\alpha_2 \phi) \cos \beta_2 \phi$$
$$- d \sinh(\alpha_2 \phi) \sin \beta_2 \phi \,] \sin kx + \overline{K}P \cos(\phi_k - \phi) \sin kx$$
$$\qquad (6.261)$$

Taking the derivative of u, v, and w with respect to x and ϕ,

$$\frac{\partial u}{\partial x} = -\frac{2PL^2}{Ehr\pi^2} \cos(\phi_k - \phi) \sin(kx) \qquad (6.275)$$

$$\frac{\partial u}{\partial \phi} = \frac{2PL^3}{Ehr\pi^3} \sin(\phi_k - \phi) \cos(kx) \qquad (6.276)$$

$$\frac{\partial v}{\partial \phi} = \frac{2P}{Eh} \left(\frac{1}{r^2 k^4} + \frac{2}{k^2} \right) \cos(\phi_k - \phi) \sin(kx) \qquad (6.277)$$

$$\frac{\partial v}{\partial x} = -\frac{2P}{Eh} \left(\frac{1}{r^2 k^3} + \frac{2}{k} \right) \sin(\phi_k - \phi) \cos(kx) \qquad (6.278)$$

$$\frac{\partial w}{\partial x} = 2\,k\cos kx\,(a\cosh\alpha_1\phi\cos\beta_1\phi - b\sinh\alpha_1\phi\sin\beta_1\phi + c\cosh\alpha_2\phi\cos\beta_2\phi$$
$$-d\sinh\alpha_2\phi\sin\beta_2\phi\,) + \overline{K}Pk\cos(\phi_k - \phi)\cos kx$$

$$(6.279)$$

$$\frac{\partial^2 w}{\partial x^2} = -2\,k^2\sin kx\,(a\cosh\alpha_1\phi\cos\beta_1\phi - b\sinh\alpha_1\phi\sin\beta_1\phi + c\cosh\alpha_2\phi\cos\beta_2\phi$$
$$-d\sinh\alpha_2\phi\sin\beta_2\phi\,) - \overline{K}Pk^2\cos(\phi_k - \phi)\sin kx$$

$$(6.280)$$

$$\frac{\partial^3 w}{\partial x^3} = -2\,k^3\cos kx\,(a\cosh\alpha_1\phi\cos\beta_1\phi - b\sinh\alpha_1\phi\sin\beta_1\phi + c\cosh\alpha_2\phi\cos\beta_2\phi$$
$$-d\sinh\alpha_2\phi\sin\beta_2\phi\,) - \overline{K}Pk^3\cos(\phi_k - \phi)\cos kx$$

$$(6.281)$$

$$\frac{\partial w}{\partial \phi} = 2\sin kx\,[(a\alpha_1 - b\beta_1)\,\sinh\alpha_1\phi\cos\beta_1\phi + (-b\alpha_1 - a\beta_1)\cosh\alpha_1\phi\sin\beta_1\phi$$
$$+(c\alpha_2 - d\beta_2)\sinh\alpha_2\phi\cos\beta_2\phi + (-d\alpha_2 - c\beta_2)\cosh\alpha_2\phi\sin\beta_2\phi\,]$$
$$+ \overline{K}P\sin(\phi_k - \phi)\sin kx$$

$$(6.282)$$

$$\frac{\partial^2 w}{\partial \phi^2} = 2\sin kx\,\big[(a\alpha_1^{\,2} - 2b\alpha_1\beta_1 - a\beta_1^{\,2})\,\cosh\alpha_1\phi\cos\beta_1\phi$$
$$+ \left(-b\alpha_1^{\,2} - 2a\alpha_1\beta_1 + b\beta_1^{\,2}\right)\sinh\alpha_1\phi\sin\beta_1\phi$$
$$+ \left(c\alpha_2^{\,2} - 2d\alpha_2\beta_2 - c\beta_2^{\,2}\right)\cosh\alpha_2\phi cos\beta_2\phi$$
$$+\left(-d\alpha_2^{\,2} - 2c\alpha_2\beta_2 + d\beta_2^{\,2}\right)\sinh\alpha_2\phi\sin\beta_2\phi\,\big]$$
$$- \overline{K}P\cos(\phi_k - \phi)\sin kx$$

$$(6.283)$$

$$\frac{\partial^3 w}{\partial \phi^3} = 2\sin kx\,[\left(a\alpha_1^{\,3} - 3b\alpha_1^{\,2}\beta_1 - 3a\alpha_1\beta_1^{\,2} + b\beta_1^{\,3}\right)\,\sinh\alpha_1\phi\cos\beta_1\phi$$
$$+ \left(-3a\alpha_1^{\,2}\beta_1 + 3b\alpha_1\beta_1^{\,2} - b\alpha_1^{\,3} + a\beta_1^{\,3}\right)\,\cosh\alpha_1\phi\sin\beta_1\phi$$
$$+ \left(c\alpha_2^{\,3} - 3d\alpha_2^{\,2}\beta_2 - 3c\alpha_2\beta_2^{\,2} + d\beta_2^{\,3}\right)\sinh\alpha_2\phi\cos\beta_2\phi$$
$$+ \left(-d\alpha_2^{\,3} - 3c\alpha_2^{\,2}\beta_2 + 3d\alpha_2\beta_2^{\,2} + c\beta_2^{\,3}\right)\cosh\alpha_2\phi\sin\beta_2\phi]$$
$$- \overline{K}P\sin(\phi_k - \phi)\sin kx$$

$$(6.284)$$

$$\frac{\partial^2 w}{\partial \phi \partial x} = 2k\cos kx[(a\alpha_1 - b\beta_1)\sinh\alpha_1\phi\cos\beta_1\phi$$
$$+ (-b\alpha_1 - a\beta_1)\cosh\alpha_1\phi\sin\beta_1\phi$$
$$+ (c\alpha_2 - d\beta_2)\sinh\alpha_2\phi\cos\beta_2\phi$$
$$+ (-d\alpha_2 - c\beta_2)\cosh\alpha_2\phi\sin\beta_2\phi] + \overline{K}Pk\,\sin(\phi_k - \phi)\cos kx \quad (6.285)$$

$$\frac{\partial^3 w}{\partial x^2 \partial \phi} = -2k^2 \sin kx[(a\alpha_1 - b\beta_1)\ \sinh \alpha_1 \phi \cos \beta_1 \phi$$

$$+ (-b\alpha_1 - a\beta_1) \cosh \alpha_1 \phi \sin \beta_1 \phi$$
$$+ (c\alpha_2 - d\beta_2) \sinh \alpha_2 \phi \cos \beta_2 \phi$$
$$+ (-d\alpha_2 - c\beta_2) \cosh \alpha_2 \phi \sin \beta_2 \phi] - \overline{K}Pk^2\ \sin(\phi_k - \phi) \sin kx \quad (6.286)$$

$$\frac{\partial^3 w}{\partial \phi^2 \partial x} = 2k\ \cos kx[(a\alpha_1^2 - 2b\alpha_1\beta_1 - a\beta_1^2)\ \cosh \alpha_1 \phi \cos \beta_1 \phi$$

$$+ (-b\alpha_1^2 - 2a\alpha_1\beta_1 - b\beta_1^2) \sinh \alpha_1 \phi \sin \beta_1 \phi$$
$$+ (c\alpha_2^2 - 2d\alpha_2\beta_2 - c\beta_2^2) \cosh \alpha_2 \phi \cos \beta_2 \phi$$
$$+ (-d\alpha_2^2 - 2c\alpha_2\beta_2 + d\beta_2^2) \sinh \alpha_2 \phi sin\beta_2 \phi]$$
$$- \overline{K}Pk\ \cos(\phi_k - \phi) \cos kx \quad (6.287)$$

Solving for each of the stresses, moments, and shears,

$$M_x = -EI\left(\frac{\partial^2 w}{\partial x^2}\right) \quad (6.266)$$

Substituting 6.280,

$$M_x = EI[2\ k^2 \sin kx\ (a \cosh \alpha_1 \phi \cos \beta_1 \phi - b \sinh \alpha_1 \phi \sin \beta_1 \phi$$
$$+ c \cosh \alpha_2 \phi \cos \beta_2 \phi - d \sinh \alpha_2 \phi \sin \beta_2 \phi)$$
$$- \overline{K}Pk^2 \cos(\phi_k - \phi) \sin kx] \quad (6.288)$$

$$M_\phi = -\frac{EI}{r^2}\left(\frac{\partial^2 w}{\partial \phi^2}\right) \quad (6.267)$$

Substituting 6.283,

$$M_\phi = -\frac{EI}{r^2}\{2 \sin kx[(a\alpha_1^2 - 2b\alpha_1\beta_1 - a\beta_1^2)\ \cosh \alpha_1 \phi \cos \beta_1 \phi$$

$$+ (-b\alpha_1^2 - 2a\alpha_1\beta_1 + b\beta_1^2) \sinh \alpha_1 \phi \sin \beta_1 \phi$$
$$+ (c\alpha_2^2 - 2d\alpha_2\beta_2 - c\beta_2^2) \cosh \alpha_2 \phi \cos \beta_2 \phi$$
$$+ (-d\alpha_2^2 - 2c\alpha_2\beta_2 + d\beta_2^2) \sinh \alpha_2 \phi \sin \beta_2 \phi]$$
$$- \overline{K}P \cos(\phi_k - \phi) \sin kx\} \quad (6.289)$$

$$M_{x\phi} = \frac{EI}{r}\left(\frac{\partial^2 w}{\partial x \partial \phi}\right) \quad (6.268)$$

Substituting 6.285,

$$M_{x\phi} = \frac{EI}{r}\{2k\cos kx[(a\alpha_1 - b\beta_1)\sinh\alpha_1\phi\cos\beta_1\phi$$

$$+ (-b\alpha_1 - a\beta_1)\cosh\alpha_1\phi\sin\beta_1\phi + (c\alpha_2 - d\beta_2)\sinh\alpha_2\phi\cos\beta_2\phi$$
$$+ (-d\alpha_2 - c\beta_2)\cosh\alpha_2\phi\sin\beta_2\phi] + \overline{K}Pk\ \sin(\phi_k - \phi)\cos kx\} \qquad (6.290)$$

$$Q_x = -EI\left(\frac{\partial^3 w}{\partial x^3}\right) + \frac{EI}{r^2}\left(\frac{\partial^3 w}{\partial x\,\partial\phi^2}\right) \qquad (6.274)$$

Substituting 6.281 and 6.287,

$$Q_x = EI\{2\,k^3\cos kx(a\cosh\alpha_1\phi\cos\beta_1\phi - b\sinh\alpha_1\phi\sin\beta_1\phi$$
$$+ c\cosh\alpha_2\phi\cos\beta_2\phi - d\sinh\alpha_2\phi\sin\beta_2\phi)$$

$$- \overline{K}Pk^3\cos(\phi_k - \phi)\cos kx\} + \frac{EI}{r^2}$$

$$\times\{2k\ \cos kx[(a\alpha_1{}^2 - 2b\alpha_1\beta_1 - a\beta_1{}^2)\ \cosh\alpha_1\phi\cos\beta_1\phi$$

$$+ (-b\alpha_1{}^2 - 2a\alpha_1\beta_1 - b\beta_1{}^2)\sinh\alpha_1\phi sin\beta_1\phi$$

$$+ (c\alpha_2{}^2 - 2d\alpha_2\beta_2 - c\beta_2{}^2)\cosh\alpha_2\phi\cos\beta_2\phi$$

$$+ (-d\alpha_2{}^2 - 2c\alpha_2\beta_2 + d\beta_2{}^2)\sinh\alpha_2\phi sin\beta_2\phi]$$
$$- \overline{K}Pk\ \cos(\phi_k - \phi)\cos kx\} \qquad (6.291)$$

$$Q_\phi = -\frac{EI}{r^3}\left(\frac{\partial^3 w}{\partial\phi^3}\right) - \frac{EI}{r}\left(\frac{\partial^3 w}{\partial x^2\,\partial\phi}\right) \qquad (6.271)$$

Substituting 6.284 and 6.286,

$$Q_\phi = -\frac{EI}{r^3}$$

$$\times\{2\sin kx[(a\alpha_1{}^3 - 3b\alpha_1{}^2\beta_1 - 3a\alpha_1\beta_1{}^2 + b\beta_1{}^3)\ \sinh\alpha_1\phi\cos\beta_1\phi$$

$$+ (-3a\alpha_1{}^2\beta_1 + 3b\alpha_1\beta_1{}^2 - b\alpha_1{}^3 + a\beta_1{}^3)\ \cosh\alpha_1\phi\sin\beta_1\phi$$

$$+ (c\alpha_2{}^3 - 3d\alpha_2{}^2\beta_2 - 3c\alpha_2\beta_2{}^2 + d\beta_2{}^3)\sinh\alpha_2\phi\cos\beta_2\phi$$

$$+ (-d\alpha_2{}^3 - 3c\alpha_2{}^2\beta_2 + 3d\alpha_2\beta_2{}^2 + c\beta_2{}^3)\cosh\alpha_2\phi\sin\beta_2\phi]$$

$$- \overline{K}P\sin(\phi_k - \phi)\sin kx\} + \frac{EI}{r}$$

$$\times\{2k^2\sin kx[(a\alpha_1 - b\beta_1)\ \sinh\alpha_1\phi\cos\beta_1\phi$$
$$+ (-b\alpha_1 - a\beta_1)\cosh\alpha_1\phi\sin\beta_1\phi + (c\alpha_2 - d\beta_2)\sinh\alpha_2\phi\cos\beta_2\phi$$

$$+ (-d\alpha_2 - c\beta_2)\cosh\alpha_2\phi\sin\beta_2\phi] - \overline{K}Pk^2\ \sin(\phi_k - \phi)\sin kx\} \qquad (6.292)$$

$$N_x = EA\left(\frac{\partial u}{\partial x}\right) \tag{6.263}$$

Substituting 6.275,

$$N_x = -\frac{2APL^2}{hr\pi^2}\cos(\phi_k - \phi)\sin(kx) \tag{6.293}$$

$$N_\phi = \frac{EA}{r}\left[\frac{\partial v}{\partial \phi} - w\right] \tag{6.264}$$

Substituting 6.277 and 6.261,

$$N_\phi = \frac{EA}{r}\left\{\frac{2P}{Eh}\left(\frac{1}{r^2k^4} + \frac{2}{k^2}\right)\cos(\phi_k - \phi)\sin(kx)\right.$$
$$- 2[a\cosh(\alpha_1\phi)\cos\beta_1\phi - b\sinh(\alpha_1\phi)\sin\beta_1\phi + c\cosh(\alpha_2\phi)\cos\beta_2\phi$$
$$\left. -d\sinh(\alpha_2\phi)\sin\beta_2\phi]\sin kx + \overline{K}P\cos(\phi_k - \phi)\sin kx\right\}$$
$$\tag{6.294}$$

$$N_{x\phi} = \frac{EA}{2}\left(\frac{\partial u}{r\partial \phi} + \frac{\partial v}{\partial x}\right) \tag{6.265}$$

Substituting 6.276 and 6.278,

$$N_{x\phi} = \frac{EA}{2}$$
$$\times\left\{\left[\frac{2PL^3}{Ehr^2\pi^3}\sin(\phi_k - \phi)\cos(kx)\right] - \frac{2P}{Eh}\left(\frac{1}{r^2k^3} + \frac{2}{k}\right)\sin(\phi_k - \phi)\cos(kx)\right\}$$
$$\tag{6.295}$$

6.5   Edge Beams

Usually, two types of edge beams are used on cylindrical barrel shells, vertical or horizontal. The edge beams that are orientated vertically are used for long shells, where stiffness is required in the vertical direction. If the shell is relatively flat, the horizontal thrusts are large and stiffness is required in the horizontal direction. The vertically orientated beam is far more common, and therefore the solution will only consider a vertical beam (see Fig. 6.17) (Billington 1982).

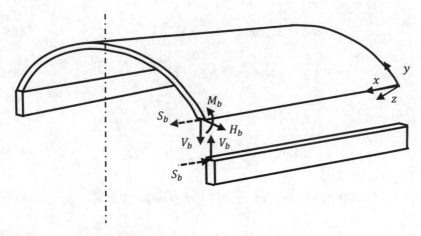

Fig. 6.17 Stresses on the edge beam

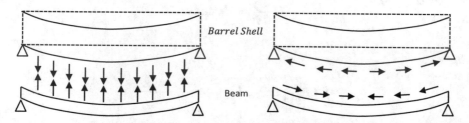

Fig. 6.18 Compatibility between the barrel shell and edge beam

6.5.1 Stress Equations

The beam is very flexible in the transverse direction, and therefore the beams offer very little twisting resistance. Furthermore, the lateral torsional stiffness is not sufficient to generate torsional reactions.

$$H_b = 0$$

$$M_b = 0$$

By introducing an edge beam, the membrane forces and deflections will not be correct. The edge beam will limit the vertical deflections and longitudinal shears. The interaction between the shell edge and beam must be the same to satisfy compatibility requirements (Fig. 6.18).

From beam flexural theory (Timoshenko 1958),

$$\text{Loading} = w = EI\frac{\partial^4 y}{\partial x^4} \tag{6.296}$$

$$\text{Shear} = -S_x = EI\frac{\partial^3 y}{\partial x^3} \tag{6.297}$$

$$\text{Moment} = -M_x = EI\frac{\partial^2 y}{\partial x^2} \tag{6.298}$$

$$\text{Slope} = \frac{\partial y}{\partial x} = -\int_0^x \frac{M_x}{EI}dx + f_1(x) \tag{6.299}$$

$$\text{Deflection} = y = -\iint_0^L \frac{M_x}{EI}dx + \int_0^L f_1(x)dx + f_2(x) \tag{6.300}$$

If we assume that the load applied from the shell onto the beam is a sinusoidal distribution,

$$EI\frac{\partial^4 y}{\partial x^4} = V_b \sin kx \tag{6.301}$$

Integrating 6.301,

$$\int EI\frac{\partial^4 y}{\partial x^4}dx = EI\frac{\partial^3 y}{\partial x^3} = \frac{-V_b \cos kx}{k} \tag{6.302}$$

$$\int\int EI\frac{\partial^4 y}{\partial x^4}dx = EI\frac{\partial^2 y}{\partial x^2} = \frac{-V_b \sin kx}{k^2} \tag{6.303}$$

$$\int\int\int EI\frac{\partial^4 y}{\partial x^4}dx = EI\frac{\partial y}{\partial x} = \frac{V_b \cos kx}{k^3} \tag{6.304}$$

$$\int\int\int\int EI\frac{\partial^4 y}{\partial x^4}dx = EIy = \frac{V_b \sin kx}{k^4} \tag{6.305}$$

or

$$y = \frac{V_b \sin kx}{EIk^4} \tag{6.306}$$

The longitudinal stress between the top of the beam and the edge of the shell is typically expressed in the following form:

$$f = \frac{M}{Z} \tag{6.307}$$

where z is the section modulus of the beam.

Substituting 6.298 and 6.303 into 6.307,

$$f = -\frac{V_b \sin kx}{Zk^2} \qquad (6.308)$$

The horizontal stress at the top of the beam is referred to as the horizontal shear. This shear varies from maximum at the ends of the beam to zero at mid-span. To represent this distribution, the stress is varied by a *cos* curve.

$$S = S_b \cos kx \qquad (6.309)$$

Summing the edge stresses from the support to any distance x,

$$T = -\int_0^x S_b \cos kx \, dx \qquad (6.310)$$

$$T = \frac{-S_b}{k} \sin kx \qquad (6.311)$$

The moment may therefore be defined as (refer to Fig. 6.19):

$$M = Te = \frac{-S_b e}{k} \sin kx \qquad (6.312)$$

From these equations, the vertical deflection caused by the bending moment may be determined. Since the bending moment is a function of the second derivative of deflection, Eqs. 6.312 and 6.300 are combined.

$$y = -\iint_0^L \frac{-S_b e}{EI \, k} \sin kx \, dx + \int_0^L f_1(x) dx + f_2(x) \qquad (6.313)$$

Integrating the above expression,

Fig. 6.19 Moment in the edge beam

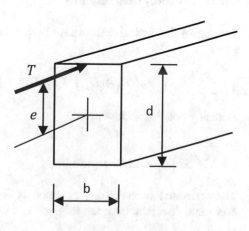

$$y = -\frac{S_b e}{EI\,k^3}\sin kx \tag{6.314}$$

where the integration constants are equal to zero, due to the boundary conditions.

$$f_1(x) = f_2(x) = 0$$

The horizontal stress between the shell and the top of the beam is the combination of two components.

$$f = \frac{T}{A} + \frac{Te}{Z} \tag{6.315}$$

where

$$z = \frac{bd^2}{6} \tag{6.316}$$

$$e = \frac{d}{2} \tag{6.317}$$

6.5.2 Compatibility Equations

With many shell solutions (e.g., cylindrical and domes), the compatibility equations are formulated in terms of the deformations. With cylindrical barrel shells, however, the compatibility equations are formulated in terms of the vertical deformations and horizontal stresses.

6.5.2.1 Vertical Compatibility

The sum of the vertical deformation between the shell edge and the beam must equal zero.

$$V_b\left(D^s_{11} + D^b_{11}\right) + S_b\left(D^s_{12} + D^b_{12}\right) + D^s_{10} + D^b_{10} = 0 \tag{6.318}$$

assuming positive is downward.

6.5.2.2 Horizontal Compatibility

The horizontal compatibility equation is formulated by summing stresses in the horizontal direction, between the shell and top of the beam.

$$V_b\left(f^s_{21} + f^b_{21}\right) + S_b\left(f^s_{22} + f^b_{22}\right) + f^s_{20} + f^b_{20} = 0 \tag{6.319}$$

Defining the sub- and postscripts

First subscript	Due to	Second subscript	Postscript
1 – vertical deflection		0 – applied load	s – shell
2 – horizontal stress		1 – vertical force	b – beam
		2 – horizontal shear	

6.5.3 Components of the Compatibility Equations

$$D^b_{11} = y = \frac{V_b \sin kx}{EIk^4} \tag{6.306}$$

where

$$I = \frac{bd^3}{12} \text{ (bean)} \tag{6.320}$$

$$k = \frac{n\pi}{L} \tag{6.189}$$

Substituting the above into 6.306,

$$D^b_{11} = y = \frac{12V_b L^4}{En^4\pi^4 bd^3} \sin kx \tag{6.321}$$

The horizontal shear also contributes to the vertical deflection.

$$D^b_{12} = y = -\frac{S_b e}{EI\,k^3} \sin kx \tag{6.314}$$

Substitute I, k, and e (6.317, 6.318 and 6.320) into the above.

$$D^b_{12} = y = -\frac{6S_b L^3}{Ebd^2 n^3\pi^3} \sin kx \tag{6.322}$$

Since,

$$f^b_{21} = \frac{M}{Z} = -\frac{V_b \sin kx}{Zk^2} \tag{6.308}$$

Substitute 6.189 and 6.316 into 6.308.

$$f_{21}^b = -\frac{6V_b L^2}{bd^2 n^2 \pi^2} \sin kx \tag{6.323}$$

The horizontal stress on top of the beam may be expressed in terms of the resultant horizontal force T.

$$f_{22}^b = \frac{T}{A} + \frac{Te}{z} \tag{6.315}$$

Substitute 6.311, 6.312, and 6.316.

$$f_{22}^b = \frac{S_b L \sin kx}{n\pi bd} + \frac{3 S_b d \, L \, \sin kx}{n\pi \, bd^2} \tag{6.324}$$

Combine the two terms.

$$f_{22}^b = \frac{4 S_b L}{n\pi bd} \sin kx \tag{6.325}$$

The displacements for the self-weight of the beam is represented by a Fourier series, where w_d is a force per unit length.

$$w = \frac{4 \, w_d}{\pi} \sum_{n=1,3,5\ldots}^{\infty} \frac{1}{n} \sin kx \tag{6.326}$$

The vertical deflection of the edge beam is determined from the well-known deflection formula,

$$D_{10}^b = \frac{5 \, wL^4}{384 \, EI} \tag{6.327}$$

Substituting 6.326 and 6.320 into 6.327,

$$D_{10}^b = \frac{5 \, w_d L^4}{8 \, \pi Ebd^3 n^4} \sin kx \tag{6.328}$$

Referring to Eq. 6.308,

$$f_{20}^b = -\frac{V_b \sin kx}{Zk^2} \tag{6.308}$$

Substituting Eqs. 6.189, 6.316 and set $V_b = \frac{4 \, w_d}{\pi}$. The load will be in the negative direction, and therefore the negative sign is canceled.

$$f_{20}^b = \frac{24\, w_d L^2}{b d^2 n^2 \pi^3} \sin kx \tag{6.329}$$

The membrane deformation was previously solved (Eq. 6.92), and comprise part of the D_{10}^s term. A component of the deformation w (Eq. 6.239) is taken in the Δ_v direction, and added to the membrane deformation equation.

$$
\begin{aligned}
D_{10}^s &= \Delta_v \\
&= \frac{2PL^4}{Ehr^2\pi^4} \sin\left(\frac{\pi x}{L}\right)\left[1 + \frac{2\pi^2 r^2}{L^2} + \left(\frac{\pi^4 r^4}{2L^4}\right)\cos^2(\phi_k - \phi)\right] \\
&\quad + w\cos(\phi_k - \phi)
\end{aligned}
\tag{6.317}
$$

Substituting Eq. 6.239 for w,

$$
\begin{aligned}
D_{10}^s &= \Delta_v \\
&= \frac{2PL^4}{Ehr^2\pi^4} \sin\left(\frac{\pi x}{L}\right)\left[1 + \frac{2\pi^2 r^2}{L^2} + \left(\frac{\pi^4 r^4}{2L^4}\right)\cos^2(\phi_k - \phi)\right] \\
&\quad + \{2\sin kx[a\cosh(\alpha_1\phi)\cos\beta_1\phi - b\sinh(\alpha_1\phi)\sin\beta_1\phi \\
&\quad + c\cosh(\alpha_2\phi)\cos\beta_2\phi - d\sinh(\alpha_2\phi)\sin\beta_2\phi] \\
&\quad + \overline{K}P\cos(\phi_k - \phi)\sin kx\}\ \cos(\phi_k - \phi)
\end{aligned}
\tag{6.330}
$$

In the above equation, the coefficients a, b, c, and d are determined by applying boundary conditions.

$$f_{20}^s = \frac{N_x}{h}\ (x = L/2) \tag{6.331}$$

The variation of V_b is assumed to have a sinusoidal variation:

$$V_b \sin kx \tag{6.332}$$

Since the loading is a function of the fourth derivative of deflection (6.296),

$$\frac{\partial^4 y}{\partial x^4} = \frac{V_b}{EI}\sin kx \tag{6.333}$$

Integrating the above expression,

$$y = \Delta_v = \frac{V_b}{EI}\left(\frac{L}{n\pi}\right)^4 \sin kx \tag{6.334}$$

For a segment of a cylindrical arch,

$$I = k'r^3h \tag{6.335}$$

where k' is a function of ϕ_k (Billington 1982).

The deflection of the shell for a sinusoidal load is equal to:

$$D_{11}^s = \frac{V_b L^4}{E k' r^3 h \pi^4} \sin kx \tag{6.336}$$

Similarly for shears,

$$D_{12}^s = \frac{S_b L^4}{E k' r^3 h} \sin kx \tag{6.337}$$

Furthermore, the stress distribution between the edge of the shell and the top of the beam is expressed in terms of a sin distribution.

$$f = \frac{My}{I} = \frac{V_b r \cos(\phi_k - \phi)}{k' r^3 h \, k^2} \sin kx \tag{6.338}$$

where

$$I = k'r^3h \tag{6.335}$$

$$y = r \cos(\phi_k - \phi) \tag{6.339}$$

$$M = \frac{V_b}{k^2} \sin kx \tag{6.340}$$

or expressed as:

$$f_{21}^s = \frac{V_b L^2 \cos(\phi_k - \phi)}{k' r^2 h \, \pi^2} \sin kx \tag{6.341}$$

Similarly,

$$f_{22}^s = \frac{S_b L^2}{r^2 h} \sin kx \tag{6.342}$$

As seen by the copious nature of the equations, the analysis is complex and does not engender a practical method of analysis. For this reason, the equations are solved and tabulated to simplify the solution.

6.6 Steps in Solving the Deformations and Stresses in the Shell

Due to the complexity of the solution when solving for the arbitrary coefficients a, b, c, and d, tables have been formulated for various ratios of $\frac{r}{L}$ and $\frac{r}{h}$. These tables are found in Appendix A and B, which are reproduced from the *ASCE Manual of Engineering Practice, Design of Shell Roofs, No. 31* (1960). The solution is based on the preceding equations, but simplified to ease calculations.

6.6.1 Barrel Shells with Free Edges

Refer to Appendix A for coefficients that are to be included in the stress equations. Similar to the deformation equations, the equations below specify the column from which the coefficients should be extracted, dependent on the ratios of $\frac{r}{h}$ and $\frac{r}{L}$. In all of the equations, only the first term of the Fourier series is considered, i.e., $n = 1$.

$$P = \frac{4P_d}{\pi}.$$

1. **Solve for the reactions V_L, H_L and S_L.**
   Initial parameters

   Span (S)
   Length (L)
   Radius (r)
   Height (H)
   Thickness (h)
   Young's modulus (E)
   Poisson's ratio (ν)

   $I = \frac{h^3}{12}$ (*moment of inertia per unit width of shell*)   6.320

   $D = \frac{EI}{(1-\nu^2)}$ 6.121

   $P = \frac{4}{\pi}P_d$ 6.343

   Edge stresses at $\phi = 0$

   N'_ϕ $N'_\phi = -Pr \cos(\phi_k)$ 6.62
   $(x = L/2)$

   $N'_{x\phi}$ $N'_{\phi x} = -\frac{2PL}{\pi} \sin(\phi_k)$ 6.67
   $(x = 0 \text{ or } x = L)$

   Solve for V_L, H_L, and S_L (correction loads along the free edge)

   $V_L = -N'_\phi \sin \phi_k$ 6.344

   $H_L = -N'_\phi \cos \phi_k$ 6.345

   $S_L = -N'_{\phi x}$ 6.346

2. **Solve for the membrane stresses**

Coefficients are taken from Appendix A. Values differ for each angle ϕ_k and ratios of $\frac{r}{h}$ and $\frac{r}{L}$

$$N'_x = -\frac{2PL^2}{r\pi^2} \cos(\phi_k - \phi) \sin\left(\frac{\pi x}{L}\right) \qquad 6.347$$
$$+V_L\left(\frac{L}{r}\right)^2 (A:col.1) \sin\left(\frac{\pi x}{L}\right)$$
$$+H_L\left(\frac{L}{r}\right)^2 (A:col.5) \sin\left(\frac{\pi x}{L}\right)$$
$$+S_L\left(\frac{L}{r}\right)^2 (A:col.9) \sin\left(\frac{\pi x}{L}\right)$$
$$N'_{\phi x} = -\frac{2PL}{\pi} \sin(\phi_k - \phi) \cos\left(\frac{\pi x}{L}\right) \qquad 6.348$$
$$+V_L\left(\frac{L}{r}\right)(A:col.2) \cos\left(\frac{\pi x}{L}\right)$$
$$+H_L\left(\frac{L}{r}\right)(A:col.6) \cos\left(\frac{\pi x}{L}\right)$$
$$+S_L\left(\frac{L}{r}\right)(A:col.10) \cos\left(\frac{\pi x}{L}\right)$$
$$N'_{\phi} = -Pr \cos(\phi_k - \phi) \sin\left(\frac{\pi x}{L}\right) \qquad 6.349$$
$$+V_L(A:col.3) \sin\left(\frac{\pi x}{L}\right)$$
$$+H_L(A:col.7) \sin\left(\frac{\pi x}{L}\right)$$
$$+S_L(A:col.11) \sin\left(\frac{\pi x}{L}\right)$$

3. **Solve for the membrane deformations in the shell.**

$$\Delta'_v = \frac{2PL^4}{Ehr^2\pi^4} \sin\left(\frac{\pi x}{L}\right)\left[1 + \frac{2\pi^2 r^2}{L^2} + \left(\frac{\pi^4 r^4}{2L^4}\right)\cos^2(\phi_k - \phi)\right] \qquad 6.92$$
$$\Delta'_H = \frac{Pr^2}{Eh} \sin(\phi_k - \phi)\cos(\phi_k - \phi)\sin\left(\frac{\pi x}{L}\right) \qquad 6.97$$

6.6.2 Barrel Shells with Edge Beams

Similar to barrel shells with free edges, the arbitrary coefficients a, b, c, and d have been solved for various ratios of $\frac{r}{L}$ and $\frac{r}{h}$. These coefficients are found in Appendix A and B. The start of the analysis is the same as the membrane solution (which is repeated for completeness).

1. **Solve for the reactions V_L, H_L and S_L.**

Initial parameters

Span (S)
Length (L)
Radius (r)
Height (H)
Thickness (h)
Young's modulus (E)
Poisson's ratio (ν)

$$I = \frac{h^3}{12} \text{ (moment of inertia per unit width of shell)} \qquad 6.320$$
$$D = \frac{EI}{(1-v^2)} \qquad 6.121$$
$$P = \frac{4}{\pi}P_d \qquad 6.343$$

Edge stresses at $\phi = 0$

N'_ϕ $N'_\phi = - Pr \cos(\phi_k)$ 6.62

$(x = L/2)$

$N'_{x\phi}$ $N'_{x\phi} = -\frac{2PL}{\pi} \sin(\phi_k)$ 6.67

$(x = 0 \text{ or } x = L)$

Solve for V_L, H_L, and S_L (correction loads along the free edge)

$$V_L = -N'_\phi \sin \phi_k$$ 6.344

$$H_L = -N'_\phi \cos \phi_k$$ 6.345

$$S_L = -N'_{\phi x}$$ 6.346

2. Solve for the deformations and stresses between the top of the beam and base of the shell

Vertical deformations
Refer to Appendix B for coefficients that are to be included in the deformation equations. The equations specify the column from which the coefficients should be extracted, dependent on the ratios of $\frac{r}{h}$ and $\frac{r}{L}$. In all of the equations, only the first term of the Fourier series is considered, and therefore $n = 1$. The equations are also expressed in terms of $\left(\frac{L^2}{E}\right) \sin\left(\frac{\pi x}{L}\right)$, which is common and therefore cancel

$$D^s_{11} = V_b \left(\frac{L^2}{E}\right)\left(\frac{L^2}{hr^3}\right)$$ 6.350

$$(B : col.1) \sin\left(\frac{\pi x}{L}\right)$$

$$D^b_{11} = V_b \left(\frac{L^2}{E}\right)$$ 6.351

$$\left(\frac{12L^2}{\pi^4 bd^3}\right) \sin\left(\frac{\pi x}{L}\right)$$

$$D^s_{12} = S_b \left(\frac{L^2}{E}\right)\left(\frac{L^2}{hr^3}\right)$$ 6.352

$$(B : col.7) \sin\left(\frac{\pi x}{L}\right)$$

$$D^b_{12} = -S_b \left(\frac{L^2}{E}\right)\left(\frac{6L}{bd^2\pi^3}\right)$$ 6.353

$$\sin\left(\frac{\pi x}{L}\right)$$

$$D^s_{10} = \left(\frac{L^2}{E}\right)\left(\frac{L^2}{hr^3}\right)$$

$$\left\{Pr\left[\frac{2}{\pi^4} + \left(\frac{2r}{\pi L}\right)^2 + \right.\right.$$ 6.354

$$\left(\frac{r}{L}\right)\cos^2(\phi_k - \phi)\right]$$

$$+[V_L(B : col.1) + H_L$$

$$(B : col.4) + S_L$$

$$\left.(B : col.7)]\right\} \sin\left(\frac{\pi x}{L}\right)$$

$$D^b_{10} = -\left(\frac{L^2}{E}\right)\left(\frac{4}{\pi}\right)$$ 6.355

$$\left(\frac{5}{384}\right)\left(\frac{12\, w^b L^2}{bd^3}\right)$$

$$\sin\left(\frac{\pi x}{L}\right)$$

$$w^b = \gamma bd$$

$$\gamma = \text{weight of beam material}$$

Horizontal shears
Refer to Appendix A for coefficients that are to be included in the stress equations. Similar to

$$f^s_{21} = V_b \left(\frac{L}{r}\right)^2 \frac{1}{h}$$ 6.356

$$(A : col.1) \sin\left(\frac{\pi x}{L}\right)$$

$$f^b_{21} = -V_b \left(\frac{L}{r}\right)^2$$ 6.357

the deformation equations, the equations below specify the column from which the coefficients should be extracted, dependent on the ratios of $\frac{r}{h}$ and $\frac{r}{L}$. In all of the equations, only the first term of the Fourier series is considered, and therefore $n = 1$

$$\left(\frac{6r^2}{bd^2\pi^2}\right)\sin\left(\frac{\pi x}{L}\right)$$

$$f_{22}^s = S_b\left(\frac{L}{r}\right)^2 \frac{1}{h}(A : col.9) \qquad 6.358$$
$$\sin\left(\frac{\pi x}{L}\right)$$

$$f_{22}^b = S_b\left(\frac{L}{r}\right)^2 \frac{4r^2}{L\pi bd} \qquad 6.359$$
$$\sin\left(\frac{\pi x}{L}\right)$$

$$f_{20}^s = \frac{N_x}{h} \qquad 6.360$$

Where,

$$N_x = \left(\frac{L}{r}\right)^2\left[-Pr\left(\frac{2}{\pi^2}\right)\cos\left(\phi_k - \phi\right) + V_L\right. \qquad 6.361$$
$$(A : col.1) + H_L(A : col.5)$$
$$\left. + S_L(A : col.9)\right]\sin\left(\frac{\pi x}{L}\right)$$

$$f_{20}^b = \left(\frac{L}{r}\right)^2\left(\frac{4}{\pi}\right) \qquad 6.362$$
$$\left(\frac{6w_d r^2}{bd^2\pi^2}\right)\sin\left(\frac{\pi x}{2}\right)$$

3. Solve the compatibility equations

$$V_b\left(D_{11}^s + D_{11}^b\right) + S_b\left(D_{12}^s + D_{12}^b\right) + D_{10}^s + D_{10}^b = 0 \qquad 6.318$$

$$V_b\left(f_{21}^s + f_{21}^b\right) + S_b\left(f_{22}^s + f_{22}^b\right) + f_{20}^s + f_{20}^b = 0 \qquad 6.319$$

$$\begin{bmatrix}\left(D_{11}^s + D_{11}^b\right) & \left(D_{12}^s + D_{12}^b\right)\\ \left(f_{21}^s + f_{21}^b\right) & \left(f_{22}^s + f_{22}^b\right)\end{bmatrix}\begin{Bmatrix}V_b\\ S_b\end{Bmatrix} = \begin{Bmatrix}-D_{10}^s - D_{10}^b\\ -f_{20}^s - f_{20}^b\end{Bmatrix} \qquad 6.363$$

$$\begin{Bmatrix}V_b\\ S_b\end{Bmatrix} = \frac{1}{\left(D_{11}^s + D_{11}^b\right)\left(f_{22}^s + f_{22}^b\right) - \left(D_{12}^s + D_{12}^b\right)\left(f_{21}^s + f_{21}^b\right)} \qquad 6.364$$

$$\begin{bmatrix}\left(f_{22}^s + f_{22}^b\right) & -\left(D_{12}^s + D_{12}^b\right)\\ -\left(f_{21}^s + f_{21}^b\right) & \left(D_{11}^s + D_{11}^b\right)\end{bmatrix}\begin{Bmatrix}-D_{10}^s - D_{10}^b\\ -f_{20}^s - f_{20}^b\end{Bmatrix}$$

4. Solve the stresses and bending moments in the shell

$$N_x = -\frac{2PL^2}{r\pi^2}\cos\left(\phi_k - \phi\right)\sin\left(\frac{\pi x}{L}\right) \qquad 6.365$$
$$+ V_L\left(\frac{L}{r}\right)^2(A : col.1)\sin\left(\frac{\pi x}{L}\right) + H_L\left(\frac{L}{r}\right)^2(A : col.5)\sin\left(\frac{\pi x}{L}\right)$$
$$+ S_L\left(\frac{L}{r}\right)^2(A : col.9)\sin\left(\frac{\pi x}{L}\right) + V_b\left(\frac{L}{r}\right)^2(A : col.1)\sin\left(\frac{\pi x}{L}\right)$$
$$+ S_b\left(\frac{L}{r}\right)^2(A : col.9)\sin\left(\frac{\pi x}{L}\right)$$

$$N_{x\phi} = -\frac{2PL}{\pi}\sin\left(\phi_k - \phi\right)\cos\left(\frac{\pi x}{L}\right) \qquad 6.366$$
$$+ V_L\left(\frac{L}{r}\right)(A : col.2)\cos\left(\frac{\pi x}{L}\right) + H_L\left(\frac{L}{r}\right)(A : col.6)\cos\left(\frac{\pi x}{L}\right)$$
$$+ S_L\left(\frac{L}{r}\right)(A : col.10)\cos\left(\frac{\pi x}{L}\right) + V_b\left(\frac{L}{r}\right)(A : col.2)\cos\left(\frac{\pi x}{L}\right)$$
$$+ S_b\left(\frac{L}{r}\right)(A : col.10)\cos\left(\frac{\pi x}{L}\right)$$

$$N_\phi = -Pr\cos\left(\phi_k - \phi\right)\sin\left(\frac{\pi x}{L}\right) \qquad 6.367$$
$$+ V_L(A : col.3)\sin\left(\frac{\pi x}{L}\right) + H_L(A : col.7)\sin\left(\frac{\pi x}{L}\right)$$
$$+ S_L(A : col.11)\sin\left(\frac{\pi x}{L}\right) + V_b(A : col.3)\sin\left(\frac{\pi x}{L}\right)$$

$$+S_b(A : col.11) \sin\left(\frac{\pi x}{L}\right)$$
$$M_\phi = V_L r(A : col.4) \sin\left(\frac{\pi x}{L}\right) \qquad\qquad 6.368$$
$$+H_L r(A : col.8) \sin\left(\frac{\pi x}{L}\right) + S_L r(A : col.12) \sin\left(\frac{\pi x}{L}\right)$$
$$+V_b r(A : col.4) \sin\left(\frac{\pi x}{L}\right) + S_b r(A : col.12) \sin\left(\frac{\pi x}{L}\right)$$

6.7 Worked Examples

6.7.1 Simply Supported Barrel Vault with Free Edges

The barrel vault illustrated in Fig. 6.20 is simply supported at each of the corners and subjected to a combined dead and live load of 3.8 KPa. The edges of the shell roof are free (no edge beam).

1. **Solve for the reactions V_L, H_L, and S_L**

   Span = 11450 mm
   Length = 20250 mm
   Radius = 8100 mm
   Height = 8100–8100 cos 45 = 2372 mm
   Thickness = 75 mm
   Young's modulus = $30(10^6)$ kN/m²
   Poisson's ratio $(\nu) = 0.15$

$$I = \frac{h^3}{12} = \frac{(1)0.075^3}{12} = 3.52\left(10^{-5}\right) \text{ m}^4 \qquad (6.320)$$

$$D = \frac{EI}{(1-\nu^2)} = \frac{30(10^6)3.52(10^{-5})}{(1-0.15^2)} = 1079 \text{ Nm}^2 \qquad (6.121)$$

$$P = \frac{4}{\pi}P_d = \frac{4}{\pi}(3.8) = 4.838 \text{ kN/m}^2 \qquad (6.343)$$

Edge stresses at $\phi = 0$

$$N'_\phi = -Pr\cos(\phi_k) = -(4.838)8.1\cos(45) = -27.71 \qquad (6.62)$$

$$N'_{x\phi} = -\frac{2PL}{\pi}\sin(\phi_k) = -\frac{2(4.838)20.25}{\pi}\sin(45) = -44.11 \qquad (6.67)$$

Solve for V_L, H_L, and S_L

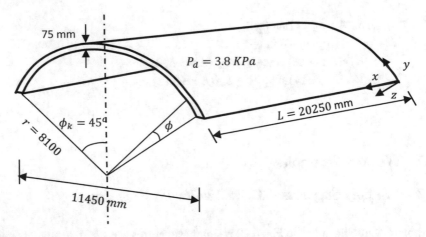

Fig. 6.20 Cylindrical barrel shell, unsupported by edge beams

$$V_L = -N'_\phi \sin \phi_k = (27.71) \sin 45 = 19.60 \text{ kN/m} \tag{6.344}$$

$$H_L = -N'_\phi \cos \phi_k = (27.71) \cos 45 = 19.60 \text{kN/m} \tag{6.345}$$

$$S_L = -N'_{x\phi} = 44.11 \text{ kN/m} \tag{6.346}$$

2. **Solve for the membrane stresses**

The membrane stress equations are expressed in term of x and ϕ.

$$N'_x = -\frac{2PL^2}{r\pi^2} \cos (\phi_k - \phi) \sin \left(\frac{\pi x}{L}\right) + V_L \left(\frac{L}{r}\right)^2 (A : col.1) \sin \left(\frac{\pi x}{L}\right)$$

$$+ H_L \left(\frac{L}{r}\right)^2 (A : col.5) \sin \left(\frac{\pi x}{L}\right) + S_L \left(\frac{L}{r}\right)^2 (A : col.9) \sin \left(\frac{\pi x}{L}\right) \tag{6.347}$$

$$N'_x = -\frac{2(4.838)20.25^2}{(8.1)\pi^2} \cos (45 - \phi) \sin \left(\frac{\pi x}{20.25}\right)$$

$$+ V_L \left(\frac{20.25}{8.1}\right)^2 (A : col.1) \sin \left(\frac{\pi x}{20.25}\right)$$

$$+ H_L \left(\frac{20.25}{8.1}\right)^2 (A : col.5) \sin \left(\frac{\pi x}{20.25}\right)$$

$$+ S_L \left(\frac{20.25}{8.1}\right)^2 (A : col.9) \sin \left(\frac{\pi x}{20.25}\right)$$

which simplifies to:

$$N'_x = -49.63 \cos{(45 - \phi)} \sin{(0.155x)} + V_L(6.25)(A : col.1) \sin{(0.155x)}$$
$$+ H_L(6.25)(A : col.5) \sin{(0.155x)} + S_L(6.25)(A : col.9) \sin{(0.155x)}$$

$$N'_{\phi x} = -\frac{2PL}{\pi} \sin{(\phi_k - \phi)} \cos{\left(\frac{\pi x}{L}\right)} + V_L\left(\frac{L}{r}\right)(A : col.2) \cos{\left(\frac{\pi x}{L}\right)}$$
$$+ H_L\left(\frac{L}{r}\right)(A : col.6) \cos{\left(\frac{\pi x}{L}\right)} + S_L\left(\frac{L}{r}\right)(A : col.10) \cos{\left(\frac{\pi x}{L}\right)} \quad (6.348)$$

$$N'_{x\phi} = -\frac{2(4.838)20.25}{\pi} \sin{(45 - \phi)} \cos{\left(\frac{\pi x}{20.25}\right)} + V_L\left(\frac{20.25}{8.1}\right)$$

$$\times (A : col.2) \cos{\left(\frac{\pi x}{20.25}\right)} + H_L\left(\frac{20.25}{8.1}\right)(A : col.6) \cos{\left(\frac{\pi x}{20.25}\right)}$$

$$+ S_L\left(\frac{20.25}{8.1}\right)(A : col.10) \cos{\left(\frac{\pi x}{20.25}\right)}$$

which simplifies to:

$$N'_{x\phi} = -62.37 \sin{(45 - \phi)} \cos{(0.155x)} + V_L(2.5)(A : col.2) \cos{(0.155x)}$$
$$+ H_L(2.5)(A : col.6) \cos{(0.155x)} + S_L(2.5)(A : col.10) \cos{(0.155x)}$$

$$N'_\phi = -Pr \cos{(\phi_k - \phi)} \sin{\left(\frac{\pi x}{L}\right)} + V_L(A : col.3) \sin{\left(\frac{\pi x}{L}\right)}$$
$$+ H_L(A : col.7) \sin{\left(\frac{\pi x}{L}\right)} + S_L(A : col.11) \sin{\left(\frac{\pi x}{L}\right)} \quad (6.349)$$

$$N'_\phi = -(4.838)8.1 \cos{(45 - \phi)} \sin{\left(\frac{\pi x}{20.25}\right)}$$

$$+ V_L(A : col.3) \sin{\left(\frac{\pi x}{20.25}\right)} + H_L(A : col.7) \sin{\left(\frac{\pi x}{20.25}\right)}$$

$$+ S_L(A : col.11) \sin{\left(\frac{\pi x}{20.25}\right)} \quad (6.348)$$

which simplifies to:

$$N'_\phi = -39.19 \cos{(45 - \phi)} \sin{(0.155x)} + V_L(A : col.3) \sin{(0.155x)}$$
$$+ H_L(A : col.7) \sin{(0.155x)} + S_L(A : col.11) \sin{(0.155x)}$$

The above stresses equations are solved in increments of $10°$, at $x = 0$ and $x = \frac{L}{2}$.

3. Solve for the membrane deformations in the shell

$$\Delta'_v = \frac{2PL^4}{Ehr^2\pi^4} \sin\left(\frac{\pi x}{L}\right)\left[1 + \frac{2\pi^2 r^2}{L^2} + \left(\frac{\pi^4 r^4}{2L^4}\right)\cos^2(\phi_k - \phi)\right]$$

$$= \frac{2(4.838)(20.25)^4}{E(0.075)(8.1)^2\pi^4} \sin\left(\frac{\pi x}{20.25}\right)$$

$$\times \left[1 + \frac{2\pi^2(8.1)^2}{(20.25)^2} + \left(\frac{\pi^4(8.1)^4}{2(20.25)^4}\right)\cos^2(45 - \phi)\right] \tag{6.92}$$

Simplified,

$$\Delta'_v = \frac{3394}{E} \sin(0.155x)\left[4.16 + 1.25\cos^2(45 - \phi)\right]$$

$$\Delta'_H = \frac{Pr^2}{Eh} \sin(\phi_k - \phi)\sin\left(\frac{\pi x}{L}\right)\cos(\phi_k - \phi)$$

$$= \frac{(4.838)8.1^2}{E(0.075)} \sin(45 - \phi)\sin\left(\frac{\pi x}{20.25}\right)\cos(45 - \phi) \tag{6.97}$$

Simplified,

$$\Delta'_H = \frac{4232}{E} \sin(45 - \phi)\sin(0.155x)\cos(45 - \phi)$$

The stresses and deformations are given at two locations in Tables 6.1 and 6.2 (at the end and mid-span). Similar to a beam, the shear stresses are maximum at the edge and zero at mid-span. Furthermore, the stresses in the x-direction are maximum at mid-span (corresponding to maximum moment), but zero at the ends. Also included in the tables are the deformations.

Table 6.1 Stresses and deformations of a barrel shell without edge beams ($x = 0$ and L)

ϕ ($x = 0$ or L)	N'_x (kN/m)	N'_ϕ (kN/m)	$N'_{x\phi}$ (kN/m)	Δ'_V (mm)	Δ'_H (mm)
0	0.00	0.00	0.00	0.00	0.00
10	0.00	0.00	−135.47	0.00	0.00
20	0.00	0.00	−89.07	0.00	0.00
30	0.00	0.00	−26.01	0.00	0.00
40	0.00	0.00	−1.56	0.00	0.00
45	0.00	0.00	0.00	0.00	0.00

Table 6.2 Stresses and deformations of a barrel shell without edge beams (x = L/2)

ϕ $(x = L/2)$	N'_x (kN/m)	N'_ϕ (kN/m)	$N'_{x\phi}$ (kN/m)	Δ'_V (mm)	Δ'_H (mm)
0	1432.59	0.00	0.00	0.54	0.07
10	29.24	−18.11	0.00	0.57	0.07
20	−328.55	−39.56	0.00	0.59	0.05
30	−207.13	−50.25	0.00	0.60	0.04
40	−32.39	−51.76	0.00	0.61	0.01
45	−4.93	−51.71	0.00	0.61	0.00

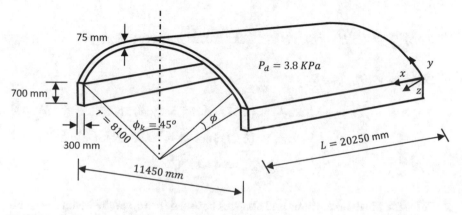

Fig. 6.21 Simply supported barrel vault with edge beams

6.7.2 Simply Supported Barrel Vault with Edge Beams

In this case, we consider the same barrel vault, but with edge beams (Fig. 6.21). Since the shell is relatively long, the edge beams are orientated in the vertical direction.

The start of the solution is identical to the barrel vault with free edges, but corrections are added for the inclusion of edge beams.

1. *Solve for the reactions V_L, H_L, and S_L*

   Span = 11450 mm
   Length = 20250 mm
   Radius = 8100 mm
   Height = 8100–8100 cos 45 = 2372 mm
   Thickness = 75 mm
   Young's modulus = $30(10^6)$ kN/m²
   Poisson's ratio $(\nu) = 0.15$

$$I = \frac{h^3}{12} = \frac{(1)0.075^3}{12} = 3.52\left(10^{-5}\right) \text{ m}^4 \qquad (6.320)$$

$$D = \frac{EI}{(1-\nu^2)} = \frac{30(10^6)3.52(10^{-5})}{(1-0.15^2)} = 1079 \text{ Nm}^2 \tag{6.121}$$

$$P = \frac{4}{\pi}P_d = \frac{4}{\pi}(3.8) = 4.838 \text{ kN/m}^2 \tag{6.343}$$

Edge stresses at $\phi = 0$

$$N'_\phi = -Pr\cos(\phi_k) = -(4.838)8.1\cos(45) = -27.71 \tag{6.62}$$

$$N'_{x\phi} = -\frac{2PL}{\pi}\sin(\phi_k) = -\frac{2(4.838)20.25}{\pi}\sin(45) = -44.11 \tag{6.67}$$

Solve for V_L, H_L, and S_L

$$V_L = -N'_\phi \sin\phi_k = (27.71)\sin 45 = 19.60 \text{ kN/m} \tag{6.344}$$

$$H_L = -N'_\phi \cos\phi_k = (27.71)\cos 45 = 19.60 \text{ kN/m} \tag{6.345}$$

$$S_L = -N'_{x\phi} = 44.1 \text{ kN/m} \tag{6.346}$$

2. **Solve for the deformations and stresses between the top of the beam and base of the shell**

Vertical deformations

$$D^s_{11} = V_b \left(\frac{L^2}{E}\right)\left(\frac{L^2}{hr^3}\right)(B:col.1)\sin\left(\frac{\pi x}{L}\right)$$

$$= V_b \left(\frac{L^2}{E}\right)\left(\frac{20.25^2}{(0.075)8.1^3}\right)(96.10)\sin\left(\frac{\pi x}{L}\right)$$

$$= V_b \left(\frac{L^2}{E}\right)(988.68)\sin\left(\frac{\pi x}{L}\right) \tag{6.350}$$

$$D^b_{11} = V_b \left(\frac{L^2}{E}\right)\left(\frac{12L^2}{\pi^4 bd^3}\right)\sin\left(\frac{\pi x}{L}\right) = V_b \left(\frac{L^2}{E}\right)\left(\frac{12(20.25)^2}{\pi^4(0.3)0.7^3}\right)\sin\left(\frac{\pi x}{L}\right)$$

$$= V_b \left(\frac{L^2}{E}\right)(490.93)\sin\left(\frac{\pi x}{L}\right) \tag{6.351}$$

$$D^s_{12} = S_b \left(\frac{L^2}{E}\right)\left(\frac{L^2}{hr^3}\right)(B:col.7)\sin\left(\frac{\pi x}{L}\right)$$

$$= S_b \left(\frac{L^2}{E}\right)\left(\frac{20.25^2}{(0.075)8.1^3}\right)(2.5)\sin\left(\frac{\pi x}{L}\right)$$

$$= S_b \left(\frac{L^2}{E}\right)(25.72)\sin\left(\frac{\pi x}{L}\right) \tag{6.352}$$

$$D_{12}^b = -S_b \left(\frac{L^2}{E}\right) \left(\frac{6L}{bd^2\pi^3}\right) \sin\left(\frac{\pi x}{L}\right)$$

$$= -S_b \left(\frac{L^2}{E}\right) \left(\frac{6(20.25)}{(0.3)0.7^2\pi^3}\right) \sin\left(\frac{\pi x}{L}\right) = -S_b \left(\frac{L^2}{E}\right)(26.66) \sin\left(\frac{\pi x}{L}\right) \quad (6.353)$$

$$D_{10}^s = \left(\frac{L^2}{E}\right)\left(\frac{L^2}{hr^3}\right)$$

$$\left\{ Pr\left[\frac{2}{\pi^4} + \left(\frac{2r}{\pi L}\right)^2 + \left(\frac{r}{L}\right)^4 \cos^2(\phi_k - \phi)\right] \right.$$

$$\left. + [V_L(B:col.1) - H_L(B:col.4) + S_L(B:col.7)] \right\} \sin\left(\frac{\pi x}{L}\right)$$

$$= \left(\frac{L^2}{E}\right)\left(\frac{20.25^2}{(0.075)8.1^3}\right)\left\{(4.838)8.1\left[\frac{2}{\pi^4} + \left(\frac{2(8.1)}{\pi(20.25)}\right)^2 + \left(\frac{8.1}{20.25}\right)^4 \cos^2(45)\right]\right.$$

$$\left. + [19.60(96.10) - 19.6(64.69) + 44.11(2.50)]\right\} \sin\left(\frac{\pi x}{L}\right)$$

$$= \left(\frac{L^2}{E}\right)(7497.61) \sin\left(\frac{\pi x}{L}\right)$$

$$(6.354)$$

$$D_{10}^b = -\left(\frac{L^2}{E}\right)\left(\frac{4}{\pi}\right)\left(\frac{5}{384}\right)\left(\frac{12\ w^b L^2}{bd^3}\right) \sin\left(\frac{\pi x}{L}\right)$$

$$= -\left(\frac{L^2}{E}\right)\left(\frac{4}{\pi}\right)\left(\frac{5}{384}\right)\left(\frac{12\ (24)\ 0.3(0.7)(20.25)^2}{(0.3)(0.7)^3}\right) \sin\left(\frac{\pi x}{L}\right)$$

$$= -\left(\frac{L^2}{E}\right)(3995.72) \sin\left(\frac{\pi x}{L}\right) \quad (6.355)$$

It should be noted that the terms $\left(\frac{L^2}{E}\right) \sin\left(\frac{\pi x}{L}\right)$ divide out of the equations.

Horizontal stress equations

x = L/2

$$f_{21}^s = V_b \left(\frac{L}{r}\right)^2 \frac{1}{h}(A:col.1) \sin\left(\frac{\pi x}{L}\right)$$

$$= V_b \left(\frac{20.25}{8.1}\right)^2 \frac{1}{0.075}(19.64) \sin\left(\frac{\pi}{2}\right) = V_b(1636.67) \quad (6.356)$$

$$f_{21}^b = -V_b \left(\frac{L}{r}\right)^2 \left(\frac{6r^2}{bd^2\pi^2}\right) \sin\left(\frac{\pi x}{L}\right)$$

$$= -V_b \left(\frac{20.25}{8.1}\right)^2 \left(\frac{6(8.1)^2}{(0.3)0.7^2\pi^2}\right) \sin\left(\frac{\pi}{2}\right) = -V_b(1695.84) \quad (6.357)$$

$$f_{22}^s = S_b \left(\frac{L}{r}\right)^2 \frac{1}{h} (A : col.9) \sin\left(\frac{\pi x}{L}\right)$$

$$= S_b \left(\frac{20.25}{8.1}\right)^2 \frac{1}{0.075} (1.388) \sin\left(\frac{\pi}{2}\right) = S_b(115.67) \qquad (6.358)$$

$$f_{22}^b = S_b \left(\frac{L}{r}\right)^2 \frac{4r^2}{L\pi bd} \sin\left(\frac{\pi x}{L}\right) = S_b \left(\frac{20.25}{8.1}\right)^2 \frac{4(8.1)^2}{20.25\,\pi\,(0.3)0.7} \sin\left(\frac{\pi}{2}\right)$$

$$= S_b(122.78) \qquad (6.359)$$

$$N_x = \left(\frac{L}{r}\right)^2 \left[-Pr\left(\frac{2}{\pi^2}\right) \cos(\phi_k - \phi) + V_L(A : col.1) + H_L(A : col.5)\right.$$
$$\left. + S_L(A : col.9)\right] \sin\left(\frac{\pi x}{L}\right)$$

$$= -\left(\frac{20.25}{8.1}\right)^2 \left[(4.838)8.1\left(\frac{2}{\pi^2}\right) \cos(45) + 19.60(19.64) + 19.60(-10.78)\right.$$
$$\left. + 44.11(1.388)\right] \sin\left(\frac{\pi}{2}\right)$$

$$= 1432.59$$

$$\qquad (6.361)$$

$$f_{20}^s = \frac{N_x}{h} = \frac{1432.59}{0.075} = 19101.22 \qquad (6.360)$$

$$f_{20}^b = \left(\frac{L}{r}\right)^2 \left(\frac{4}{\pi}\right) \left(\frac{6\,w^b r^2}{bd^2 \pi^2}\right) \sin\left(\frac{\pi}{2}\right)$$

$$= \left(\frac{20.25}{8.1}\right)^2 \left(\frac{4}{\pi}\right) \left(\frac{(6)24\,(0.3)0.7\,(8.1)^2}{(0.3)0.7^2\pi^2}\right) \sin\left(\frac{\pi}{2}\right) = 10882.41 \qquad (6.362)$$

3. Solve the compatibility equations

$$V_b(988.68 + 490.93) + S_b(25.72 - 26.66) + 7497.61 - 3995.72 = 0 \qquad (6.318)$$

$$V_b(1636.67 - 1695.84) + S_b(115.67 + 122.78) + 19101.22$$
$$+ 10882.40$$
$$= 0 \qquad (6.319)$$

or

$$V_b(1479.61) + S_b(-0.94) + 3501.88 = 0$$
$$V_b(-59.17) + S_b(238.44) + 29983.60 = 0$$

Set in matrix form,

$$\begin{bmatrix} 1479.61 & -0.94 \\ -59.17 & 238.44 \end{bmatrix} \begin{Bmatrix} V_b \\ S_b \end{Bmatrix} = \begin{Bmatrix} -3501.88 \\ -29983.60 \end{Bmatrix} \tag{6.363}$$

$$\begin{Bmatrix} V_b \\ S_b \end{Bmatrix} = \begin{bmatrix} 1479.61 & -0.94 \\ -59.17 & 238.44 \end{bmatrix}^{-1} \begin{Bmatrix} -3501.88 \\ -29983.60 \end{Bmatrix}$$

Solving the matrix, the vertical load and horizontal shears are determined.

$$\begin{Bmatrix} V_b \\ S_b \end{Bmatrix} = \frac{1}{(1479.61)(238.44) - (-0.94)(-59.17)} \begin{bmatrix} 238.44 & 0.94 \\ 59.17 & 1479.61 \end{bmatrix}$$

$$\times \begin{Bmatrix} -3501.88 \\ -29983.60 \end{Bmatrix} \tag{6.364}$$

Therefore,

$$\begin{Bmatrix} V_b \\ S_b \end{Bmatrix} = \begin{Bmatrix} -2.45 \\ -126.35 \end{Bmatrix}$$

4. **Solve for stresses and bending moments in the shell**

These stresses are both membrane and boundary effects stresses.

$$N_x = -\frac{2PL^2}{r\pi^2} \cos\left(\phi_k - \phi\right) \sin\left(\frac{\pi x}{L}\right) + V_L\left(\frac{L}{r}\right)^2 (A : col.1) \sin\left(\frac{\pi x}{L}\right)$$

$$+ H_L\left(\frac{L}{r}\right)^2 (A : col.5) \sin\left(\frac{\pi x}{L}\right) + S_L\left(\frac{L}{r}\right)^2 (A : col.9) \sin\left(\frac{\pi x}{L}\right)$$

$$+ V_b\left(\frac{L}{r}\right)^2 (A : col.1) \sin\left(\frac{\pi x}{L}\right) + S_b\left(\frac{L}{r}\right)^2 (A : col.9) \sin\left(\frac{\pi x}{L}\right)$$

$$= -\frac{2(4.838)20.25^2}{8.1\pi^2} \cos\left(45 - \phi\right) \sin\left(\frac{\pi x}{20.25}\right)$$

$$+ 19.60\left(\frac{20.25}{8.1}\right)^2 (A : col.1) \sin\left(\frac{\pi x}{20.25}\right)$$

$$+ 19.60\left(\frac{20.25}{8.1}\right)^2 (A : col.5) \sin\left(\frac{\pi x}{20.25}\right)$$

$$+ 44.11\left(\frac{20.25}{8.1}\right)^2 (A : col.9) \sin\left(\frac{\pi x}{20.25}\right)$$

$$- 2.45\left(\frac{20.25}{8.1}\right)^2 (A : col.1) \sin\left(\frac{\pi x}{20.25}\right)$$

$$- 126.35\left(\frac{20.25}{8.1}\right)^2 (A : col.9) \sin\left(\frac{\pi x}{20.25}\right) \tag{6.365}$$

which simplifies to:

$$N_x = -49.63 \cos{(45 - \phi)} \sin{(0.155x)} + 122.50 \, (A : col.1) \sin{(0.155x)}$$
$$+ 122.50 \, (A : col.5) \sin{(0.155x)} + 275.69 \, (A : col.9) \sin{(0.155x)}$$
$$- 21.75 \, (A : col.1) \sin{(0.155x)} - 789.69(A : col.9) \sin{(0.155x)}$$

$$N_{x\phi} = -\frac{2PL}{\pi} \sin{(\phi_k - \phi)} \cos{\left(\frac{\pi x}{L}\right)} + V_L\left(\frac{L}{r}\right)(A : col.2) \cos{\left(\frac{\pi x}{L}\right)}$$
$$+ H_L\left(\frac{L}{r}\right)(A : col.6) \cos{\left(\frac{\pi x}{L}\right)} + S_L\left(\frac{L}{r}\right)(A : col.10) \cos{\left(\frac{\pi x}{L}\right)}$$
$$+ V_b\left(\frac{L}{r}\right)(A : col.2) \cos{\left(\frac{\pi x}{L}\right)} + S_b\left(\frac{L}{r}\right)(A : col.10) \cos{\left(\frac{\pi x}{L}\right)}$$
$$= -\frac{2(4.838)20.25}{\pi} \sin{(45 - \phi)} \cos{\left(\frac{\pi x}{20.25}\right)} + 19.60\left(\frac{20.25}{8.1}\right)$$
$$\times (A : col.2) \cos{\left(\frac{\pi x}{20.25}\right)} + 19.60\left(\frac{20.25}{8.1}\right)$$
$$\times (A : col.6) \cos{\left(\frac{\pi x}{20.25}\right)} + 44.11\left(\frac{20.25}{8.1}\right)$$
$$\times (A : col.10) \cos{\left(\frac{\pi x}{20.25}\right)} - 2.45\left(\frac{20.25}{8.1}\right)$$
$$\times (A : col.2) \cos{\left(\frac{\pi x}{20.25}\right)} - 126.35\left(\frac{20.25}{8.1}\right)$$
$$\times (A : col.10) \cos{\left(\frac{\pi x}{20.25}\right)} \tag{6.366}$$

which simplifies to:

$$N_{x\phi} = -62.37 \sin{(45 - \phi)} \cos{(0.155x)} + 49.00 \, (A : col.2) \cos{(0.155x)}$$
$$+ 49.00 \, (A : col.6) \cos{(0.155x)} + 110.28 \, (A : col.10) \cos{(0.155x)}$$
$$- 6.13 \, (A : col.2) \cos{(0.155x)} - 315.88 \, (A : col.10) \cos{(0.155x)}$$

$$N_\phi = -Pr\cos(\phi_k - \phi)\sin\left(\frac{\pi x}{L}\right) + V_L(A : col.3)\sin\left(\frac{\pi x}{L}\right)$$

$$+ H_L(A : col.7)\sin\left(\frac{\pi x}{L}\right) + S_L(A : col.11)\sin\left(\frac{\pi x}{L}\right)$$

$$+ V_b(A : col.3)\sin\left(\frac{\pi x}{L}\right) + S_b(A : col.11)\sin\left(\frac{\pi x}{L}\right)$$

$$= -(4.838)8.1\cos(45 - \phi)\sin\left(\frac{\pi x}{20.25}\right)$$

$$+ 19.60(A : col.3)\sin\left(\frac{\pi x}{20.25}\right) + 19.60(A : col.7)\sin\left(\frac{\pi x}{20.25}\right)$$

$$+ 44.11(A : col.11)\sin\left(\frac{\pi x}{20.25}\right) - 2.45(A : col.3)\sin\left(\frac{\pi x}{20.25}\right)$$

$$- 126.35\,(A : col.11)\sin\left(\frac{\pi x}{20.25}\right) \tag{6.367}$$

which simplifies to:

$$N_\phi = -39.15\cos(45 - \phi)\sin(0.155x) + 19.60(A : col.3)\sin(0.155x)$$
$$+ 19.60(A : col.7)\sin(0.155x) + 44.11(A : col.11)\sin(0.155x)$$
$$- 2.45(A : col.3)\sin(0.155x) - 126.35\,(A : col.11)\sin(0.155x)$$

$$M_\phi = V_Lr(A : col.4)\sin\left(\frac{\pi x}{L}\right) + H_Lr(A : col.8)\sin\left(\frac{\pi x}{L}\right)$$

$$+ S_Lr(A : col.12)\sin\left(\frac{\pi x}{L}\right) + V_br(A : col.4)\sin\left(\frac{\pi x}{L}\right)$$

$$+ S_br(A : col.12)\sin\left(\frac{\pi x}{L}\right)$$

$$= (19.60)8.1(A : col.4)\sin\left(\frac{\pi x}{20.25}\right)$$

$$+ (19.60)8.1(A : col.8)\sin\left(\frac{\pi x}{20.25}\right)$$

$$+ (44.11)8.1(A : col.12)\sin\left(\frac{\pi x}{20.25}\right)$$

$$- (2.45)8.1(A : col.4)\sin\left(\frac{\pi x}{20.25}\right)$$

$$- (126.35)8.1(A : col.12)\sin\left(\frac{\pi x}{20.25}\right) \tag{6.368}$$

which simplifies to:

$$M_\phi = 158.76\,(A : col.4)\sin(0.155x) + 158.76\,(A : col.8)\sin(0.155x)$$
$$+ 357.29(A : col.12)\sin(0.155x) - 19.85\,(A : col.4)\sin(0.155x)$$
$$- 1023.44\,(A : col.12)\sin(0.155x)$$

The above stresses and bending moments are given in increments of $10°$ and at $x = 0$ and $x = \frac{L}{2}$.

6.8　Exercises

6.8.1. Using a spreadsheet or alternative programing language, resolve the worked example 6.7.1 for a barrel vault with free edges. Reproduce Tables 6.1 and 6.2 for the resultant stresses and deformations at x = 0 and x = L/2. In addition, solve for the stresses and deformations at x = L/4.

6.8.2. Resolve for the worked example 6.7.1, but change the length of the barrel vault to L = 13.500 m. Solve for the resultant stresses at x = 0 and x = L/2.

6.8.3. Graph the membrane stresses and correction stresses for exercises 6.8.1 and 6.8.2. Describe your observations concerning the relationship between the membrane stresses, and correcting for errors along the free edge. Only consider N_x and N_ϕ at x = L/2 and $N_{x\phi}$ at x = 0.

6.8.4. Using a spreadsheet or alternative programing language, resolve for the worked example 6.7.2 for a barrel vault supported by an edge beam. Reproduce Tables 6.3 and 6.4 for resultant stresses at x = 0 and x = L/2. In addition, solve for the stresses at x = L/4.

6.8.5. If a larger edge beam is used in worked example 6.7.2, how would this affect the stress distribution? Use beam size b = 500 mm and d = 1000 mm. Reproduce Tables 6.3 and 6.4 for resultant stresses at x = 0 and x = L/2.

Table 6.3 Stresses and bending moments in a shell with edge beams (x = L/2)

ϕ (deg.)	N_x (kN/m)	N_ϕ (kN/m)	$N_{x\phi}$ (kN/m)	M_ϕ (kN/m)
0	36.13	−1.73	0.00	0.00
10	−207.78	−23.30	0.00	0.57
20	−209.11	−36.96	0.00	−1.07
30	−111.75	−42.34	0.00	−2.57
40	−31.31	−43.12	0.00	−3.28
45	−19.56	−43.12	0.00	−3.35

Table 6.4 Stresses and bending moments in a shell with edge beams (x = 0)

ϕ (deg.)	N_x (kN/m)	N_ϕ (kN/m)	$N_{x\phi}$ (kN/m)	M_ϕ (kN/m)
0	0.00	0.00	−126.35	0.00
10	0.00	0.00	−101.75	0.00
20	0.00	0.00	−52.94	0.00
30	0.00	0.00	−17.17	0.00
40	0.00	0.00	−2.57	0.00
45	0.00	0.00	0.00	0.00

References

ASCE Manual of Engineering Practice, N. 3., 1960. *Design of Cylindrical Shell Roofs.* s.l.:ASCE publications.

Billington, D. P., 1982. *Thin Shell Concrete Structures.* 2 ed. New York: McGraw-Hill.

Chatterjee, B. K., 1971. *Theory and Design of Concrete Structures.* Calcutta: Oxford and IBH Publishing Co..

Flugge, W., 1960. *Stresses in Shells.* Berlin: Springer-Verag.

Jaeger, L. G., 1969. *Elementary Theory of Elastic Plates.* Oxford: Thomson.

Szilard, R., 1974. *Theory and Analysis of Plates.* New Jersey: Prentice-Hall.

Timoshenko, S., 1958. *Strength of Materials.* New York: Van Nostrand Reinhold Co.

Timoshenko, S., 1959. *Theory of Plates and Shells.* 2 ed. New York: McGraw-Hill Book Co..

Chapter 7
Catenary Arches and Domes

Abstract This chapter develops theory to determine the thrust-line forces in a catenary arch for a uniform distributed load (gravity). An iterative point load method (funicular iterative method) is also given to solve the pure compression shape for any number of point loads applied at any angle. This method successively adds segments to the arch model, allowing the arch to grow and form naturally into a pure compression arch. Finally, the membrane solution for a catenary dome is derived, including a solution for a catenary dome with an oculus or opening for a skylight.

7.1 Introduction

The terms funicular and catenary are anglicized words from Latin, meaning rope and chain respectively. From a force-flow perspective, a rope and a chain react identically. However, each term has taken on a separate classification of shapes. The catenary is the shape of a hanging chain under its own weight, or a uniform load. The term funicular is a collection of shapes, where any number of weights are hung from a chain, creating unique shape in tension. Thus, funicular shapes are infinite in number, whereas the catenary shape is a single shape of a chain under its own self-weight. The catenary and a various funicular shapes are given in Fig. 7.1.

Both catenary and funicular shapes are in pure tension (without bending or shear forces). If the links are locked and the shape is flipped upright, the forces in the chain are reversed, and in pure compression. The structural characteristics of catenary and funicular shapes are therefore highly favored in the construction of masonry arches and domes, which do not perform well under conditions of tension and bending forces. In addition, as seen in Chapter 2, pure compression structures are more durable and have greater material strengths (Allen and Zalewski 2010).

Although catenary and funicular shapes are not popular in architecture, they are the most optimal shapes, in terms of stress economy. Most importantly, the thrust-line in catenary and funicular shapes are designed to match the axis of the shell or arch, thus eliminating bending and shears. If bending is eliminated, the stresses in the shell are significantly reduced. Furthermore, many designers fail to realize that a thrust-line will tend to flow in a natural pattern in all structural members (i.e., arches,

© The Author(s), under exclusive license to Springer Nature Switzerland AG 2022   293
M. Gohnert, *Shell Structures*, https://doi.org/10.1007/978-3-030-84807-1_7

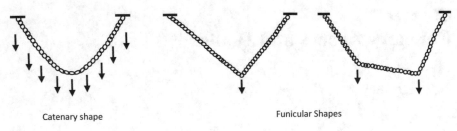

Catenary shape   Funicular Shapes

Fig 7.1 The catenary and various funicular shapes

Fig. 7.2 Hanging chain
forms a catenary shape

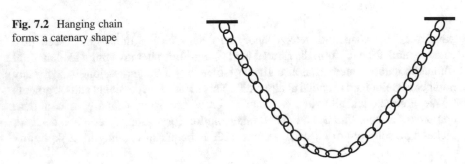

domes, beams, and slabs), resembling a catenary (Heyman 1995) (Ockleston 1958). Optimization is only achieved if the thrust-line matches the structure.

7.2 The Catenary Arch

The catenary shape is defined by the shape of a hanging chain, supported at the two ends and allowed to drape naturally under its own self-weight (see Fig. 7.2).

The catenary is very similar to a parabolic shape, but the catenary has its own distinct form, which is defined by Eq. 7.1.

$$y = a \cosh(x/a) \tag{7.1}$$

Equation 7.1 is illustrated as the lower curve in Fig. 7.3. The catenary arch, however, is illustrated as the upper curve in the same figure. To solve for the arch equation (Gohnert and Bradley 2020a), the height (H) and the base length (L) should be selected beforehand to define the proportions of the arch or vault. Equation 7.2 is then iterated to determine the value of "a" (scaling factor).

$$H = a \cosh(L/2a) - a \tag{7.2}$$

Once "a" is determined, the arch equation is defined, and expressed by Eq. 7.3.

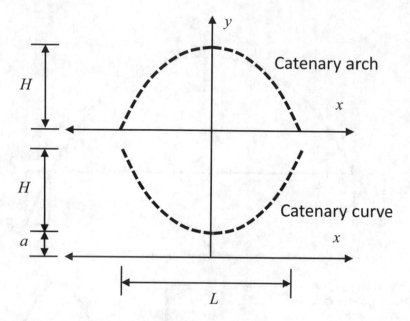

Fig. 7.3 Catenary arch for uniform load (i.e., self-weight)

$$y = H - a\ \cosh\left(\frac{x}{a}\right) + a \tag{7.3}$$

The length of the catenary curve (S) is solved by integrating Eq. 7.1 from 0 to $L/2$, and multiplying by 2.

$$S = 2a\ \sinh\left(L/2a\right) \tag{7.4}$$

The total weight (W) of the arch is equal to the unit weight of the arch material (γ_m) times the thickness (h), length (S), and width (b).

$$W = \gamma_m Shb \tag{7.5}$$

The vertical reactions at each base are equal to the total weight divided by 2.

$$R_v = W/2 \tag{7.6}$$

The angle of the base reaction, which is also the angle of the base, is determined by first taking the derivative of Eq. 7.1.

$$\frac{dy}{dx} = \sinh\left(x/a\right) \tag{7.7}$$

Since,

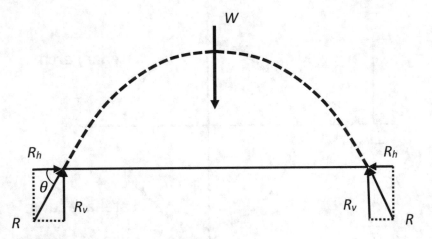

Fig. 7.4 Reactions of a catenary arch subjected to self-weight

Fig. 7.5 Distribution of the
horizontal and axial forces
in a catenary arch

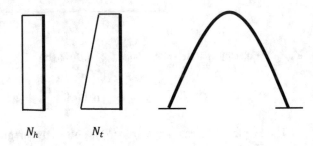

$$\frac{dy}{dx} = \tan\theta \tag{7.8}$$

The angle at the base of the arch is solved by equating Eqs. 7.7 and 7.8, and setting $x = L/2$.

$$\theta = \tan^{-1}[\sinh L/2a] \tag{7.9}$$

The diagonal and the horizontal reactions are then determined (see Fig. 7.4).

$$R = R_v/\sin\theta \tag{7.10}$$

$$R_h = R\cos\theta \tag{7.11}$$

The forces in a catenary arch due to self-weight are predictable. The horizontal forces are constant. Therefore, at the top of the arch, the axial force (N_t) is equal to the horizontal force (N_h). The axial force varies linearly from the base to apex of the arch (see Fig. 7.5, which illustrates the distribution of horizontal and axial forces).

From these known force characteristics, the axial force in the arch may be determined.

Knowing the axial force distribution, the force at the top of the arch is equal to:

$$N_{t(y=H)} = R_h \qquad (7.12)$$

Substituting Eqs. 7.10 and 7.11 into 7.12,

$$N_{t(y=H)} = R\cos\theta = \frac{R_v\cos\theta}{\sin\theta} \qquad (7.13)$$

where θ is the angle of the reaction at the base of the arch.

Expressing the above equation in terms of the weight of the arch:

$$N_{t(y=H)} = \frac{W\,\cos\theta}{2\,\sin\theta} \qquad (7.14)$$

Similarly, the axial force at the base of the arch is equal to the reaction.

$$N_{t(y=0)} = R \qquad (7.15)$$

Substituting 7.10 and 7.6,

$$N_{t(y=0)} = \frac{R_v}{\sin\theta} = \frac{W}{2\sin\theta} \qquad (7.16)$$

The forces at the top and bottom of the arch (Eqs. 7.14 and 7.16) only differ by a $\cos\theta$ term in the numerator. Therefore, a general equation may be expressed in terms of k, which distributes the forces from the bottom to the top of the arch.

$$N_t = \frac{W}{2\,\sin\theta}k \qquad (7.17)$$

The expression k is determined by applying the following boundary conditions:

$$y = 0, k = 1$$
$$y = H, k = \cos\theta$$

Since the distribution of force is linear, k takes the form:

$$k = my + c \qquad (7.18)$$

Substituting the first boundary condition:

$$1 = m(0) + c$$
$$c = 1 \qquad\qquad (7.19)$$

Substituting the second boundary condition:

$$\cos\theta = m\,(H) + 1$$
$$m = \frac{\cos\theta - 1}{H} \qquad\qquad (7.20)$$

Therefore,

$$k = \left(\frac{\cos\theta - 1}{H}\right) y + 1 \qquad\qquad (7.21)$$

Substituting 7.21 into 7.17, the axial force is defined at all points y in the arch.

$$N_t = \frac{W}{2\sin\theta}\left[\left(\frac{\cos\theta - 1}{H}\right) y + 1\right] \qquad\qquad (7.22)$$

Since the catenary arch is a pure compression form, and free of bending and shears, the axial force is the only force determined, where

$$\sigma_t = \frac{N_t}{bh}$$

7.3   The Funicular Arch

As previously stated, the pure compression shape for a uniform load (i.e., self-weight) is a catenary arch. The analogy of a chain is a powerful tool to understand the structural properties of a catenary. Likewise, the analogy of a chain may also be used to define pure compression shapes of arches which are subjected to point loads. If a chain is suspended between two points, and a point load is applied at some location along the chain, the chain will form a triangle (assuming no self-weight or if the self-weight is small compared to the magnitude of the point load), and the two legs of the triangle will be in pure tension (see the lower triangle of Fig. 7.6). If the links are locked and the chain is flipped upright, the stresses are reversed and in pure compression (see upper triangle of Fig. 7.6). Since the chain is in pure compression, the angles of the reactions are the same as the angles of the legs of the triangle.

The method used to define the shape of a funicular arch is a numerical solution and is similar to the finite element method, where a series of elements are connected together to form and analyze a structural shape. However, there are distinct

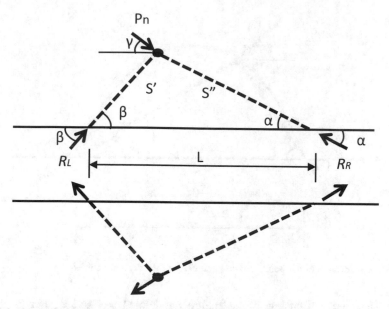

Fig. 7.6 Triangular shape to support a point load in pure compression

differences between the two methods. Unlike the finite element method, the funicular shape is not predetermined or known beforehand, nor is the number of elements/segments known. At the start of the analysis, only the length of the first, or base segment, is selected beforehand. As the analysis proceeds, segments are continuously added to the profile. The catenary shape literally grows, or forms into a pure compressive shape. During the analysis, the catenary will ascend from the initial support and then descend to the second support, creating a unique pure compression shape (Gohnert and Bradley 2020a, b).

7.3.1   Derivation of the Funicular Arch Theory, Using an Iterative Approach

The solution is composed of two parts: The first part solves for a combined reaction for the self-weight and all of the point loads applied to the catenary arch. The second part of the analysis solves for the pure compressive shape of the catenary. This is achieved by consecutively adding segments to the model, allowing the catenary to grow naturally into a pure compression arch.

All arches will have self-weight and therefore a catenary arch will be initially assumed, with point loads applied to the exterior. As depicted in Fig. 7.7, the location of externally applied point loads (P_n) is assumed to be located along the profile of the catenary arch. Any number of point loads may be applied to an arch, thus the subscript (n), which is numbered according to the number of applied point

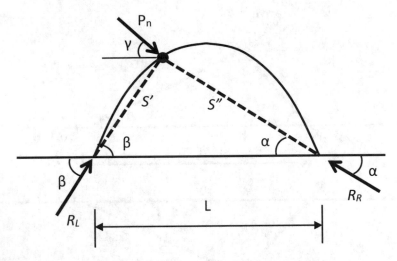

Fig. 7.7 The location of the point load and geometry of the catenary curve

loads. The lengths of the legs of the triangle are defined as S' and S'', and the angles of the legs are respectively β and α. The left and right reactions (R_L and R_R) are also depicted in Fig. 7.6, which are tangent to the angle of the leg of the triangle.

Referring to Fig. 7.7, and using the law of cosines for oblique triangles, the angles of the reactions are determined.

$$\alpha = \cos^{-1}\left[\frac{S''^2 + L^2 - S'^2}{2S''L}\right] \tag{7.23}$$

$$\beta = \cos^{-1}\left[\frac{S'^2 + L^2 - S''^2}{2S'L}\right] \tag{7.24}$$

The reactions are determined by formulating equilibrium equations in the x and y directions.

$$\sum F_x : P_n \cos\gamma + R_L \cos\beta - R_R \cos\alpha = 0 \tag{7.25}$$

$$\sum F_y : P_n \sin\gamma - R_L \sin\beta - R_R \sin\alpha = 0 \tag{7.26}$$

Although we can form the arch from the left or right supports, the left support will be assumed as the location of the start of the arch. Substituting Eqs. 7.26 into 7.25, the reaction (R_R) is determined.

$$R_R = \frac{P_n(\cos\gamma + \sin\gamma \cot\beta)}{\cos\alpha\,(1 + \tan\alpha \cot\beta)} \tag{7.27}$$

Substituting R_R into Eq. (7.26), the reaction (R_L) is solved.

$$R_L = \frac{P_n \sin\gamma - R_R \sin\alpha}{\sin\beta} \tag{7.28}$$

The base reactions, defined by Eqs. 7.27 and 7.28, are determined for every point load applied to the arch.

It should be noted that this method is also applicable to arches only subjected to self-weight, or continuous loads. A continuous load is modeled by dividing the load into a series of point loads.

Once the reactions for each of the point loads are determined, the next step is to sum the reactions, to solve for a single reaction. However, to enable the summation, the reactions are split into vertical and horizontal components. Considering only the left side of the arch,

$$R_{Lv} = \sum_1^n R_L^n \sin\beta_n \tag{7.29}$$

$$R_{Lh} = \sum_1^n R_L^n \cos\beta_n \tag{7.30}$$

where R_L^n are the reactions for each of the point loads applied to the arch. The number of reactions will equal the number of point loads.

Equations 7.29 and 7.30 are then resolved back into a single reaction and angle.

$$R_L = \sqrt{R_{Lv}^2 + R_{Lh}^2} \tag{7.31}$$

$$\beta_L = \tan^{-1}(R_{Lv}/R_{Lh}) \tag{7.32}$$

If point loads are applied to a catenary arch, the pure compression shape will be a distorted catenary, referred to as a funicular shape. The formation of the funicular arch begins with the base segment. In this segment (see Fig. 7.8), all of the terms are known except for T_1, which is the magnitude of the axial thrust between the first and second segments. Since the shape must be in pure compression, the segments have only one degree-of-freedom, which is the axial thrust in the arch. Since the arch is in pure compression, the angle of the axial thrust must be the same as the angle of the segment.

The equilibrium equations of the base element are expressed in terms the components of T_1 (i.e., T_{1x} and T_{1y}).

Summing forces in the x-direction (refer to base segment Fig.7.8),

$$\sum F_x : P_1 \cos\gamma_1 + R_L \cos\beta - T_{1x} = 0$$

or

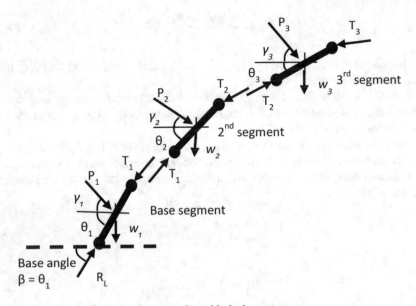

Fig. 7.8 Forming the funicular shape, starting with the base segment

$$T_{1x} = P_1 \cos \gamma_1 + R_L \cos \beta \qquad (7.33)$$

Summing forces in the y-direction,

$$\sum F_y : \ P_1 \sin \gamma_1 - R_L \sin \beta + w_1 + T_{1y} = 0$$

or

$$T_{1y} = -P_1 \sin \gamma_1 + R_L \sin \beta - w_1 \qquad (7.34)$$

where w_1 is the weight of the segment and P_1 is an external load, if applied to the segment.

The axial thrust and angle are given by 7.35 and 7.36.

$$T_1 = \sqrt{T_{1x}^2 + T_{1y}^2} \qquad (7.35)$$

$$\Theta_2 = \tan^{-1}\left(T_{1y}/T_{1x}\right) \qquad (7.36)$$

Equations 7.33 and 7.34 are the equilibrium equations for the base segment. For subsequent segments (i.e., $n = 2, 3, 4, \ldots$), the equations are identical, but expressed in a general form:

$$T_{nx} = P_n \cos \gamma_n + T_{n-1} \cos \theta_n \qquad (7.37)$$

$$T_{ny} = -P_n \sin \gamma_n + T_{n-1} \sin \theta_n - w_n \qquad (7.38)$$

where

$$T_n = \sqrt{T_{nx}^2 + T_{ny}^2} \qquad (7.39)$$

$$\theta_{n+1} = \tan^{-1}\left(T_{ny}/T_{nx}\right) \qquad (7.40)$$

The growth of the funicular shape begins with the base segment. Subsequent segments are added to the ends of the previous segments. Each subsequent segment is solved to determine the axial thrust and angle of that segment. As arch segments are added, the curve will grow upwards and then downward, forming an arch. The adding of segments is only terminated when the segments reach the second support.

7.4   Membrane Theory of Catenary Domes

Catenary domes have gained in popularity over the last decade, but the shape is far from new. Catenary shapes have been applied to dome design for at least 400 years. The designs, however, were based on the study of hanging chains or graphical statics (Block et al. 2006). Solutions are readily available for many dome shapes, such as circular, elliptical, conical, and conoidal, but the derivation for the catenary shape is lacking. Therefore, a complete derivation of the catenary dome is given for a uniformly distributed load (gravity load) (Gohnert and Bradley 2020a, b). The solution also includes a derivation for catenaries with a circular hole located at the apex, frequently referred to as a skylight or oculus.

The theory presented is a membrane solution and does not include boundary effects (Farshad 1992). A membrane solution is where all of the stresses are in-plane, which includes meridian and hoop stresses. Boundary effects occur at the boundaries and are caused by fixity at the base. This fixity causes bending and out-of-plane shears, which may, or may not, significantly affect the stresses in the shell. However, boundary effects at the base are minimal in catenary shells, compared to other dome shapes.

7.4.1   Catenary Relationships

Shell stress equations are traditionally derived by formulating equilibrium equations, i.e., summing forces in the meridian and hoop directions. These equations, however, are usually derived in terms of angles (Billington 1982), but the catenary equation is

Fig. 7.9 The catenary curve
and arch

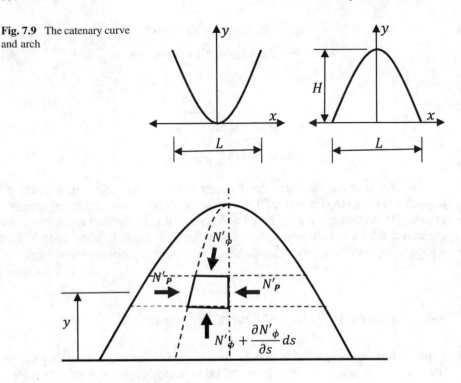

Fig. 7.10 Differential element in a catenary dome

in terms of Cartesian coordinates. Similar to the derivation of elliptical domes, the
relationship between the angles and coordinates must first be established.

The equation for a catenary arch (see Fig. 7.9) is given by Eq. 7.3.

$$y = H - a \cosh (k) + a \tag{7.3}$$

where

$$k = x/a \tag{7.41}$$

Taking the derivative of the curve equation 7.3,

$$\frac{dy}{dx} = - \sinh (k) \tag{7.42}$$

The differential element on the catenary dome is illustrated in Fig. 7.10, and a side
view of the element is given in Fig. 7.11. Referring to Fig. 7.11,

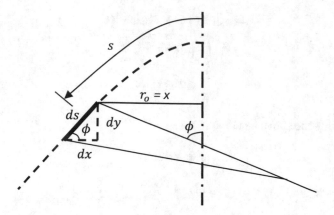

Fig. 7.11 Side view of the differential element

$$\frac{dy}{dx} = -\tan\phi \tag{7.43}$$

Equating Eqs. 7.42 and 7.43,

$$\tan\phi = \sinh(k)$$

or

$$\phi = \tan^{-1}[\sinh(k)] \tag{7.44}$$

which defines the relationship between the angle ϕ and the x-coordinate.

7.4.2   Arch Length of the Catenary Curve

Using the arch length equation,

$$s = \int_0^x \sqrt{1 + \left(\frac{dy}{dx}\right)^2}\, dx \tag{7.45}$$

Substituting Eq. (7.42),

$$s = \int_0^x \sqrt{1 + [\sinh(k)]^2}\, dx$$

The above equation is simplified by substituting the hyperbolic identity,

$$\cosh^2(k) = 1 + \sinh^2(k)$$

$$s = \int_0^x \sqrt{\cosh^2(k)}dx$$

$$s = \int_0^x \cosh(k)\, dx$$

Integrating the above equation,

$$s = a\sinh(k) \tag{7.46}$$

Evaluating from $x = 0$ to $x = L/2$,

$$s = a\, \sinh\left(\frac{L}{2a}\right) \tag{7.47}$$

However, Eq. 7.47 is only half of the arch length. The total length is double.

$$s = 2a\, \sinh\left(\frac{L}{2a}\right) \tag{7.48}$$

7.4.3   Membrane Stress Equations

Several assumptions are applied to the solution of the catenary stress equations.

1. *Only membrane stresses are considered.* The stress distribution is assumed to be in the plane of the shell and boundary effects are ignored. Bending and out-of-plane shear stresses occur at the base of shells, which may be significant in some dome shapes. However, boundary effects in catenary domes are minimal.
2. *The loading is symmetrical and uniformly distributed.* A membrane solution initially includes the meridian, hoop, and in-plane shear stresses. Along the meridian and hoop directions, the in-plane shears are zero for symmetrical loading (Billington 1982) and drop from the equilibrium equations. To reduce the complexity of the derivation, the self-weight of the shell is the only load case considered.
3. *The meridian stresses are independent of the hoop stresses, but the hoop stresses are dependent on the meridian stresses.* The meridian stresses are determined from the weight and geometry of the dome, varies from apex to base, and are independent of the hoop stresses. The hoop stresses, however, are dependent on the meridian stresses, also varies from apex to base, but are constant around the shell at y-coordinates.

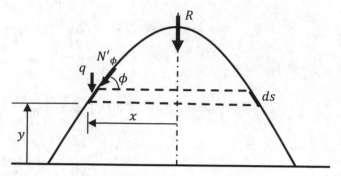

Fig. 7.12 Weight of the dome cap above y

The meridian stress (N'_ϕ) is solved using the same approach used in the derivation of circular domes (Billington 1982). Referring to Fig. 7.12, the meridian resultant stress is equal to the total load R of a portion of the dome above point y, divided by the perimeter length. The $\sin \phi$ term aligns the load in the direction of the meridian stress.

From classical circular dome theory,

$$N'_\phi = \frac{-R}{2\pi r_o \sin \phi} \tag{7.49}$$

or

$$N'_\phi = \frac{-R}{2\pi x} \csc\phi \tag{7.50}$$

where $r_o = x$ (see Fig. 7.11)

Solving for R, which is achieved by summing the differential strips from y to the apex of the dome (see Fig. 7.12).

$$R = \int_0^x q(2\pi x)ds \tag{7.51}$$

where q is the self-weight of the dome per unit area.

Referring to Fig. 7.11,

$$ds = dx/\cos \phi$$
$$ds = \sec \phi\, dx \tag{7.52}$$

Substituting 7.52 into 7.51, and rearranging,

$$R = 2\pi q \int_0^x x \sec \phi \, dx$$

Replacing ϕ with Eq. 7.44,

$$R = 2\pi q \int_0^x x \sec \left\{ \tan^{-1}[\sinh(k)] \right\} dx$$

Since,

$$\cosh(k) = \sec \left\{ \tan^{-1}[\sinh(k)] \right\} \tag{7.53}$$

$$R = 2\pi q \int_0^x x \cosh(k) \, dx$$

Integrating the above expression, and evaluating the results from 0 to x,

$$R = 2\pi q \left[ax \sinh(k) - a^2 \cosh(k) + a^2 + c \right]$$

The constant c is determined from boundary conditions.
At $x = 0$, $R = 0$ and $c = 0$.
Therefore,

$$R = 2\pi q a^2 [k \sinh(k) - \cosh(k) + 1] \tag{7.54}$$

where $k = x/a$, as defined previously.

Substituting the total weight R of the dome cap into Eq. 7.50, and replace ϕ with Eq. 7.44,

$$N'_\phi = \frac{-2\pi q a^2 [k \sinh(k) - \cosh(k) + 1]}{2\pi x} \csc \left\{ \tan^{-1}[\sinh(k)] \right\}$$

Since,

$$\coth(k) = \csc \left\{ \tan^{-1}[\sinh(k)] \right\} \tag{7.55}$$

$$N'_\phi = -q a^2 \left[\left(\frac{k}{x}\right) \sinh(k) - \left(\frac{1}{x}\right) \cosh(k) + \left(\frac{1}{x}\right) \right] \coth(k)$$

$$N'_\phi = -\frac{qa}{k} \coth(k) [k \sinh(k) - \cosh(k) + 1] \tag{7.56}$$

The k value in the denominator makes the solution undefined at $x = 0$ (i.e., at the apex of the dome). However, a solution at the apex is possible by setting x equal to a small value.

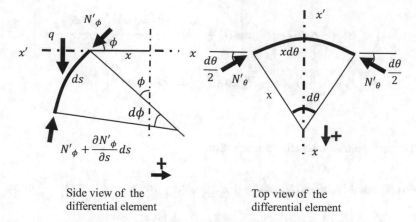

Side view of the
differential element

Top view of the
differential element

Fig. 7.13 Resultant stresses and geometry of a differential element

Fig. 7.14 Edge lengths of
the differential element

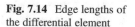

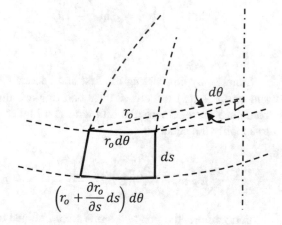

The solution for the hoop stresses are determined by summing forces in the x-direction, i.e., along the line x-x' of Fig. 7.13. The equilibrium equation is formulated by multiplying the resultant stresses by the lengths of the differential element (Billington 1982). The edge lengths of the differential element are given in Fig. 7.14.

Referring to Figs. 7.13 and 7.14, sum forces in the x-direction.

$$-\left(N'_\phi \cos\phi\right)r_o d\theta + \left[N'_\phi \cos\left(\phi+d\phi\right) + \frac{\partial\left(N'_\phi \cos\left(\phi+d\phi\right)\right)}{\partial s}ds\right]$$

$$\times\left(r_o + \frac{\partial r_o}{\partial s}ds\right)d\theta - N'_\theta \sin\left(\frac{d\theta}{2}\right)ds - N'_\theta \sin\left(\frac{d\theta}{2}\right)ds$$

$$= 0$$

Expand the above equilibrium equation.

$$-\left(N'_\phi \cos\phi\right)r_o d\theta + \left[N'_\phi \cos\left(\phi+d\phi\right)\right]r_o d\theta + \left[N'_\phi \cos\left(\phi+d\phi\right)\right]\frac{\partial r_o}{\partial s}ds\, d\theta$$

$$+\frac{\partial\left[N'_\phi \cos\left(\phi+d\phi\right)\right]}{\partial s}ds\, r_o d\theta + \frac{\partial\left[N'_\phi \cos\left(\phi+d\phi\right)\right]}{\partial s}ds\frac{\partial r_o}{\partial s}ds\, d\theta$$

$$-N'_\theta \sin\left(\frac{d\theta}{2}\right)ds - N'_\theta \sin\left(\frac{d\theta}{2}\right)ds$$

$$= 0$$

Simplify by eliminating the first and second terms and assume $sin\left(\frac{d\theta}{2}\right) \approx \left(\frac{d\theta}{2}\right)$ and $cos(\phi + d\phi) \approx \cos\phi$ for small angles, allowing the last two terms to be combined. Furthermore, the fifth term is a higher order term, and may be dropped to simplify the equation.

$$\left(N'_\phi \cos\phi\right)\frac{\partial r_o}{\partial s}ds\, d\theta + \frac{\partial\left(N'_\phi \cos\phi\right)}{\partial s}r_o ds\, d\theta - N'_\theta\, ds\, d\theta = 0$$

Furthermore, the first two terms are combined using the product rule.

$$\frac{\partial\left(N'_\phi \cos\phi\, r_0\right)}{\partial s}ds\, d\theta - N'_\theta\, ds\, d\theta = 0$$

Divide by $ds\, d\theta$ and rearrange.

$$N'_\theta = \frac{\partial\left(N'_\phi \cos\phi\, r_0\right)}{\partial s} \tag{7.57}$$

The above expression indicates the dependency of the hoop on the meridian stress. In addition, the formation of the relationship signifies that the solution is statically determinate (Heyman 1967).

Replace,

$$r_0 = x$$
$$\phi = \tan^{-1}[\sinh(k)] \tag{7.44}$$

Therefore,

$$N'_\theta = \frac{\partial\left(N'_\phi \, x\cos\left\{\tan^{-1}[\sinh(k)]\right\}\right)}{\partial s}$$

Since,

$$\text{sech}(k) = \cos\left\{\tan^{-1}[\sinh(k)]\right\} \tag{7.58}$$

$$N'_\theta = \frac{\partial\left[N'_\phi x\,\text{sech}(k)\right]}{\partial s} \tag{7.59}$$

Substituting the meridian stress Eqs. 7.56 into 7.59,

$$N'_\theta = \frac{\partial\left[-\frac{qa}{k}\coth(k)[k\sinh(k) - \cosh(k) + 1]x\,\text{sech}(k)\right]}{\partial s}$$

Simplify using the following hyperbolic identities:

$$\coth(k)\text{sech}(k) = \text{csch}(k)$$
$$\text{csch}(k)\sinh(k) = 1$$
$$\text{csch}(k)\cosh(k) = \coth(k)$$

$$N'_\theta = \frac{\partial\left\{-qa^2[k - \coth(k) + \text{csch}(k)]\right\}}{\partial s}$$

In the above equation a disparity exists, since k is in terms of x and the derivative is with respect to the arch length s. However, the relationship between s and x was previously defined,

$$s = a\sin h(k) \tag{7.46}$$

or

$$k = \sin h^{-1}\left(\frac{s}{a}\right) \tag{7.60}$$

Taking the derivative of Eq. 7.60,

$$\frac{dk}{ds} = \frac{1/a}{\sqrt{1 + \left(\frac{s}{a}\right)^2}} \tag{7.61}$$

where

$$\frac{s}{a} = \sinh(k) \tag{7.62}$$

Replacing the value of k and pulling the constants out of the derivative,

$$N_\theta' = -qa^2 \frac{\partial \left\{ \sinh^{-1}\left(\frac{s}{a}\right) - \coth\left[\sinh^{-1}\left(\frac{s}{a}\right)\right] + \operatorname{csch}\left[\sinh^{-1}\left(\frac{s}{a}\right)\right] \right\}}{\partial s}$$

Take the derivative of the above expression using the technique of derivative substitution.

$$N_\theta' = -qa^2 \left\{ \frac{1/a}{\sqrt{1 + \left(\frac{s}{a}\right)^2}} + \frac{\frac{1}{a}}{\sqrt{1 + \left(\frac{s}{a}\right)^2}} \operatorname{csch}^2\left[\sinh^{-1}\left(\frac{s}{a}\right)\right] \right.$$
$$\left. - \frac{\frac{1}{a}}{\sqrt{1 + \left(\frac{s}{a}\right)^2}} \operatorname{csch}\left[\sinh^{-1}\left(\frac{s}{a}\right)\right] \coth\left[\sinh^{-1}\left(\frac{s}{a}\right)\right] \right\}$$

Since,

$$k = \sinh^{-1}\left(\frac{s}{a}\right) \tag{7.60}$$

$$\frac{s}{a} = \sinh(k) \tag{7.62}$$

Make this substitution and rearrange,

$$N_\theta' = -\frac{qa}{\sqrt{1 + \sinh^2(k)}} \left[1 + \operatorname{csch}^2(k) - \operatorname{csch}(k)\coth(k)\right]$$

Using the hyperbolic identities,

$$\cosh^2(k) = 1 + \sinh^2(k)$$

$$\frac{1}{\cosh(k)} = \operatorname{sech}(k)$$

$$N'_\theta = -qa\text{sech}(k)\left[1 + \text{csch}^2(k) - \text{csch}(k)\text{coth}(k)\right]$$

Using the additional identities,

$$\text{csch}^2(k) + 1 = \text{coth}^2(k)$$

$$\text{sech}(k)\text{coth}(k) = \text{csch}(k)$$

$$\text{coth}(k) = \cosh(k)\text{csch}(k)$$

$$N'_\theta = -qa\text{sech}(k)\left[\text{coth}^2(k) - \text{csch}(k)\text{coth}(k)\right]$$

$$N'_\theta = -qa\text{csch}(k)\left[\text{coth}(k) - \text{csch}(k)\right]$$

$$N'_\theta = -qa\,\textbf{csch}^2(k)\left[\cosh(k) - 1\right] \tag{7.63}$$

7.4.4 *Membrane Stress Equation for a Dome with an Oculus*

The theory assumes that the oculus or skylight is concentric, circular, and located at the apex of the dome. The meridian stress is derived by calculating the weight of the dome cap, and distributing this load over the perimeter. The same stress equations are used, but the stress will be calculated by subtracting the weight of the removed dome cap (see Fig. 7.15) from the total weight. The x-coordinate, defining opening of the oculus, is equal to x_o. Therefore, the weight of the removed dome cap is equal to a constant.

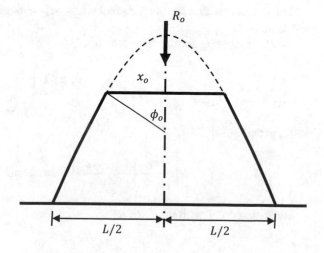

Fig. 7.15 Catenary dome with an oculus

$$R_o = 2\pi q a^2 [k_o \sinh(k_o) - \cosh(k_o) + 1] \qquad (7.64)$$

where

$$k_o = \frac{x_o}{a} \qquad (7.65)$$

The meridian stress for a catenary with an oculus may therefore be defined.

$$N'_\phi = \frac{-(R - R_o)}{2\pi x} \csc\phi \qquad (7.66)$$

Substituting the total weight R (Eq. 7.54) and replacing ϕ with Eq. 7.44,

$$N'_\phi = \frac{-2\pi q a^2 [k \sinh(k) - \cosh(k) + 1] + R_o}{2\pi x} \csc\{\tan^{-1}[\sinh(k)]\}$$

Since,

$$\coth(k) = \csc\{\tan^{-1}[\sinh(k)]\} \qquad (7.67)$$

$$N'_\phi = -q a^2 \left[\frac{k}{x} \sinh(k)\coth(k) - \frac{1}{x} \cosh(k)\coth(k) + \frac{1}{x}\coth(k) \right] + \frac{R_o\coth(k)}{2\pi x}$$

$$N'_\phi = -q \left[a \sinh(k)\coth(k) - \frac{a}{k} \cosh(k)\coth(k) + \frac{a}{k}\coth(k) \right] + \frac{R_o\coth(k)}{2\pi x}$$

$$N'_\phi = -\frac{qa}{k}\coth(k) \left[k \sinh(k) - \cosh(k) + 1 - \frac{R_o}{2\pi a^2 q} \right] \qquad (7.68)$$

The above equation for the meridian stress only changes by the inclusion of the last term.

Similarly, solving for the hoop stresses,

$$N'_\theta = \frac{\partial \left[N'_\phi x \operatorname{sech}(k) \right]}{\partial s} \qquad (7.59)$$

Substituting Eq. 7.66,

$$N'_\theta = \frac{\partial}{\partial s} \left\{ \frac{-(R - R_o)}{2\pi x} \csc\phi x \operatorname{sech}(k) \right\}$$

Replacing csc ϕ with Eq. (7.67),

$$N'_\theta = \frac{\partial}{\partial s}\left[\frac{-(R - R_o)}{2\pi x}\coth(k)x\mathrm{sech}(k)\right]$$

Using the hyperbolic identity and simplify,

$$\mathrm{csch}(k) = \coth(k)\mathrm{sech}(k)$$

$$N'_\theta = \frac{\partial}{\partial s}\left[\frac{-(R - R_o)\mathrm{csch}(k)}{2\pi}\right]$$

Substitute the equation for R.

$$N'_\theta = \frac{\partial}{\partial s}\left[-\frac{2\pi qa^2[k\sinh(k) - \cosh(k) + 1]\mathrm{csch}(k)}{2\pi} + \frac{R_o\mathrm{csch}(k)}{2\pi}\right]$$

Since,

$$\sinh(k)\mathrm{csch}(k) = 1$$

$$\cosh(k)\mathrm{csch}(k) = \coth(k)$$

$$N'_\theta = -qa^2\frac{\partial}{\partial s}\left[k - \coth(k) + \mathrm{csch}(k) - \frac{R_o}{2\pi qa^2}\mathrm{csch}(k)\right]$$

Replacing k in terms of the arch length s,

$$k = \sinh^{-1}\left(\frac{s}{a}\right) \tag{7.60}$$

$$N'_\theta = -qa^2\frac{\partial}{\partial s}\left\{\sinh^{-1}\left(\frac{s}{a}\right) - \coth\left[\sinh^{-1}\left(\frac{s}{a}\right)\right] + \mathrm{csch}\left[\sinh^{-1}\left(\frac{s}{a}\right)\right]\right.$$
$$\left.\frac{R_o}{2\pi qa^2}\mathrm{csch}\left(\sinh^{-1}\left(\frac{s}{a}\right)\right)\right\}$$

Taking the derivative,

$$N'_\theta = -qa^2 \left\{ \frac{\frac{1}{a}}{\sqrt{1+\left(\frac{s}{a}\right)^2}} + \frac{\frac{1}{a}}{\sqrt{1+\left(\frac{s}{a}\right)^2}} \mathrm{csch}^2\left[\sinh^{-1}\left(\frac{s}{a}\right)\right] - \right.$$

$$\frac{\frac{1}{a}}{\sqrt{1+\left(\frac{s}{a}\right)^2}} \mathrm{csch}\left[\sinh^{-1}\left(\frac{s}{a}\right)\right]\coth\left[\sinh^{-1}\left(\frac{s}{a}\right)\right] +$$

$$\left. \frac{\frac{1}{a}}{\sqrt{1+\left(\frac{s}{a}\right)^2}} \frac{R_o}{2\pi qa^2}\mathrm{csch}\left[\sinh^{-1}\left(\frac{s}{a}\right)\right]\coth\left[\sinh^{-1}\left(\frac{s}{a}\right)\right] \right\}$$

where

$$\frac{dk}{ds} = \frac{1/a}{\sqrt{1+\left(\frac{s}{a}\right)^2}} \tag{7.61}$$

Expressing the equation in terms of x and k and rearranging,

$$k = \sinh^{-1}\left(\frac{s}{a}\right) \tag{7.60}$$

$$\frac{s}{a} = \sinh(k) \tag{7.62}$$

$$N'_\theta = -\frac{qa}{\sqrt{1+\sinh^2(k)}}\left\{1+\mathrm{csch}^2(k) - \right.$$

$$\left. \mathrm{csch}(k)\coth(k) + \frac{R_o}{2\pi qa^2}\mathrm{csch}(k)\coth(k)\right\}$$

Simplifying using the identities,

$$\cosh^2(k) = 1 + \sinh^2(k)$$

$$\frac{1}{\cosh(k)} = \mathrm{sech}(k)$$

$$\mathrm{sech}(k)\coth(k) = \mathrm{csch}(k)$$

$$N'_\theta = -qa\left[\mathrm{sech}(k) + \mathrm{csch}^2(k)\mathrm{sech}(k) - \mathrm{csch}^2(k) + \frac{R_o}{2\pi qa^2}\mathrm{csch}^2(k)\right]$$

$$N'_\theta = -qa\left\{\mathrm{sech}(k)\left[1+\mathrm{csch}^2(k)\right] - \mathrm{csch}^2(k)\left[1-\frac{R_o}{2\pi qa^2}\right]\right\}$$

Substituting the identity,

$$csch^2(k) + 1 = \coth^2(k)$$

$$N'_\theta = -qa \left\{ \text{sech}(k) \coth^2(k) - csch^2(k) \left[1 - \frac{R_o}{2\pi qa^2} \right] \right\}$$

Since,

$$\text{sech}(k) \coth^2(k) = \cosh(k) \, csch^2(k)$$

$$N'_\theta = -qa \, csch^2(k) \left[\cosh(k) - 1 + \frac{R_o}{2\pi qa^2} \right] \qquad (7.69)$$

7.5 Worked Examples

7.5.1 Arch Subjected to a Uniform Gravity Load

The optimum pure compressive shape for an arch or vault, subjected to a uniform load, is a catenary curve. The accuracy of the catenary solution (Sect. 7.2) and the funicular analysis method (Sect. 7.3) are verified by comparing the solutions with the hyperbolic catenary equation (Eq. 7.22).

The geometry and material density of the arch:

H = 3 m
L = 3 m
b = 1 m (unit width)
a = 0.608149905
γ_m = 1900 kg/m^3 or 18.639 kN/m^3
h = 0.250 m

The value of "a" was solved by iterating Eq. 7.2.
Solving for the length of the arch.

$$S = 2a \, \sinh(L/2a) = 2 \, (0.60815) \sinh \left[\frac{3}{2 \, (0.60815)} \right] = 7.113 \, m \qquad (7.4)$$

From the length, the total weight of the arch (per m width) is determined.

$$W = \gamma_m S h b = 18.639 \, (7.113) \, 0.25 \, (1) = 33.145 \, kN \qquad (7.5)$$

The vertical reactions are half of the total weight.

Table 7.1 Stress results of the catenary and funicular arch methods

x (m)	y (m)	N_t (Catenary Method Eqn. 7.22) (kN/m)	N_t (Funicular method) (74 segments) (kN/m)	N_t (Funicular method) (714 segments) (kN/m)
0	0	16.81	16.81	16.81
0.73	0.5	14.48	14.52	14.48
0.99	1.0	12.15	12.25	12.15
1.16	1.5	9.82	9.99	9.81
1.30	2.0	7.49	7.35	7.48
1.41	2.5	5.16	5.28	5.16
1.50	3.0	2.83	2.84	2.83
1.59	2.5	5.16	5.23	5.19
1.70	2.0	7.49	7.73	7.51
1.84	1.5	9.82	9.93	9.84
2.01	1.0	12.15	12.19	12.19
2.27	0.5	14.48	14.92	14.51
3.00	0	16.81	17.21	16.84

$$R_v = \frac{W}{2} = \frac{33.145}{2} = 16.573 \text{ kN} \tag{7.6}$$

Solving for the angle at the base of the arch,

$$\theta = \tan^{-1}[\sinh L/2a] = \tan^{-1}[\sinh 3/2(0.60815)] = 80.297° \tag{7.9}$$

With the angle at the base solved, the reaction is determined.

$$R = \frac{R_v}{\sin \theta} = \frac{16.573}{\sin 80.297} = 16.813 \text{ kN} \tag{7.10}$$

The remaining term is the horizontal component of the reaction.

$$R_h = R \cos \theta = 16.813 \cos 80.297 = 2.834 \text{ kN} \tag{7.11}$$

From the reactions, the internal stress is solved, as a function of height.

$$N_t = \frac{W}{2 \sin \theta} \left[\left(\frac{\cos \theta - 1}{H} \right) y + 1 \right]$$

$$= \frac{33.15}{2 \sin 80.30} \left[\left(\frac{\cos 80.30 - 1}{3} \right) y + 1 \right] \tag{7.22}$$

The above equation is solved in increments of 0.5 m, and given in Table 7.1 under the heading catenary method.

Fig. 7.16 Base segment geometry of stresses

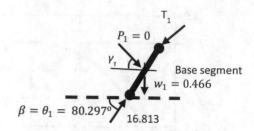

$P_1 = 0$

γ_1

T_1

Base segment
$w_1 = 0.466$

$\beta = \theta_1 = 80.297°$

16.813

The funicular arch method is also applied to the same arch. The accuracy of this method is dependent on the size of each segment, or element—the smaller the segments, the higher the accuracy. The method requires that the base segment is calculated first. The result of the calculation is then used to calculate the axial thrust and angular orientation of the next segment. This process is repeated again and again, allowing the shape to form, rising and descending to the second support. In this example, the same geometry and material density is used. The length of each segment is 100 mm in length, corresponding to a course of bricks.

Base segment, n = 1

Similarly to the catenary arch method (Sect. 7.2), the reaction (R) and angle of the reaction (θ_1) is determined using Eqs. 7.9 and 7.10 (see Fig. 7.16). Since we are solving the same arch, the results will be similar, or identical.

$$\theta_1 = 80.297° \tag{7.9}$$

$$R = 16.813 \text{ kN} \tag{7.10}$$

Since the arch is only subjected to self-weight, the external load (P_1) is zero. The value w_1 is the self-weight of the segment.

$$w_1 = \gamma_m S_1 h = 18.630 \,(0.1)0.25 = 0.466 \, kN$$

Solving for the reaction T_1,

$$T_{1x} = P_1 \cos \gamma_1 + R_L \cos \beta = 0 \cos \gamma_1 + 16.813 \, \cos 80.297 = 2.834 \, kN \tag{7.33}$$

$$T_{1y} = -P_1 \sin \gamma_1 + R_L \sin \beta - w_1$$
$$= -0 \sin \gamma_1 + 16.813 \sin 80.297 - 0.466 = 16.107 \, kN \tag{7.34}$$

Solving for the resultant reaction and angle,

$$T_1 = \sqrt{T_{1x}^2 + T_{1y}^2} = \sqrt{(2.834)^2 + (16.107)^2} = 16.354 \, kN \tag{7.35}$$

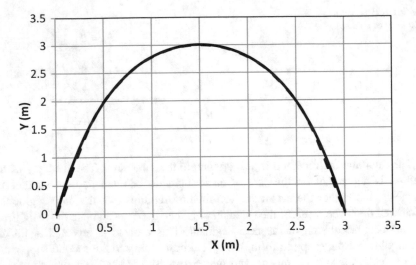

Fig. 7.17 Comparison of the catenary and funicular solutions

$$\Theta_2 = \tan^{-1}(T_{1y}/T_{1x}) = \tan^{-1}\left(\frac{16.107}{2.834}\right) = 80.021° \qquad (7.36)$$

Second element, n = 2

The above equations complete the analysis of the first segment, or base element. The second segment is then added to the first segment, and the solution is given by solving for Eqs. 7.37, 7.38, 7.39, and 7.40.

$$w_2 = \gamma_m S_2 h = 18.630\,(0.1)0.25 = 0.466\ kN$$

$$T_{2x} = 0\,\cos\gamma_n + 16.354\cos 80.021 = 2.834\ kN \qquad (7.37)$$

$$T_{2y} = 0\,\sin\gamma_n + 16.354\sin 80.021 - 0.466 = 15.641\ kN \qquad (7.38)$$

$$T_2 = \sqrt{(2.834)^2 + (15.641)^2} = 15.895\ kN \qquad (7.39)$$

$$\theta_3 = \tan^{-1}\left(\frac{15.641}{2.834}\right) = 79.730° \qquad (7.40)$$

The above calculation sequence is repeated, adding elements and thus allowing the growth of the arch. The arch will automatically rise and fall to the second support. The results of the calculations are given in Table 7.1, applying both the catenary and funicular methods.

As seen in Table 7.1, the funicular method provides a good approximation of the catenary curve. It is also evident that the funicular method converges to the exact solution, as the number of segments is increased. When plotted (Fig. 7.17), both methods produce an almost identical curve. From Table 7.1, it is also apparent that the solution improves as the size of the segment decreases.

7.5.2   *Arch Subjected to a Uniform Gravity Load and a Point Load*

The power of funicular arch method is its application to arches with point loads. Pure compression shapes may be determined for any number, or orientation, of point loads.

To demonstrate this method, 5 kN load is applied to the catenary arch example (see Fig. 7.18). The load is applied 500 mm from the center, and orientated at an angle of 60° from the horizontal.

The first step is to determine the height of the point load.

$$y = H - a \, \cosh\left(\frac{x}{a}\right) + a = 3 - 0.60815 \cosh\left(\frac{0.5}{0.60815}\right) + 0.60815$$

$$= 2.783 \text{ m} \tag{7.3}$$

Next, length S' and S'' are determined using Pythagoras theorem.

$$S' = \sqrt{1^2 + 2.783^2} = 2.957 \text{ m}$$

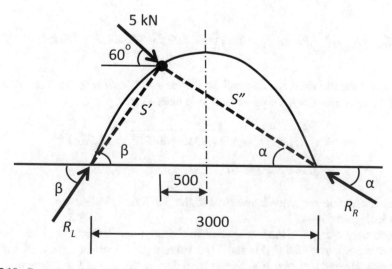

Fig. 7.18 Geometry of the point load

$$S'' = \sqrt{2^2 + 2.783^2} = 3.427 \text{ m}$$

Solving for the angles α and β,

$$\alpha = \cos^{-1}\left[\frac{S''^2 + L^2 - S'^2}{2S''L}\right] = \cos^{-1}\left[\frac{(3.427)^2 + 3^2 - (2.957)^2}{2\,(3.427)\,3}\right]$$

$$= 54.294° \tag{7.23}$$

$$\beta = \cos^{-1}\left[\frac{S'^2 + L^2 - S''^2}{2S'L}\right] = \cos^{-1}\left[\frac{(2.957)^2 + 3^2 - (3.427)^2}{2\,(2.957)\,3}\right]$$

$$= 70.233° \tag{7.24}$$

Solving for the reaction of the point loads,

$$R_R = \frac{P_1(\cos\gamma + \sin\gamma\cot\beta)}{\cos\alpha\,(1 + \tan\alpha\cot\beta)} = \frac{5\,(\cos 60 + \sin 60\cot 70.233)}{\cos 54.294\,(1 + \tan 54.294\cot 70.233)}$$

$$= 4.633 \text{ kN} \tag{7.27}$$

$$R_L = \frac{P_n\sin\gamma - R_R\sin\alpha}{\sin\beta} = \frac{5\sin 60 - 5.111\sin 54.294}{\sin 70.233} = 0.699 \text{ kN} \tag{7.28}$$

Using the left reaction, the horizontal and vertical components of the point load are solved and added to the self-weight reactions of the catenary arch.

$$R_{Lv} = R_v + \sum_1^1 R_L^1\sin\beta_1 = 16.573 + 0.699\sin 70.233 = 17.231 \text{ kN} \tag{7.29}$$

$$R_{Lh} = R_h + \sum_1^1 R_L^1\cos\beta_1 = 2.834 + 0.699\cos 70.233 = 3.070 \text{ kN} \tag{7.30}$$

After all the reactions are summed (in this case only one point load reaction and the self-weight reaction), a single reaction is determined.

$$R = \sqrt{R_{Lv}^2 + R_{Lh}^2} = \sqrt{17.231^2 + 3.070^2} = 17.502 \text{ kN} \tag{7.31}$$

$$\beta = \theta_1 = \tan^{-1}(17.231/3.070) = 79.896° \tag{7.32}$$

The calculation sequence is identical to the previous calculation.
Base segment, n = 1
Since only one point load is applied to the arch and located 1m from the left support, the external load (P_1) is zero. The value w_1 is calculated assuming a height of 10 mm. Higher accuracy is achieved by reducing the size of the segment.

$$w_1 = \gamma_m S_1 h = 18.639\,(0.01)0.25 = 0.0466\ kN$$

Solving for the reaction T_1,

$$T_{1x} = P_1\,\cos\,\gamma_1 + R_L\cos\beta = 0\cos\gamma_1 + 17.502\cos 79.896 = 3.070\ kN \quad (7.33)$$

$$\begin{aligned} T_{1y} &= -P_1\sin\gamma_1 + R_L\sin\beta - w_1 \\ &= -0\sin\gamma_1 + 17.502\sin 79.896 - 0.0466 = 17.408\ kN \end{aligned} \quad (7.34)$$

$$T_1 = \sqrt{T_{1x}^2 + T_{1y}^2} = \sqrt{(3.070)^2 + (17.184)^2} = 17.408\ kN \quad (7.35)$$

$$\Theta_1 = \tan^{-1}(T_{1y}/T_{1x}) = \tan^{-1}\left(\frac{17.184}{3.070}\right) = 79.870^\circ \quad (7.36)$$

The subsequent elements (i.e., $n = 2, 3, 4, \ldots$) are solved in the same manner as set out in the previous example. However, the point load is reached after the 300 segment is added to the arch, i.e., $x = 1$ m.

$$w_{300} = \gamma_{300} S_{300} h = 18.639\,(0.01)0.25 = 0.0466\ kN$$

$$T_{300x} = 5\,\cos\,60 + 4.534\cos\,46.79 = 5.604\ kN \quad (7.37)$$

$$T_{300y} = 5\,\sin 60 + 4.534\sin 46.79 - 0.0466 = -1.072\ kN \quad (7.38)$$

$$T_{300} = \sqrt{(5.427)^2 + (-1.235)^2} = 5.566\ kN \quad (7.39)$$

$$\theta_{301} = \tan^{-1}\left(\frac{-1.072}{5.604}\right) = -10.829^\circ \quad (7.40)$$

On close inspection, at the point load, the arch takes a sudden change in direction, and curves downward. The second base is reached after 703 segments are added to the base segment. The solution is illustrated in Fig. 7.19, together with the catenary arch without the 5 kN point load (dashed line). If the arch is only subjected to a point load, the shape would be triangular. However, the arch has a self-weight and therefore the shape is a pointed catenary arch.

In Fig. 7.19, it is observed that the curve leans into point load, which is necessary to ensure that the arch is in pure compression.

An important aspect of the funicular solution is that the growth of the curve ascended and descended to a second support. When this occurs, the arch will be in pure compression. However, a divergent solution (i.e., the arch never descends to the second support) indicates that a pure compression arch is not possible for the given set of loads.

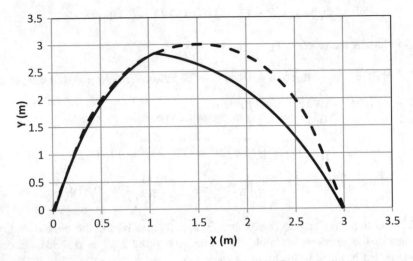

Fig 7.19 The funicular shape of a catenary arch subjected to a single point load

7.5.3   Membrane Solution of a Catenary Dome Subjected to Gravity Load

The theory for a catenary dome is solved using only two membrane equations. Since catenary domes are pure compression shapes and the boundary effects are minimal, only the membrane solution is usually required.

h = 0.1 m
H = 3 m
L = 6 m
γ = 19.5 kN/m^3
q = 19.5(0.1) = 1.95 kN/m^2
a = 1.85627768331

The scaling factor "a" is determined using Eq. 7.2, and the y coordinate is determined using 7.3.

Using Eq. 7.56 to solve the meridian stresses,

$$N'_\phi = -\frac{1.95\,(1.85628)^2}{x}\coth\left(\frac{x}{1.85628}\right)$$
$$\times\left[\frac{x}{1.85628}\sinh\left(\frac{x}{1.85628}\right) - \cosh\left(\frac{x}{1.85628}\right) + 1\right] \quad (7.56)$$

and Eq. 7.63 for the hoop stresses,

Table 7.2 Resultant stresses in a catenary dome (from apex to base)

x (m)	y (m)	N'_ϕ (kN/m)	N'_θ (kN/m)
0.00	3.00	−1.81	−1.81
0.25	2.98	−1.83	−1.80
0.50	2.93	−1.89	−1.78
0.75	2.45	−1.99	−1.74
1.00	2.72	−2.13	−1.68
1.25	2.56	−2.32	−1.62
1.50	2.36	−2.56	−1.54
1.75	2.11	−2.86	−1.46
2.00	1.81	−3.22	−1.37
2.25	1.46	−3.66	−1.28
2.50	1.05	−4.19	−1.19
2.75	0.56	−4.82	−1.09
3.00	0.00	−5.55	−1.00

Fig. 7.20 Meridian stresses in the catenary dome

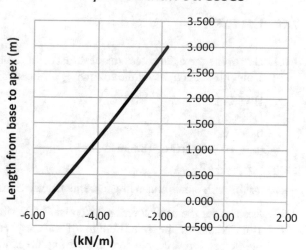

$$N'_\theta = -1.95 \ (1.85628) \ \operatorname{csch}^2\left(\frac{x}{1.85628}\right)\left[\cosh\left(\frac{x}{1.85628}\right) - 1\right] \qquad (7.63)$$

The stresses are determined (see Table 7.2) and the graphs are given in Figs. 7.20 and 7.21.

As seen from the solution, the meridian and hoop stresses are entirely compressive. The stresses in the catenary shape differs significantly from the circular hemisphere, where hoop tensions occur in the lower portion of the shell. Although a boundary solution was not derived, the boundary effect are small in catenary domes, compared to other dome shapes.

Fig. 7.21 Hoop stresses in
the catenary dome

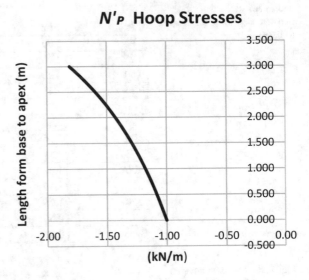

7.6   Exercises

7.6.1. Determine the axial forces (axial thrust) of a catenary vault. Only consider a
       1 m wide section, using the following parameters in the analysis:

       $H = 3$ m
       $L = 6$ m
       $b = 1$ m
       $\gamma_m = 1900$ kg/m3 or 18.639 kN/m3
       $h = 0.125$ m

Solve for the arch axial thrusts in y increments of 0.5 m.

7.6.2. Resolve for the axial forces of exercise 7.6.1, with a reduced height of $H =$
       1 m. Maintain the same length of catenary curve (S is equal for both
       problems). Draw conclusions concerning the effect of reducing the height.
7.6.3. Using the funicular iteration method for solving arches, program the theory in
       a spreadsheet or alternative programing language. Resolve exercise 7.6.1. Use
       a segment length of 500 mm, compare the funicular iteration method with the
       catenary equations. Graphically compare the geometric configurations and
       tabulate the axial thrusts.
7.6.4. Repeat exercise 7.6.3, but with a segment length of 50 mm.
7.6.5. Repeat exercise 7.6.4, but with a vertical point load of 5 kN placed at the apex
       of the catenary curve (pointing down). Graph the predicted geometry of the
       pure compression shape, and tabulate the axial forces.
7.6.6. Solve for the membrane stresses of a concrete catenary dome, based on the
       following parameters:

$$H = 2.5 \text{ m}$$
$$L = 6 \text{ m}$$
$$q = 4.69 \text{ kN/m}^2$$
$$h = 100 \text{ mm}$$

Compare the results with the worked example 4.8 for a circular dome. Tabulate the membrane resultant stresses, and draw conclusions concerning the efficiency of catenary domes.

7.6.7. Resolve exercise 7.6.6, but with an oculus radius of 500 mm.

References

Allen, E. & Zalewski, W., 2010. *Form and Forces, Expressive Structures.* New Jersey: John Wiley and Son.

Billington, D. P., 1982. *Thin Shell Concrete Structures.* 2 ed. New York: McGraw-Hill.

Block, P., Dejong, M. & Ochsendorf, J., 2006. As Hangs the Flexible Line: Equilibrium of Masonry Structures. *Nexus Network Journal.*

Farshad, M., 1992. *Design and Analysis of Shell Structures.* s.l.:Springer, Kluwer Academic Publishers.

Gohnert, M. & Bradley, R., 2020a. Catenary Solution for Arches and Vaults. *Journal of Architectural Engineering,* 26(2).

Gohnert, M. & Bradley, R., 2020b. Membrane Solution of Catenary Domes. *International Association for Shell and Spatial Structures,* 61(2).

Heyman, J., 1967. On Shell Solutions of Catenary Domes. *International Journal of Solids and Structures,* Volume 3, pp. 227–241.

Heyman, J., 1995. *The Stone Skeleton, Structural Engineering of Masonry Architecture.* Cambridge: Cambridge University Press.

Ockleston, A. J., 1958. Arching Action in Reinforced Concrete. *The Structural Engineer,* 36(6), pp. 197–201.

Appendixes

Appendix A

Appendix B

M. Gohnert, *Shell Structures*, https://doi.org/10.1007/978-3-030-84807-1

Table A1 r/t = 100 and r/L = 0.1

φ	Vertical edge load				Horizontal edge load				Shear edge load				Edge moment			
	N_x	$N_{\phi x}$	N_ϕ	M_ϕ	N_x	$N_{\phi x}$	N_ϕ	M_ϕ	N_x	$N_{\phi x}$	N_ϕ	M_ϕ	N_x	$N_{\phi x}$	N_ϕ	M_ϕ
$\phi_k = 30$																
30	−5.382	0	−3.471	−0.3278	−0.0013	0	+0.9997	+0.1310	−0.0904	0	−0.0207	−0.0010	−0.0038	0	+0.0008	+0.9738
20	−3.575	−2.619	−2.656	−0.2779	+0.0017	−0.0001	+0.9848	+0.1162	−0.0396	−0.0403	−0.0081	−0.0007	+0.0172	+0.0021	+0.0022	+0.9767
10	+1.809	−3.265	−0.775	−0.1512	+0.0034	+0.0017	+0.9399	+0.0722	+0.1118	−0.0250	+0.0142	+0.0002	+0.0264	+0.0173	+0.0014	+0.9854
0	+10.69	0	+0.500	0	−0.0151	0	+0.8660	0	+0.3600	+0.1000	0	0	−0.1370	0	0	+1.0000
$\phi_k = 35$																
35	−3.430	0	−2.959	−0.3703	−0.0061	0	+0.9986	+0.1754	−0.0774	0	−0.0243	−0.0015	−0.0310	0	−0.0051	+0.9642
30	−3.220	−0.921	−2.820	−0.3594	−0.0043	−0.0068	+0.9951	+0.1717	−0.0694	−0.0205	−0.0213	−0.0014	−0.0211	−0.0076	−0.0037	+0.9650
20	−1.538	−2.302	−1.812	−0.2781	+0.0060	−0.0136	+0.9663	+0.1425	−0.0054	−0.0439	−0.0015	−0.0008	+0.0373	−0.0043	+0.0040	+0.9710
10	+1.822	−2.301	−0.330	−0.1444	+0.0080	+0.0041	+0.9071	+0.0848	+0.1212	−0.0149	+0.0185	−0.0002	+0.0469	+0.0251	+0.0045	+0.9825
0	+6.845	0	+0.574	0	−0.0316	0	+0.8192	0	+0.3076	+0.1000	0	0	−0.2051	0	0	+1.0000
$\phi_k = 40$																
40	−2.210	0	−2.534	−0.4048	−0.0165	0	+0.9954	+0.2247	−0.0675	0	−0.0279	−0.0022	−0.0712	0	−0.0184	+0.9530
30	−1.806	−1.138	−2.168	−0.3683	−0.0049	−0.0068	+0.9836	+0.2041	−0.0461	−0.0331	−0.0179	−0.0018	−0.0196	−0.0293	−0.0074	+0.9570
20	−0.578	−1.830	−1.205	−0.2691	+0.0176	−0.0032	+0.9420	+0.1672	+0.0181	−0.0423	+0.0054	−0.0009	+0.0815	−0.0118	+0.0112	+0.9653
10	+1.516	−1.614	−0.045	−0.1352	+0.0156	+0.0081	+0.8685	+0.1047	+0.1236	−0.0057	+0.0221	−0.0002	+0.0724	+0.0998	+0.0106	+0.9798
0	+4.534	0	+0.643	0	−0.0631	0	+0.7660	0	+0.2682	+0.1000	0	0	−0.3099	0	0	+1.0000
$\phi_k = 45$																
45	−1.602	0	−2.231	−0.4355	−0.0359	0	+0.9873	+0.2779	−0.0596	0	−0.0315	−0.0031	−0.1280	0	−0.0437	+0.9392
40	−1.546	−0.434	−2.163	−0.4274	−0.0309	−0.0094	+0.9849	+0.2743	−0.0559	−0.0160	−0.0292	−0.0030	−0.1102	−0.0325	−0.0385	+0.9402
30	−1.098	−1.180	−1.646	−0.3654	+0.0033	−0.0181	+0.9642	+0.2461	−0.0258	−0.0398	−0.0123	−0.0022	+0.0128	−0.0642	−0.0047	−0.9469
20	−0.166	−1.550	−0.763	−0.2562	+0.0412	−0.0050	+0.9138	+0.1899	+0.0338	−0.0389	+0.0120	−0.0010	+0.1523	−0.0158	+0.0277	+0.9595
10	+1.310	−1.263	+0.179	−0.1255	+0.0260	+0.1175	+0.8251	+0.1072	+0.1222	+0.0025	+0.0250	−0.0002	+0.0981	+0.0680	+0.0212	+0.9771
0	+3.407	0	+0.707	0	−0.1185	0	+0.7071	0	+0.2378	+0.1000	0	0	−0.4613	0	0	+1.0000
$\phi_k = 50$																
50	−1.051	0	−1.931	−0.4541	−0.0693	0	+0.9696	+0.3336	−0.0531	0	−0.0350	−0.0042	−0.2048	0	−0.0885	+0.9218
40	−0.948	−0.558	−1.741	−0.4265	−0.0213	−0.0326	+0.9650	+0.3198	−0.0423	−0.0271	−0.0268	−0.0037	−0.1192	−0.0964	−0.0573	+0.9258
30	−0.619	−0.999	−1.208	−0.3492	+0.0284	−0.0366	+0.9430	+0.2786	−0.0098	−0.0424	−0.0057	−0.0025	+0.0849	−0.1086	+0.0111	+0.9371
20	−0.001	−1.184	−0.451	−0.2379	+0.0820	−0.0039	+0.8851	+0.2104	+0.0440	−0.0340	+0.0181	−0.0010	+0.2503	−0.0099	+0.0576	+0.9537
10	+0.998	−0.930	+0.317	−0.1141	+0.0374	+0.0357	+0.7738	+0.1167	+0.1188	+0.0077	+0.0275	−0.0002	+0.1176	+0.1119	+0.0373	+0.9745
0	+2.488	0	+0.766	0	−0.2047	0	+0.6428	0	+0.2138	+0.1000	0	0	−0.6695	0	0	+1.0000

Table A2 r/t = 100 and r/L = 0.2

φ	Vertical edge load				Horizontal edge load				Shear edge load				Edge moment			
	N_x	$N_{\phi x}$	N_ϕ	M_ϕ	N_x	$N_{\phi x}$	N_ϕ	M_ϕ	N_x	$N_{\phi x}$	N_ϕ	M_ϕ	N_x	$N_{\phi x}$	N_ϕ	M_ϕ
$\phi_k = 30$																
30	−5.160	0	−3.420	−0.3028	−0.0742	0	+0.9815	+0.1219	−0.1946	0	−0.0414	−0.0005	−0.549	0	−0.0813	+0.9009
20	−3.440	−2.531	−2.632	−0.2565	−0.0123	−0.0292	+0.9757	+0.1084	−0.0755	−0.0776	−0.0145		−0.073	−0.2142	−0.0153	+0.9124
10	−1.763	−3.228	−0.787	−0.1395	+0.0700	−0.0087	+0.9427	+0.0678	+0.2208	−0.0469	+0.0287	+0.0009	+0.557	−0.0429	+0.0568	+0.9456
0	+10.92	0	−0.500	0	−0.1140	0	+0.8660	0	+0.7119	+0.2000	0	0	−1.067	0	0	+1.0000
$\phi_k = 35$																
35	−3.087	0	−2.871	−0.3347	−0.1552	0	+0.9603	+0.1591	−0.1487	0	−0.0458	−0.0024	−0.912	0	−0.1914	+0.8637
30	−2.939	−0.833	−2.744	−0.3248	−0.1260	−0.0399	+0.9622	+0.1559	−0.1334	−0.0394	−0.0401	−0.0022	−0.742	−0.2343	−0.1585	+0.8669
20	−1.639	−2.156	−1.807	−0.2514	+0.0561	−0.0634	+0.9649	+0.1301	−0.0106	−0.0845	−0.0018	−0.0012	+0.327	−0.3747	+0.0315	+0.8914
10	+1.501	−2.296	−0.365	−0.1302	+0.1526	+0.0074	+0.9224	+0.0781	+0.2378	−0.0280	+0.0369	−0.0001	+0.928	+0.0486	+0.1212	+0.9362
0	+7.434	0	+0.574	0	−0.3244	0	+0.8192	0	+0.6159	+0.2000	0	0	−2.045	0	0	+1.0000
$\phi_k = 40$																
40	−1.764	0	−2.398	−0.3556	−0.3096	0	+0.9087	+0.1962	−0.1273	0	−0.0525	−0.0035	−1.421	0	−0.3885	+0.8156
30	−1.640	−0.947	−2.105	−0.3239	−0.1336	−0.1389	+0.9351	+0.1845	−0.0880	−0.0626	−0.0335	−0.0029	−0.631	−0.6461	−0.2007	+0.8306
20	−0.965	−1.697	−1.251	−0.2374	+0.2243	−0.1133	+0.9680	+0.1487	+0.0323	−0.0817	+0.0109	−0.0013	+1.015	−0.5396	+0.1574	+0.8707
10	+1.111	−1.753	−0.109	−0.1188	+0.2753	+0.0557	+0.9101	+0.0870	+0.2408	−0.0111	+0.0434	−0.0007	+1.316	+0.2523	+0.2259	+0.9271
0	+5.836	0	+0.643	0	−0.7210	0	+0.7660	0	+0.5460	+0.2000	0	0	−3.452	0	0	+1.0000
$\phi_k = 45$																
45	−0.775	0	−1.934	−0.3619	−0.5679	0	+0.7941	+0.2293	−0.1067	0	−0.0577	−0.0050	−2.074	0	−0.7365	+0.7509
40	−0.820	−0.217	−1.897	−0.3556	−0.4987	−0.1493	+0.8118	+0.2269	−0.0988	−0.0287	−0.0335	−0.0048	−1.828	−0.5460	−0.6582	+0.7558
30	−1.062	−0.731	−1.584	−0.3068	−0.0210	−0.3078	+0.9229	+0.2071	−0.0506	−0.0726	−0.0228	−0.0034	−0.116	−1.1380	−0.1410	+0.7920
20	−0.934	−1.317	−0.910	−0.2175	+0.5381	−0.1552	+1.0080	+0.1641	+0.0570	−0.0737	+0.0224	−0.0014	+1.958	−0.5987	+0.3956	+0.8510
10	+0.700	−1.478	+0.039	−0.1073	+0.4245	+0.1611	+0.9154	+0.0946	+0.2353	+0.0028	+0.0482	0	+1.641	+0.5811	+0.3778	+0.9196
0	+5.344	0	+0.707	0	−1.3920	0	+0.7071	0	+0.4995	+0.2000	0	0	−5.299	0	0	+1.0000
$\phi_k = 50$																
50	+0.052	0	−1.437	−0.3493	−0.9409	0	+0.5813	+0.2514	−0.0848	0	−0.0601	−0.0065	−2.814	0	−1.245	+0.6613
40	−0.291	−0.028	−1.413	−0.3315	−0.5827	−0.4585	+0.7040	+0.2452	−0.0714	−0.0439	−0.0468	−0.0056	−1.762	−1.381	−0.8336	+0.6869
30	−0.016	−0.383	−1.246	−0.2786	+0.2885	−0.5526	+0.9656	+0.2228	−0.0258	−0.0723	−0.0113	−0.0035	+0.797	−1.690	+0.0859	+0.7523
20	−1.244	−1.059	−0.745	−0.1953	+0.1015	−0.1620	+1.1120	+0.1764	+0.0664	−0.0639	+0.0312	−0.0013	+3.038	−0.536	+0.7607	+0.8338
10	+0.381	−1.427	+0.110	−0.0943	+0.5624	+0.3610	+0.9471	+0.1015	+0.2257	+0.0125	+0.0513	0	+1.811	+1.083	+0.5718	+0.9154
0	+5.670	0	+0.766	0	−2.398	0	+0.0428	0	+0.4748	+0.2000	0	0	−7.473	0	0	+1.0000

Table A3 $r/t = 100$ and $r/L = 0.3$

ϕ	Vertical edge load				Horizontal edge load				Shear edge load				Edge moment			
	N_x	$N_{\phi x}$	N_ϕ	M_ϕ	N_x	$N_{\phi x}$	N_ϕ	M_ϕ	N_x	$N_{\phi x}$	N_ϕ	M_ϕ	N_x	$N_{\phi x}$	N_ϕ	M_ϕ
$\phi_k = 30$																
30	-4.709	0	-3.362	-0.2769	-0.2662	0	+0.9571	+0.1099	-0.2618	0	-0.0596	-0.0024	-2.101	0	-0.3444	-0.7859
20	-3.568	-2.381	-2.632	-0.2347	-0.0094	-0.0961	+0.9752	+0.0082	-0.1182	-0.1174	-0.0231	-0.0016	-0.108	-0.7669	-0.0802	+0.8119
10	+1.053	-3.290	-0.837	-0.1275	+0.2942	+0.0035	+0.9639	-0.0622	+0.3251	-0.0750	+0.0421	-0.0003	+2.372	+0.0141	+0.1945	+0.8845
0	+12.43	0	+0.500	0	-0.6606	0	+0.8660	0	+1.1020	+0.3000		0	-5.456	0	0	+1.000
$\phi_k = 35$																
35	-2.156	0	-2.682	-0.2903	-0.6039	0	+0.8706	+0.1362	-0.2140	0	-0.0685	-0.0036	-3.493	0	-0.7613	+0.7084
30	-2.203	-0.595	-2.591	-0.2821	-0.4823	-0.1544	+0.8892	+0.1338	-0.1937	-0.0568	-0.0603	-0.0033	-2.808	-0.8945	-0.6319	+0.7161
20	-2.106	-1.817	-1.830	-0.2201	+0.2698	-0.2300	+0.9839	+0.1137	-0.0244	-0.1248	-0.0046	-0.0018	+1.506	-1.3580	+0.1098	+0.7726
10	+0.513	-2.456	-0.482	-0.1148	+0.6242	+0.0755	+0.9800	+0.0698	+0.3456	-0.0471	+0.0535	-0.0002	+3.764	+0.4236	+0.4372	+0.8689
0	+10.03	0	+0.574	0	-1.543	0	+0.8192	0	+0.9676	+0.3000		0	-9.403	0	0	+1.0000
$\phi_k = 40$																
40	-0.216	0	-1.977	-0.2853	-1.1760	0	+0.6733	+0.1553	-0.1668	0	-0.0736	-0.0048	-5.193	0	-1.469	+0.6071
30	-1.020	-0.275	-1.892	-0.2631	-0.1805	-0.5149	+0.8208	+0.1488	-0.1238	-0.0837	-0.0485	-0.0039	-2.222	-2.288	-0.7475	+0.6444
20	-2.210	-1.203	-1.430	-0.1982	+0.9141	-0.3958	+1.0690	+0.1252	+0.0263	-0.1164	+0.0125	-0.0018	+3.993	-1.809	+0.5716	+0.7382
10	-0.295	-2.141	-0.338	-0.1018	+1.0730	+0.2676	+1.0400	+0.0762	+0.3397	-0.0252	+0.0611	-0.0001	+5.004	+1.180	+0.7966	+0.8572
0	+10.01	0	+0.643	0	-3.0680	0	+0.7660	0	+0.8034	+0.3000		0	-14.380	0	0	+1.0000
$\phi_k = 45$																
45	+1.400	0	-1.178	-0.2565	-1.970	0	+0.3055	+0.15091	-0.1142	0	-0.0719	-0.0059	-6.814	0	-2.463	+0.4746
40	+1.083	+0.354	-1.224	-0.2537	-1.728	-0.5178	+0.3766	+0.1589	-0.1112	-0.0310	-0.0674	-0.0056	-6.032	-1.802	-2.202	+0.4871
30	-1.040	+0.432	-1.454	-0.2296	-0.0472	-1.061	+0.8385	+0.1550	-0.0776	-0.0851	-0.0328	-0.0040	-0.316	-0.3743	-0.4779	+0.5768
20	-3.110	-0.788	-1.326	-0.1743	+1.941	-0.5051	+1.279	+0.1338	+0.0382	-0.1013	+0.0241	-0.0016	+6.770	-1.867	+1.312	+0.7124
10	-0.946	-2.214	-0.330	-0.0899	+1.513	+0.6328	+1.158	+0.0821	+0.3209	-0.0128	+0.0645	-0.0001	+5.696	+2.230	+1.229	+0.8510
0	+11.24	0	+0.707	0	-5.257	0	+0.7071	0	+0.8743	+0.3000		0	-19.610	0	0	+1.000
$\phi_k = 50$																
50	+2.605	0	-0.326	-0.2300	-2.821	0	-0.2407	+0.1409	-0.0565	0	-0.0611	-0.0063	-7.871	0	-3.553	+0.3153
40	+1.276	+1.180	-0.676	-0.2032	-1.744	-1.346	+0.1577	+0.1474	-0.0651	-0.0327	-0.0514	-0.0056	-4.994	-3.782	-2.394	+0.3754
30	-1.837	+1.062	-1.342	-0.1926	+0.8732	-1.624	+1.0330	+0.1561	-0.0643	-0.0700	-0.0208	-0.0036	+2.207	-4.666	+0.2204	+0.5257
20	-4.099	-0.690	-1.432	-0.1527	+3.129	-0.4486	+1.6230	+0.1422	+0.0214	-0.0879	+0.0276	-0.0013	+8.917	-1.422	+2.171	+0.6995
10	-1.187	-2.497	-0.379	-0.0801	+1.766	+1.155	+1.3150	+0.0885	+0.3005	-0.0113	+0.0639	+0.0002	+5.588	+3.317	+1.625	+0.8508
0	+12.66	0	+0.766	0	-7.773	0	+0.6428	0	+0.8992	+0.3000		0	-23.76	0	0	+1.0000

Table A4 r/t = 100 and r/L = 0.4

φ	Vertical edge load				Horizontal edge load				Shear edge load				Edge moment			
	N_x	$N_{\phi x}$	N_ϕ	M_ϕ	N_x	$N_{\phi x}$	N_ϕ	M_ϕ	N_x	$N_{\phi x}$	N_ϕ	M_ϕ	N_x	$N_{\phi x}$	N_ϕ	M_ϕ
$\phi_k = 30$																
30	−3.508	0	−3.175	−0.2404	−0.736	0	+0.881	+0.0947	−0.3318	0	−0.0772	−0.0028	−5.33	0	−0.879	+0.6558
20	−3.500	−1.942	−2.592	−0.2048	−0.038	−0.2682	+0.958	+0.0856	−0.1568	−0.1503	−0.0304	−0.0019	−0.14	−1.920	−0.196	+0.6995
10	−0.244	−3.281	−0.947	−0.1121	+0.804	−0.0008	+1.008	+0.0544	+0.4153	−0.0998	+0.0545	−0.0003	+6.27	+0.187	+0.502	+0.8169
0	+15.23	0	+0.500	0	−1.772	0	+0.866	0	+1.5080	+0.4000	0	0	−15.27	0	0	+1.0000
$\phi_k = 35$																
35	−0.217	0	−2.261	−0.2351	−1.518	0	+0.670	+0.1091	−0.2518	0	−0.0842	−0.0040	−8.48	0	−1.890	+0.5364
30	−0.656	−0.102	−2.241	−0.2293	−1.214	−0.3878	+0.723	+0.1078	−0.2316	−0.0672	−0.0746	−0.0037	−6.82	−2.173	−1.574	+0.5500
20	−2.982	−1.086	−1.911	−0.1833	+0.680	−0.5777	+1.014	+0.0952	−0.0480	−0.1540	−0.0073	−0.0019	+3.69	−3.300	+0.248	+0.6463
10	−1.512	−2.715	−0.712	−0.0982	+1.596	+0.1982	+1.092	+0.0612	+0.4268	−0.0671	+0.0673	−0.0001	+9.37	+1.106	+1.064	+0.8000
0	+15.11	0	+0.574	0	−3.990	0	+0.819	0	+1.3730	+0.4000	0	0	−23.99	0	0	+1.0000
$\phi_k = 40$																
40	+2.485	0	−1.221	−0.2044	−2.664	0	+0.250	+0.1088	−0.1542	0	−0.0818	−0.0049	−11.15	0	−3.211	+0.3896
30	+0.117	+0.913	−1.512	−0.1948	−1.123	−1.171	+0.606	+0.1093	−0.1315	−0.0851	−0.0567	−0.0039	−4.83	−4.928	−0.164	+0.4541
20	−4.296	−0.278	−1.730	−0.1580	+2.058	−0.9139	+1.236	+0.1013	+0.0023	−0.1351	+0.0099	−0.0017	+8.71	−3.904	+1.250	+0.6098
10	−2.814	−2.749	−0.749	−0.0862	+2.495	+0.6036	+1.272	+0.0665	+0.4228	−0.0471	+0.0728	0	+11.30	+2.733	+1.764	+0.7918
0	+17.20	0	+0.643	0	−7.122	0	+0.766	0	+1.368	+0.4000	0	0	−33.39	0	0	+1.000
$\phi_k = 45$																
45	+4.329	0	−0.119	−0.1513	−3.817	0	−0.372	+0.0899	−0.0653	0	−0.0657	−0.0051	−12.50	0	−4.647	+0.2210
40	+3.663	+1.126	−0.277	−0.1526	−3.361	−1.005	−0.230	+0.0923	−0.0723	−0.0186	−0.0630	−0.0049	−11.09	−3.299	−4.168	+0.2412
30	−0.918	+2.209	−1.249	−0.1564	−0.150	−2.079	+0.703	+0.1060	−0.1030	−0.0673	−0.0384	−0.0035	−0.92	−6.970	−0.971	+0.3846
20	−6.034	−0.008	−1.889	−0.1377	+3.796	−1.015	+1.638	+0.1083	−0.0344	−0.1142	+0.0158	−0.0013	+12.56	−3.637	+2.441	+0.5941
10	−3.355	−3.194	−0.854	−0.0780	+3.104	+1.260	+1.497	+0.0731	+0.3651	−0.0449	+0.0716	+0.0003	+11.51	+4.262	+2.363	+0.7936
0	+19.64	0	+0.707	0	−10.78	0	+0.707	0	+1.388	+0.4000	0	0	−39.59	0	0	+1.0000
$\phi_k = 50$																
50	+4.973	0	+0.784	−0.0881	−4.477	0	−1.043	+0.0545	+0.0216	0	−0.0395	−0.0045	−11.45	0	−5.493	+0.0676
40	+2.829	+2.328	+0.079	−0.1028	−2.858	−2.155	−0.394	+0.0716	−0.0276	+0.0025	−0.0408	−0.0041	−7.65	−4.182	−3.773	+0.1539
30	−2.433	+2.516	−1.399	−0.1279	+1.235	−2.673	+1.065	+0.1057	−0.1165	−0.0382	−0.0313	−0.0029	+2.61	−3.745	+0.216	+0.3594
20	−6.881	−0.213	−2.118	−0.1244	+5.125	−0.8202	+2.120	+0.1184	−0.0779	−0.1025	+0.0109	−0.0010	+13.82	−1.157	+3.391	+0.5981
10	−3.049	−3.540	−0.904	−0.0724	+3.212	+1.909	+1.707	+0.0810	+0.3428	−0.0541	+0.0667	+0.0003	+10.19	+5.359	+2.668	+0.7998
0	+20.35	0	+0.766	0	−13.59	0	+0.643	0	+1.459	+0.4000	0	0	−41.80	0	0	+1.0000

Table A5 $r/t = 100$ and $r/L = 0.5$

ϕ	Vertical edge load				Horizontal edge load				Shear edge load				Edge moment			
	N_x	$N_{\phi x}$	N_ϕ	M_ϕ	N_x	$N_{\phi x}$	N_ϕ	M_ϕ	N_x	$N_{\phi x}$	N_ϕ	M_ϕ	N_x	$N_{\phi x}$	N_ϕ	M_ϕ
$\phi_k = 30$																
30	−1.757	0	−2.879	−0.2019	−1.420	0	+0.767	+0.0789	−0.3849	0	−0.0912	−0.0030	−10.64	0	−1.779	+0.5185
20	−3.508	−1.323	−2.524	−0.1738	−0.037	−0.5108	+0.932	+0.0727	−0.1963	−0.1773	−0.0368	−0.0020	−0.48	−3.877	−0.421	+0.5825
10	−2.292	−3.378	−1.108	−0.0969	+1.616	+0.0381	+1.071	+0.0489	+0.4854	−0.1263	+0.0654	−0.0002	+12.53	+0.236	+0.998	+0.7480
0	+20.22	0	+0.500	0	−3.767	0	+0.866	0	+1.964	+0.5000	0	0	−30.35	0	0	+1.0000
$\phi_k = 35$																
35	+2.471	0	−1.677	−0.1788	−2.776	0	+0.395	+0.0816	−0.2558	0	−0.0962	−0.0039	−14.99	0	−3.345	+0.3692
30	+1.493	+0.587	−1.758	−0.1758	−2.223	−0.7101	+0.495	+0.0816	−0.2428	−0.0837	−0.0875	−0.0037	−12.12	−4.668	−2.786	+0.3888
20	−4.209	−0.060	−2.002	−0.1480	+1.246	−1.062	+1.056	+0.0775	−0.0879	−0.1706	−0.0227	−0.0018	+6.40	−5.907	+0.453	+0.5266
10	−4.426	−3.090	−1.055	−0.0837	+2.986	+0.3737	+1.252	+0.0535	+0.4733	−0.0921	+0.0629	+0.0000	+17.12	+1.978	+1.931	+0.7345
0	+22.54	0	+0.574	0	−7.555	0	+0.819	0	+1.867	+0.5000	0	0	−44.56	0	0	+1.0000
$\phi_k = 40$																
40	+5.472	0	−0.358	−0.1324	−4.282	0	−0.220	+0.0682	−0.1143	0	−0.0775	−0.0034	−17.17	0	−5.060	+0.2033
30	+1.404	+2.232	−1.069	−0.1354	−1.839	−1.889	+0.363	+0.0754	−0.1408	−0.0683	−0.0585	−0.0034	−7.86	−7.678	−2.645	+0.2912
20	−6.661	+0.757	−2.062	−0.1254	+3.336	−1.488	+1.415	+0.0819	−0.0853	−0.1408	+0.0027	−0.0014	+13.14	−6.372	+1.904	+0.5001
10	−5.826	−3.503	−1.214	−0.0751	+4.196	+1.017	+1.528	+0.0593	+0.4218	−0.0802	+0.0780	+0.0003	+18.72	+4.129	+2.821	+0.7347
0	+26.12	0	+0.643	0	−12.17	0	+0.766	0	+1.908	+0.5000	0	0	−54.72	0	0	+1.0000
$\phi_k = 45$																
45	+6.655	0	+0.771	−0.0750	−4.823	0	−0.923	+0.0405	+0.0153	0	−0.0477	−0.0038	−15.73	0	−6.128	+0.0523
40	+5.749	+1.741	+0.524	−0.0793	−4.625	−1.375	−0.727	+0.0448	−0.0054	+0.0023	−0.0480	−0.0037	−14.17	−4.167	−5.525	+0.0765
30	−0.652	+3.364	−1.053	−0.1039	−0.365	−2.902	+0.573	+0.0709	−0.1266	−0.0318	−0.0417	−0.0028	−2.11	−9.122	−1.418	+0.2487
20	−8.443	+0.730	−2.362	−0.1125	+5.266	−1.497	+1.929	+0.0901	−0.1395	−0.1180	+0.0012	−0.0011	+16.12	−5.253	+3.200	+0.5015
10	−5.758	−4.033	−1.326	−0.0708	+4.705	+1.781	+1.798	+0.0670	+0.3807	−0.0882	+0.0726	+0.0003	+17.30	+5.658	+3.294	+0.7426
0	+27.75	0	+0.707	0	−16.12	0	+0.707	0	+2.0020	+0.5000	0	0	−58.57	0	0	+1.0000
$\phi_k = 50$																
50	+6.074	0	+1.416	−0.0253	−5.122	0	−1.462	+0.0088	+0.1028	0	−0.0139	−0.0027	−11.38	0	−6.094	−0.9520
40	+3.738	+2.900	+0.542	−0.0469	−3.478	−2.507	−0.713	+0.0302	+0.0147	−0.0398	−0.0269	−0.0027	−8.59	−5.742	−4.363	−0.0339
30	−2.427	+3.394	−1.371	−0.0900	+1.103	−3.275	+1.025	+0.0759	−0.1664	−0.0015	−0.0410	−0.0022	+0.92	−8.235	−0.093	+0.2533
20	−8.536	+0.212	−2.512	−0.1065	+6.245	−1.164	+2.385	+0.1024	−0.1868	−0.1144	−0.0079	−0.0010	+15.24	−3.756	+3.776	+0.5154
10	−4.760	−4.221	−1.261	−0.0672	+4.596	+2.387	+1.966	+0.0752	+0.3719	−0.1018	+0.0671	+0.0003	+14.96	+6.079	+3.305	+0.7497
0	+26.68	0	+0.766	0	−18.51	0	+0.643	0	+2.065	+0.5000	0	0	−56.51	0	0	+1.0000

Table A6 $r/t = 100$ and $r/L = 0.6$

ϕ	Vertical edge load				Horizontal edge load				Shear edge load				Edge moment			
	N_x	$N_{\phi x}$	N_ϕ	M_ϕ	N_x	$N_{\phi x}$	N_ϕ	M_ϕ	N_x	$N_{\phi x}$	N_ϕ	M_ϕ	N_x	$N_{\phi x}$	N_ϕ	M_ϕ
$\phi_k = 30$																
30	+0.690	0	−2.477	−0.1646	−2.375	0	+0.609	+0.0636	−0.4130	0	−0.1015	−0.0031	−17.24	0	−2.909	+0.3912
20	−3.149	−0.438	−2.432	−0.1447	−0.079	−0.8580	+0.895	+0.0605	−0.2349	−0.1448	−0.0424	−0.0010	−0.990	−6.327	−0.707	+0.4746
10	−5.088	−3.429	−1.334	−0.0837	+2.720	+0.0560	+1.160	+0.0431	+0.5284	−0.1530	+0.0739	−0.0001	+16.89	+0.297	+1.631	+0.6844
0	+26.76	0	+0.500	0	−6.379	0	+0.866	0	+2.481	+0.6000	0	0	−50.48	0	0	+1.0000
$\phi_k = 35$																
35	+5.530	0	+0.999	−0.1294	−4.187	0	+0.085	+0.0577	−0.2245	0	−0.0937	−0.0036	−21.54	0	−4.900	+0.2286
30	+3.964	+1.371	−1.194	−0.1291	−3.368	−1.072	+0.231	+0.0589	−0.2258	−0.0617	−0.0849	−0.0034	−17.56	−5.539	−4.097	+0.2534
20	−5.521	+1.142	−2.096	−0.1187	+1.835	−1.625	+1.098	+0.0628	−0.1434	−0.1738	−0.0163	−0.0014	+8.78	−8.709	+0.609	+0.4257
10	−7.862	−3.492	−1.452	−0.0731	+4.617	+0.5579	+1.438	+0.0477	+0.4825	−0.1215	+0.0821	+0.0005	+25.70	+2.778	+2.864	+0.6789
0	+31.29	0	+0.574	0	−11.74	0	+0.891	0	+2.433	+0.6000	0	0	−67.87	0	0	+1.0000
$\phi_k = 40$																
40	+8.029	0	+0.412	−0.0781	−5.611	0	+0.632	+0.0378	−0.0378	0	−0.0653	−0.0034	−20.91	0	−6.413	+0.0741
30	+2.640	+3.389	−0.660	−0.0907	−2.518	−2.498	+0.140	+0.0500	−0.1275	−0.0385	−0.0561	−0.0020	−10.34	−9.508	−3.449	+0.1748
20	−8.598	+1.754	−2.347	−0.1015	+4.328	−2.036	+1.563	+0.0673	−0.1799	−0.1354	−0.0078	−0.0012	+16.02	−8.383	+2.316	+0.4147
10	−8.783	−4.118	−1.651	−0.0674	+5.859	+1.341	+1.421	+0.0538	+0.4136	−0.1192	+0.0787	+0.0004	+25.53	+5.071	−3.686	+0.6842
0	+34.71	0	+0.643	0	−17.02	0	+0.766	0	+2.5580	+0.6000	0	0	−75.81	0	0	+1.0000
$\phi_k = 45$																
45	+7.904	0	+1.346	−0.0293	−5.855	0	−1.258	+0.0115	+0.1009	0	−0.0267	−0.0026	−15.56	0	−6.655	−0.0364
40	+6.935	+2.078	+1.051	−0.0352	−5.259	−1.550	−1.038	+0.0166	+0.0669	+0.0245	−0.0302	−0.0025	−14.36	−4.156	−6.057	−0.0122
30	−0.196	+4.195	−0.884	−0.0712	−0.727	−3.373	−0.459	+0.0490	−0.1460	+0.0082	−0.0439	−0.0022	−3.96	−9.685	−1.817	+0.1633
20	−9.841	+1.352	−2.657	−0.0959	+6.025	−1.905	+2.094	+0.0775	−0.2510	−0.1174	−0.0148	−0.0010	+16.43	−6.537	+3.383	+0.4285
10	−7.893	−4.563	−1.682	−0.0659	+6.150	+2.089	+2.020	+0.0623	+0.3813	−0.1339	+0.0714	+0.0004	+22.53	+6.087	+3.873	+0.6947
0	+34.56	0	+0.707	0	−20.64	0	+0.707	0	+2.6590	+0.6000	0	0	−74.68	0	0	+1.0000
$\phi_k = 50$																
50	+6.029	0 + 2.955	+1.647	+0.0038	−4.847	0	−1.571	−0.0114	+0.1678	0	+0.0073	−0.0015	−8.26	0	−5.666	−0.0950
40	+4.096	+3.764	+0.283	−0.0192	−3.656	−2.446	−0.851	+0.0100	+0.0523	+0.0699	−0.0151	−0.0017	−8.00	−4.483	−4.344	−0.0164
30	−1.803	+0.659	−1.264	−0.0627	+0.458	−3.469	−0.919	+0.0574	−0.2030	+0.0305	−0.0491	−0.0018	−1.93	−7.812	−0.623	+0.1835
20	−9.145		−2.736	−0.0932	+6.530	−1.539	+2.469	+0.0897	−0.2862	−0.1225	−0.0245	−0.0010	+13.86	−5.044	+3.549	+0.4425

(continued)

Table A6 (continued)

ϕ	Vertical edge load				Horizontal edge load				Shear edge load				Edge moment			
	N_x	$N_{\phi x}$	N_ϕ	M_ϕ	N_x	$N_{\phi x}$	N_ϕ	M_ϕ	N_x	$N_{\phi x}$	N_ϕ	M_ϕ	N_x	$N_{\phi x}$	N_ϕ	M_ϕ
10	−6.445	−4.528	−1.501	−0.0624	+6.010	+2.570	+2.141	+0.0696	+0.3899	−0.1463	+0.0675	+0.0003	+20.05	+5.942	+3.658	+0.6986
0	+31.83	0	+0.766	0	−22.64	0	+0.643	0	+2.6880	+0.6000	0	0	−70.37	0	0	+1.0000

Table A7 r/t = 200 and r/L = 0.1

φ	Vertical edge load				Horizontal edge load				Shear edge load				Edge moment			
	N_x	$N_{\phi x}$	N_ϕ	M_ϕ	N_x	$N_{\phi x}$	N_ϕ	M_ϕ	N_x	$N_{\phi x}$	N_ϕ	M_ϕ	N_x	$N_{\phi x}$	N_ϕ	M_ϕ
$\phi_k = 30$																
30	−5.388	0	−3.482	−0.3278	−0.0129	0	+0.9979	+0.1310	−0.0905	0	−0.0207	−0.0009	−0.1005	0	−0.0164	+0.9730
20	−3.600	−2.628	−2.666	−0.2779	+0.0014	−0.0043	+0.9845	+0.1162	−0.0397	−0.0403	−0.0080	−0.0006	+0.0123	−0.0332	−0.0024	+0.9761
10	+1.787	−3.290	−0.781	−0.1511	+0.0164	+0.0020	+0.9410	+0.0722	+0.1117	−0.0251	+0.0142	−0.0002	+0.1340	+0.0177	+0.0099	+0.9852
0	+10.83	0	+0.500	0	−0.0447	0	+0.8660	0	+0.3604	+1.000	0	0	−0.3807	0	0	+1.0000
$\phi_k = 35$																
35	−3.373	0	−2.947	−0.3692	−0.0349	0	+0.9926	+0.1751	−0.0773	0	−0.0243	−0.0015	−0.2044	0	−0.0435	+0.9623
30	−3.175	−0.907	−2.810	−0.3584	−0.0272	−0.0089	+0.9901	+0.1715	−0.0693	−0.0205	−0.0212	−0.0014	−0.1601	−0.0520	−0.0356	+0.9632
20	−1.566	−2.282	−1.814	−0.2774	+0.0192	−0.0121	+0.9675	+0.1424	−0.0055	−0.0439	−0.0015	−0.0008	+0.1134	−0.0710	+0.0090	+0.9700
10	+1.763	−2.310	−0.337	−0.1441	+0.0378	+0.0075	+0.9107	+0.0832	+0.1211	−0.0150	+0.0185	−0.0002	+0.2305	+0.0464	+0.0256	+0.9823
0	+6.999	0	+0.574	0	−0.1053	0	+0.8192	0	+0.3082	+0.1000	0	0	−0.6605	0	0	+1.0000
$\phi_k = 40$																
40	−2.193	0	−2.526	−0.4041	−0.0791	0	+0.9782	+0.2237	−0.0670	0	−0.0278	−0.0022	−0.3607	0	−0.1000	+0.9479
30	−1.834	−1.137	−2.168	−0.3677	−0.0303	−0.0341	+0.9742	+0.2095	−0.0458	−0.0329	−0.0178	−0.0018	−0.1407	−0.1563	−0.0487	+0.9521
20	−0.678	−1.860	−1.220	−0.2690	+0.0667	−0.0239	+0.9491	+0.1669	+0.0177	−0.0425	+0.0054	−0.0009	+0.3060	−0.1102	+0.0428	+0.9635
10	+1.483	−1.699	−0.048	−0.1334	+0.0718	+0.0226	+8.777	+0.0965	+0.1231	−0.0058	−0.0220	−0.0002	+0.3419	+0.1068	+0.0540	+0.9795
0	+4.960	0	+0.643	0	−0.2243	0	+0.7660	0	+0.2696	+0.1000	0	0	−1.0830	0	0	+1.0000
$\phi_k = 45$																
45	−1.415	0	−2.167	−0.4305	−0.1591	0	+0.9446	+0.2747	−0.0585	0	−0.0311	−0.0031	−0.5815	0	−0.2039	+0.9270
40	−1.384	−0.385	−2.106	−0.4226	−0.1386	−0.0418	+0.9469	+0.2713	−0.0549	−0.0127	−0.0288	−0.0030	−0.5076	−0.1526	−0.1817	+0.9284
30	−1.097	−1.081	−1.635	−0.3622	+0.0019	−0.0838	+0.9561	+0.2441	−0.0259	−0.0392	−0.0122	−0.0021	+0.0014	−0.3085	−0.0344	+0.9392
20	−0.350	−1.505	−0.798	−0.2550	+0.1623	−0.0353	+0.9371	+0.1892	+0.0326	−0.0387	+0.0118	−0.0010	+0.5987	−0.1323	+0.1135	+0.9566
10	+1.175	−1.323	+0.148	−0.1254	+0.1167	+0.0570	+0.8460	+0.1071	+0.1213	+0.0021	+0.0248	−0.0002	+0.4491	+0.2141	+0.0996	+0.9768
0	+3.887	0	+0.707	0	−0.4371	0	+0.7071	0	+0.2407	+0.1000	0	0	−1.6840	0	0	+1.0000
$\phi_k = 50$																
50	−0.838	0	−1.831	−0.4457	−0.2911	0	+0.8743	+0.3250	−0.0506	0	−0.0340	−0.0041	−0.8725	0	−0.3803	+0.8948
40	−0.842	−0.461	−1.678	−0.4195	−0.1746	−0.1378	+0.9021	+0.3125	−0.0408	−0.0260	−0.0262	−0.0036	−0.5294	−0.4145	−0.2512	+0.9030
30	−0.759	−0.908	−1.225	−0.3455	+0.1012	−0.1615	+0.9516	+0.2745	−0.0107	−0.0411	−0.0176	−0.0024	+0.2947	−0.4901	+0.0344	+0.9238
20	−0.314	−1.227	−0.517	−0.2369	+0.3240	−0.0358	+0.9439	+0.2092	+0.0369	−0.0337	+0.0174	−0.0010	+0.9856	−0.1131	+0.2362	+0.9497
10	+0.907	−1.111	+0.281	−0.1142	+0.1645	+0.1245	+0.8204	+0.1167	+0.1175	+0.0087	+0.0271	−0.0001	+0.5268	+0.3830	+0.1664	+0.9741
0	+3.402	0	+0.766	0	−0.7874	0	+0.6428	0	+0.2199	+0.1000	0	0	−2.4870	0	0	+1.0000

Table A8 r/t = 200 and r/L = 0.2

φ	Vertical edge load				Horizontal edge load				Shear edge load				Edge moment			
	N_x	$N_{\phi x}$	N_ϕ	M_ϕ	N_x	$N_{\phi x}$	N_ϕ	M_ϕ	N_x	$N_{\phi x}$	N_ϕ	M_ϕ	N_x	$N_{\phi x}$	N_ϕ	M_ϕ
$\phi_k = 30$																
30	−4.804	0	−3.394	−0.3054	−0.2444	0	+0.9630	+0.1216	−0.1770	0	−0.0407	−0.0017	−1.972	0	−0.3043	+0.8904
20	−3.563	−2.414	−2.649	−0.2592	−0.0144	−0.0894	+0.9776	+0.1084	−0.0792	−0.0792	−0.0160	−0.0012	−0.154	−0.7298	−0.0618	+0.99051
10	+1.170	−3.280	−0.836	−0.1412	+0.2619	−0.0028	+0.9629	+0.0679	+0.2192	−0.0502	+0.0279	−0.0003	+2.142	+0.0427	+0.1925	+0.9435
0	+12.12	0	+0.500	0	−0.5594	0	+0.8660	0	+0.7297	+0.2000	0	0	−4.639	0	0	+1.0000
$\phi_k = 35$																
35	−2.269	0	−2.720	−0.3304	−0.5645	0	+0.8832	+0.1560	−0.1455	0	−0.0466	−0.0027	−3.344	0	−0.7035	+0.8373
30	−2.290	−0.624	−2.624	−0.3211	−0.4523	−0.1446	+0.9000	+0.1531	−0.1313	−0.0386	−0.0410	−0.0025	−2.696	−0.8582	−0.5812	+0.8425
20	−2.035	−1.854	−1.855	−0.2507	+0.2416	−0.2183	+0.9868	+0.1291	−0.0150	−0.0842	−0.0033	−0.0014	+1.373	−1.3190	+0.1200	+0.8787
10	+0.648	−2.423	−0.474	−0.1313	+0.5770	−0.0604	+0.9767	+0.0783	+0.2331	−0.0309	+0.0358	−0.0002	+3.519	+0.3350	+0.4303	+0.9335
0	+9.620	0	+0.574	0	−1.373	0	+0.8191	0	+0.6376	+0.2000	0	0	−8.444	0	0	+1.0000
$\phi_k = 40$																
40	−0.324	0	−2.021	−0.3363	−1.1330	0	+0.6921	+0.1847	−0.1144	0	−0.0503	−0.0037	−5.147	0	−1.4220	+0.7549
30	−1.057	−0.319	−1.919	−0.3098	−0.4724	−0.4965	+0.8324	+0.1760	−0.0838	−0.0572	−0.0331	−0.0030	−2.213	−2.270	−0.7127	+0.7838
20	−2.100	−1.229	−1.425	−0.2331	+0.8653	−0.3869	+1.0670	+0.1458	+0.0202	−0.0785	+0.0084	−0.0015	+3.880	−1.816	+0.5816	+0.8519
10	−0.178	−2.090	−0.327	−0.1202	+1.0210	+0.2424	+1.0360	+0.0873	+0.2295	−0.0158	+0.0411	−0.0002	+4.849	+1.085	+0.8003	+0.9255
0	+9.537	0	+0.643	0	−2.8510	0	+0.7660	0	+0.5858	+0.2000	0	0	−13.55	0	0	+1.0000
$\phi_k = 45$																
45	+1.358	0	−1.210	−0.3140	−1.967	0	+0.3169	+0.1973	−0.0793	0	−0.0494	−0.0047	−7.076	0	−2.500	+0.6266
40	+1.046	+0.344	−1.254	−0.3103	−1.725	−0.5171	+0.3877	+0.1967	−0.0769	−0.0215	−0.0462	−0.0045	−6.231	−1.862	−2.231	+0.6378
30	−1.033	−0.270	−1.469	−0.2789	−0.050	−0.6113	+0.8459	+0.1884	−0.0516	−0.0642	−0.0223	−0.0032	−0.316	−2.536	−0.5827	+0.7167
20	−3.035	−0.788	−1.324	−0.2104	+1.916	−0.5110	+1.2810	+0.1588	+0.0289	−0.0681	+0.0165	−0.0014	+6.897	−1.941	+1.367	+0.8267
10	−0.885	−2.169	−0.327	−0.1087	+1.486	+0.6094	+1.1590	+0.0955	+0.2164	−0.0071	+0.0436	−0.0001	+5.708	+2.201	+1.273	+0.9200
0	+10.91	0	+0.707	0	−5.085	0	+0.7071	0	+0.5724	+0.2000	0	0	−19.32	0	0	+1.0000
$\phi_k = 50$																
50	+2.686	0	−0.312	−0.2588	−2.918	0	−0.2665	+0.1807	−0.0395	0	−0.0419	−0.0052	−8.502	0	−3.7700	+0.4461
40	+1.320	+1.218	−0.671	−0.2562	−1.798	−1.391	+0.1445	+0.1856	−0.0445	−0.0226	−0.0351	−0.0046	−5.350	−4.060	−2.5230	+0.5106
30	−1.860	+1.105	−1.355	−0.2385	+0.904	−1.675	+1.0460	+0.1889	−0.0420	−0.0476	−0.0140	−0.0031	+2.433	−4.983	+0.2750	+0.6671
20	−4.145	−0.669	−1.455	−0.1867	+3.198	−0.4671	+1.6490	+0.1655	+0.0168	−0.0585	+0.0187	−0.0013	+9.459	−1.527	+2.3360	+0.8314
10	−1.190	−2.490	−0.395	−0.0979	−1.779	+1.160	+1.3320	+0.0992	+0.2016	−0.0062	+0.0428	−0.0001	+5.734	+3.392	+1.7370	+0.9393
0	+12.58	0	+0.766	0	−7.777	0	+0.6428	0	+0.5914	+0.2000	0	0	−24.30	0	0	+1.0000

Table A9 $r/t = 200$ and $r/L = 0.3$

ϕ	Vertical edge load				Horizontal edge load				Shear edge load				Edge moment			
	N_x	$N_{\phi x}$	N_ϕ	M_ϕ	N_x	$N_{\phi x}$	N_ϕ	M_ϕ	N_x	$N_{\phi x}$	N_ϕ	M_ϕ	N_x	$N_{\phi x}$	N_ϕ	M_ϕ
$\phi_k = 30$																
30	−2.875	0	−3.097	−0.2688	−1.018	0	+0.8433	+0.1065	−0.2481	0	−0.0585	−0.0023	−8.01	0	−1.251	+0.7583
20	−3.544	−1.727	−2.595	−0.2300	−0.025	−0.3675	+0.9558	+0.0961	−0.1187	−0.0759	−0.0236	−0.0016	−0.32	−2.906	−0.242	+0.7953
10	−1.006	−3.351	−1.021	−0.1270	+1.142	+0.0251	+1.0380	+0.0619	+0.3098	−0.1126	+0.0402	−0.0003	+9.20	+0.158	+0.787	+0.8828
0	+17.18	0	+0.500	0	−2.618	0	+0.8660	0	+1.1380	+0.3000	0	0	−21.66	0	0	+1.0000
$\phi_k = 35$																
35	+1.157	0	−2.003	−0.2639	−3.540	0	+0.5377	+0.1233	−0.1777	0	−0.0616	−0.0033	−12.73	0	−2.708	+0.6320 + 0.6464
30	+0.434	+0.250	−2.034	−0.2580	−1.763	−0.5643	+0.6157	+0.1220	−0.1650	−0.0475	−0.0548	−0.0031	−10.23	−3.260	−2.236	+0.7427
20	−3.630	−0.571	−1.989	−0.2098	+1.001	−0.8379	+1.0510	+0.1087	−0.0418	−0.1115	−0.0066	−0.0017	+5.57	−4.934	+0.474	+0.8689
10	−2.937	−2.907	−0.903	−0.1149	+2.318	+0.2927	+1.1840	+0.0698	+0.3091	−0.0522	+0.0487	−0.0002	+13.92	+1.650	+1.650	+1.0000
0	+18.66	0	+0.574	0	−5.780	0	+0.8192	0	+1.0580	+0.3000	0	0	−35.32	0	0	+1.0000
$\phi_k = 40$																
40	+4.491	0	−0.695	−0.2230	−4.689	0	−0.0553	+0.1198	−0.0958	0	−0.0550	−0.0039	−16.80	0	−4.723	+0.4470
30	+0.927	+1.787	−1.263	−0.2170	−1.599	−1.680	+0.4612	+0.1225	−0.0958	−0.0529	−0.0397	−0.0032	−7.300 + 12.97	−7.430	−2.380	+0.5253
20	−5.904	+0.374	−1.979	−0.1836	+2.992	−1.296	+1.3780	+0.1168	−0.0298	−0.0936	+0.0040	−0.0016	+16.75	−5.929	+1.932	+0.6997
10	−4.643	−3.248	−1.057	−0.1038	+3.576	+0.8937	+1.4510	+0.0774	+0.2786	−0.0416	+0.0506	−0.0001		+3.932	+2.673	+0.8621
0	+22.55	0	+0.643	0	−10.290	0	+0.7660	0	+1.0690	+0.3000	0	0	−48.57	0	0	+1.0000
$\phi_k = 45$																
45	+6.463	0	+0.608	−0.1511	−5.443	0	−0.8614	+0.0899	−0.0114	0	−0.0370	−0.0038	−17.92	0	−6.551	+0.2280
40	+5.541	+1.688	+0.369	−0.1543	−4.617	−1.379	−0.6652	+0.0937	−0.0214	−0.0039	−0.0364	−0.0037	−15.90	−4.718	−5.866	+0.2542
30	−0.843	+3.190	−1.138	−0.1700	−0.203	−2.859	+0.6257	+0.1158	−0.0772	−0.0304	−0.0273 + 0.0042	−0.0028	−1.340	−9.994	−1.292	+0.4354
20	−8.160	+0.555	−2.324	−0.1610	+5.220	−1.401	+1.9290	+0.1250	−0.0627	−0.0765	+0.0471	−0.0013	+17.97	−5.233	+3.575	+0.6795
10	−5.055	−3.891	−1.245	−0.0955	+4.276	+1.728	+1.7620	+0.0858	+0.2474	−0.0456		−0.0000	+16.41	+6.050	+3.437	+0.8640
0	+25.47	0	+0.707	0	−14.77	0	+0.7071	0	+1.1340	+0.3000	0	0	−56.01	0	0	+1.0000
$\phi_k = 50$																
50	+6.626	0	+1.510	−0.0742	−5.740	0	−1.603	+0.0441	+0.0544	0	−0.0135	−0.0031	−15.39	0	−7.343	+0.0368
40	+3.900	+3.127	+0.564	−0.0969	−3.696	−2.769	−0.7707	+0.0676	+0.0026	+0.0201	−0.0201	−0.0029	−10.44	−7.529	−5.048	+0.1442
30	−2.878	+3.516	−1.463	−0.1393	+1.552	−3.453	+1.111	+0.1158	−0.1000	−0.0070	−0.0254	−0.0023	+3.25	−9.817	+0.304	+0.4060
20	−8.823	+0.094	−2.598	+0.1467	+6.637	−1.072	+2.491	+0.1375	−0.1008	−0.0715	−0.0025	−0.0011	+18.60	−3.576	+4.612	+0.6832
10	−4.273	−4.274	−1.262	−0.0891	+4.214	+2.483	+1.987	+0.0953	+0.2354	−0.0563	+0.0424	−0.0000	+14.04	+7.094	+3.666	+0.8700
0	+25.53	0	+0.766	0	−17.75	0	+0.6428	0	+1.1930	+0.3000	0	0	−56.30	0	0	+1.0000

Table A10 r/t = 200 and r/L = 0.4

ϕ	Vertical edge load				Horizontal edge load				Shear edge load				Edge moment			
	N_x	$N_{\phi x}$	N_ϕ	M_ϕ	N_x	$N_{\phi x}$	N_ϕ	M_ϕ	N_x	$N_{\phi x}$	N_ϕ	M_ϕ	N_x	$N_{\phi x}$	N_ϕ	M_ϕ
$\phi_k = 30$																
30	+1.124	0	−2.479	−0.2211	−2.625	0	+0.5934	+0.0868	−0.2847	0	−0.0710	−0.0026	−20.19	0	−3.173	+0.5913
20	−3.463	−0.295	−2.481	−0.1937	−0.066	−0.9430	+0.9092	+0.0810	−0.1583	−0.1339	−0.0302	−0.0018	−0.87	−7.343	−0.619	+0.6619
10	−5.528	−3.468	−1.407	−0.1117	+2.974	+0.0700	+1.1940	+0.0553	+0.3622	−0.1029	+0.0493	−0.0003	+23.57	+0.435	+1.999	+0.8155
0	+27.68	0	+0.500	0	−6.899	0	+0.8660	0	+1.6340	+0.4000	0	0	−56.52	0	0	+1.0000
$\phi_k = 35$																
35	+6.873	0	−0.795	−0.1830	−4.932	0	−0.0417	+0.0837	−0.1462	0	−0.0635	−0.0032	−27.50	0	−5.935	+0.3911
30	+5.012	+1.712	−1.039	−0.1820	−3.951	−1.261	−0.1374	+0.0848	−0.1482	−0.0403	−0.0577	−0.0030	−22.24	−7.055	−4.913	+0.4179
20	−6.203	+1.614	−2.206	−0.1647	−2.208	−1.889	+1.153	+0.0862	−0.0986	−0.152	−0.0124	−0.0017	+11.73	−10.85	+1.001	+0.5937
10	−9.087	−3.682	−1.633	−0.1001	+5.303	+0.6597	+1.536	+0.0621	+0.3204	−0.0822	+0.0539	−0.0001	+31.27	+3.583	+3.651	+0.8049
0	+34.19	0	+0.574	0	−13.36	0	+0.8192	0	+1.639	+0.4000	0	0	−80.78	0	0	+1.0000
$\phi_k = 40$																
40	+11.34	0	+0.768	−0.1216	−7.637	0	−0.8602	+0.0620	+0.0319	0	−0.0448	−0.0032	−32.54	0	−8.301	+0.2027
30	+3.392	+4.247	−0.465	−0.1293	−2.992	−3.037	+0.0122	+0.0723	−0.0781	−0.0171	−0.0365	−0.0026	−13.32	−12.80	−4.292	+0.2888
20	−10.20	+2.361	−2.620	−0.1409	+5.333	−2.419	+1.7310	+0.0915	−0.1367	−0.0848	−0.0086	−0.0014	+21.73	−10.76	+3.315	+0.5642
10	−10.34	−4.580	−1.955	−0.0931	+6.841	+1.629	+1.9570	+0.0704	+0.2620	−0.0845	+0.0492	−0.0000	+31.74	+6.881	+4.858	+0.8083
0	+39.09	0	+0.643	0	−19.78	0	+0.7660	0	+1.7440	+0.4000	0	0	−92.98	0	0	+1.0000
$\phi_k = 45$																
45	+9.787	0	+1.934	−0.0436	−7.229	0	−1.692	+0.0199	+0.0957	0	−0.0099	−0.0065	−22.45	0	−8.938	−0.0169
40	+8.573	+2.572	+1.567	−0.0513	−6.456	−1.911	−1.419	+0.0264	+0.0691	+0.0238	−0.0133	−0.0043	−20.42	−5.970	−8.074	+0.0165
30	−0.227	+5.188	−0.825	−0.0982	−0.709	−4.101	+0.4128	+0.0678	−0.0987	+0.0199	−0.0292	−0.0020	−4.080	−13.43	−2.116	+0.2506
20	−11.70	+1.769	−3.045	−0.1296	+7.382	−2.215	+2.378	+0.1027	−0.1953	−0.0728	−0.0161	−0.0012	+23.35	−8.260	+4.819	+0.5714
10	−9.060	−5.153	−2.000	−0.0887	+7.017	+2.520	+2.255	+0.0800	+0.2396	−0.0981	+0.0429	−0.0000	+27.40	+8.293	+5.115	+0.8168
0	+38.68	0	+0.707	0	−23.69	0	+0.7071	0	+1.8260	+0.4000	0	0	−91.16	0	0	+1.0000
$\phi_k = 50$																
50	+7.414	0	+2.207	+0.0042	−6.022	0	−2.055	−0.0134	+0.1395	0	+0.0147	−0.0013	−13.14	0	−7.852	−0.1157
40	+4.947	+3.616	+1.124	−0.0267	−4.389	−3.008	−1.161	+0.0151	+0.0505	+0.0598	−0.0042	−0.0015	−11.30	−6.932	−5.801	−0.0061
30	−2.266	+4.540	−1.356	−0.0910	+0.8355	−4.146	+0.9927	+0.0776	−0.1446	+0.0347	−0.0352	−0.0017	−1.18	−10.95	−0.390	+0.2725
20	−10.69	−0.849	−3.073	−0.1245	+7.857	−1.711	+2.811	+0.1179	−0.2195	−0.0793	−0.0240	−0.0011	+19.84	−6.154	+5.079	+0.5892
10	−7.142	−5.081	−1.774	−0.0835	+6.637	+3.045	+2.378	+0.0892	+0.2513	−0.1076	+0.0402	−0.0001	+23.56	+7.944	+4.801	+0.8219
0	+35.00	0	+0.766	0	−25.44	0	+0.6428	0	+1.8400	+0.4000	0	0	−84.42	0	0	+1.0000

Table A11 $r/t = 200$ and $r/L = 0.5$

ϕ	Vertical edge load				Horizontal edge load				Shear edge load				Edge moment			
	N_x	$N_{\phi x}$	N_ϕ	M_ϕ	N_x	$N_{\phi x}$	N_ϕ	M_ϕ	N_x	$N_{\phi x}$	N_ϕ	M_ϕ	N_x	$N_{\phi x}$	N_ϕ	M_ϕ
$\phi_k = 30$																
30	+6.908	0	−1.578	−0.1695	−4.932	0	+0.2325	+0.0656	−0.2715	0	−0.0759	−0.0026	−36.86	0	−5.858	+0.4164
20	−3.091	+1.791	−2.323	−0.1564	−2.320	−1.780	+0.8452	+0.0654	−0.1974	−0.1375	−0.0355	−0.0018	−2.000	−13.49	−1.174	+0.5258
10	−12.18	−3.626	−1.972	−0.0957	+5.663	+0.128	+1.4220	+0.0492	+0.3600	−0.1314	+0.0537	−0.0002	+43.95	+0.71	+3.711	+0.7505
0	+43.35	0	+0.500	0	−13.29	0	−0.8660	0	+2.2530	+0.5000	0	0	−107.5		0	+1.0000
$\phi_k = 35$																
35	+12.92	0	+0.520	−0.1106	−7.760	0	−0.6687	+0.0486	+0.0566	0	−0.0528	−0.0027	−41.17	0	−9.181	+0.1851
30	+10.14	+3.193	+0.037	−0.1151	−6.371	−1.952	−0.3803	+0.0518	−0.0847	−0.0181	−0.0505	−0.0025	−33.79	−10.62	−7.611	+0.2222
20	−8.763	+4.022	−2.425	−0.1266	+3.367	−3.036	+1.253	+0.0670	−0.1818	−0.0966	−0.0204	−0.0015	+16.53	−16.90	+1.429	+0.4668
10	−16.09	−4.484	−2.447	−0.0897	+8.692	+1.030	+1.926	+0.0564	+0.2700	−0.1200	+0.0516	−0.0001	+50.25	+5.26	+5.747	+0.7504
0	+56.21	0	+0.574	0	−22.16	0	+0.8192	0	+2.3740	+0.5000	0	0	−132.8		0	+1.0000
$\phi_k = 40$																
40	+13.93	0	+1.963	−0.0443	−8.898	0	−1.520	+0.0185	+0.1093	0	−0.0175	−0.0020	−32.96	0	−10.30	−0.0072
30	+5.361	+5.888	+0.146	−0.0720	−4.097	−3.897	−0.3259	+0.0398	−0.0475	+0.0298	−0.0294	−0.0019	−17.70	−15.29	−5.586	+0.1389
20	−12.79	+3.938	−3.042	−0.1129	+6.622	−3.303	+1.951	+0.0737	−0.2613	−0.0652	−0.0242	−0.0012	+24.15	−14.33	+3.905	+0.4658
10	−15.31	−5.476	−2.659	−0.0866	+9.721	+2.099	+2.350	+0.0645	+0.2136	−0.1319	+0.0433	−0.0000	+44.95	+8.13	+6.422	+0.7611
0	+53.53	0	+0.643	0	−28.17	0	+0.7660	0	+2.5090	+0.5000	0	0	−133.2		0	+1.0000
$\phi_k = 45$																
45	+10.22	0	+2.435	+0.0045	−7.196	0	−1.959	−0.0106	+0.1830	0	+0.0132	−0.0011	−18.31	0	−8.981	−0.1089
40	+9.385	+2.771	+2.035	−0.0052	−6.757	−1.967	−1.675	−0.0034	+0.1446	+0.0466	+0.0066	−0.0012	−17.78	−4.98	−8.263	−0.0771
30	+0.957	+5.868	−0.565	−0.0616	−1.640	−4.438	+0.229	+0.0431	−0.1100	−0.0637	−0.0303	−0.0014	−8.900	−13.02	−2.879	+0.1523
20	−12.61	+2.686	−3.298	−0.1102	+7.723	−2.850	+2.505	+0.0881	−0.3155	−0.0676	−0.0341	−0.0011	+20.73	−10.69	+4.696	+0.4876
10	−12.52	−5.653	−2.457	−0.0832	+9.498	+2.791	+2.554	+0.0750	+0.2211	−0.1467	+0.0389	−0.0001	+38.31	+8.33	+6.068	+0.7681
0	+48.96	0	+0.707	0	−30.94	0	+0.707	0	+2.5390	+0.5000	0	0	−123.1		0	+1.0000
$\phi_k = 50$																
50	+5.720	0	+2.151	+0.0279	−4.333	0	−1.931	−0.0295	+0.1739	0	+0.0295	+0.0002	−4.66	0	−6.575	−0.1404
40	+5.105	+3.070	+1.265	−0.0015	−4.431	−2.438	−1.238	−0.0037	+0.0786	+0.0777	+0.0054	−0.0007	−9.77	−3.60	−5.629	−0.0518
30	−0.025	+4.652	−1.014	−0.0656	−1.252	−4.161	−0.6846	+0.0569	−0.1578	+0.0593	−0.0387	−0.0013	−10.31	−9.96	−1.682	+0.1945
20	−10.40	+1.897	−3.112	−0.1093	+7.442	−2.639	+2.813	+0.1043	−0.2949	−0.0794	−0.0364	−0.0011	+16.18	−9.65	+4.675	+0.5157
10	−12.65	−5.039	−2.643	−0.0852	+11.50	+3.115	+3.155	+0.0913	+0.2141	−0.1221	+0.0366	−0.0000	+41.78	+9.30	+7.575	+0.8127
0	+42.10	0	+0.766	0	−31.89	0	+0.6428	0	+2.4410	+0.5000	0	0	−116.1		0	+1.0000

Table A12 r/t = 200 and r/L = 0.6

φ	Vertical edge load				Horizontal edge load				Shear edge load				Edge moment			
	N_x	$N_{\phi x}$	N_ϕ	M_ϕ	N_x	$N_{\phi x}$	N_ϕ	M_ϕ	N_x	$N_{\phi x}$	N_ϕ	M_ϕ	N_x	$N_{\phi x}$	N_ϕ	M_ϕ
$\phi_k = 30$																
30	+13.41	0	−0.550	−0.1215	−7.494	0	−0.177	+0.0459	−0.2086	0	−0.0732	−0.0024	−53.95	0	−8.721	+0.2589
20	−2.95	+4.174	−2.106	−0.1221	−0.317	−2.725	+0.756	+0.0512	−0.2355	−0.1233	−0.0394	−0.0016	−4.01	−20.04	−1.828	+0.4042
10	−19.88	−3.751	−2.624	−0.0867	+8.756	+0.171	+1.682	+0.0443	+0.3019	−0.1613	+0.0532	−0.0001	+66.27	+0.71	+5.559	+0.6931
0	+61.86	0	+0.500	0	−20.84	0	+0.866	0	+3.003	+0.6000	0	0	−166.4	0	0	+1.0000
$\phi_k = 35$																
35	+17.48	0	+1.581	−0.0576	−9.794	0	−1.155	+0.0231	+0.0614	0	−0.0363	−0.0020	−48.68	0	−11.31	+0.0433
30	+13.78	+4.442	+0.942	−0.0648	−8.004	−2.520	−0.794	+0.0279	+0.0019	+0.0113	−0.0381	−0.0020	−40.77	−12.62	−9.488	+0.0867
20	−10.37	+6.000	−2.570	−0.0989	+4.010	−3.973	+1.313	+0.0531	−0.2759	−0.0663	−0.0294	−0.0012	+17.11	−21.22	+1.522	+0.3710
10	−22.33	−5.090	−3.136	−0.0827	+11.70	+1.270	+2.252	+0.0523	+0.1839	−0.1613	+0.0454	−0.0000	+66.22	+5.83	+7.338	+0.7029
0	+68.85	0	+0.574	0	−30.32	0	+0.819	0	+3.2040	+0.6000	0	0	−180.9	0	0	+1.000
$\phi_k = 40$																
40	+14.46	0	+2.530	−0.0052	−8.900	0	−1.788	−0.0029	+0.2096	0	+0.0030	−0.0012	−29.22	0	−10.53	−0.0850
30	+6.70	+6.504	+0.519	−0.0397	−4.840	−4.143	−0.527	+0.0213	−0.0157	+0.0719	−0.0224	−0.0013	−20.21	−14.50	−6.191	+0.0578
20	−13.24	+4.992	−3.187	−0.0946	+6.608	−3.889	+2.003	+0.0630	−0.3733	−0.0458	−0.0384	−0.0010	+19.37	−16.37	+3.594	+0.3909
10	−19.32	−5.810	−3.114	−0.0812	+12.10	+2.231	+2.601	+0.0616	+0.1569	−0.1797	+0.0373	−0.0001	+56.09	+7.56	+7.285	+0.7139
0	+65.23	0	+0.643	0	−35.15	0	+0.766	0	+3.2970	+0.6000	0	0	−169.2	0	0	+1.0000
$\phi_k = 45$																
45	+8.60	0	+2.397	+0.0217	−5.729	0	−1.858	−0.0205	+0.2347	0	+0.0286	−0.0004	−8.38	0	−7.554	−0.1241
40	+8.15	+2.319	+2.068	+0.0133	−5.622	−1.562	−1.636	−0.0140	+0.1924	+0.0605	+0.0199	−0.0005	−10.38	−2.49	−7.198	−0.0988
30	+2.68	+5.666	−0.311	−0.0418	−2.980	−4.165	+0.041	+0.0293	−0.1065	+0.0944	−0.0306	−0.0011	−15.36	−10.11	−3.686	+0.0939
20	−11.68	+3.500	−3.237	−0.0948	+6.764	−3.438	+2.416	+0.0759	−0.4030	−0.0596	−0.0476	−0.0011	+12.42	−13.03	+3.667	+0.4113
10	−16.02	−5.641	−2.752	−0.0772	+12.12	+2.709	+2.752	+0.0695	+0.1906	−0.1895	+0.0361	−0.0001	+50.54	+6.80	+6.668	+0.7160
0	+58.53	0	+0.707	0	−38.04	0	+0.707	0	+3.2680	+0.5000	0	0	−158.3	0	0	+1.000
$\phi_k = 50$																
50	+3.09	0	+1.709	+0.0284	−1.816	0	−1.467	−0.0273	+0.1854	0	+0.0382	+0.0002	+7.18	0	−3.972	−0.1083
40	+4.42	+1.981	+1.179	+0.0063	−3.693	−1.374	−1.120	−0.0084	+0.1051	+0.0872	+0.0118	−0.0003	−5.77	+1.40	−4.634	−0.0572
30	+2.57	+4.287	−0.638	−0.0476	−3.650	−3.722	+0.325	+0.0406	−0.1481	+0.0826	−0.0419	−0.0012	−20.51	−6.83	−3.347	+0.1206
20	−9.53	+2.775	−2.920	−0.0908	+6.372	−3.425	+2.581	+0.0862	−0.3819	−0.0757	−0.0497	−0.0012	+7.78	−12.83	+3.041	+0.4149
10	−14.34	−5.142	−2.383	−0.0706	+13.11	+2.955	+2.877	+0.0757	+0.2199	−0.1883	+0.0384	−0.0001	+51.00	+5.85	+6.371	+0.7134
0	+53.46	0	+0.766	0	−41.80	0	+0.643	0	+3.218	+0.6000	0	0	−156.6	0	0	+1.0000

Table B1 r/t = 100

φ	Vertical edge load			Horizontal edge load			Shear edge load			Edge moment		
	Δ_V	Δ_H	φ	Δ_V	Δ_H	φ	Δ_V	Δ_H	φ	Δ_V	Δ_H	φ
r/L = 0.1												
30	+12.53	+0.1445	+0.1108	−0.1445	−0.0585	−0.04556	+0.3401	+0.000480	+0.000363	−1.330	−0.5469	−0.5144
35	+6.265	+0.2508	+0.1420	−0.2508	−0.1231	−0.07108	+0.2178	+0.001006	+0.000544	−1.704	−0.8531	−0.5964
40	+3.611	+0.4009	+0.1752	−0.4009	−0.2325	−0.1039	+0.1443	+0.002009	+0.000822	−2.102	−1.247	−0.6749
45	+2.591	+0.5987	+0.2088	−0.5981	−0.4040	−0.1444	+0.1084	+0.003771	+0.001223	−2.505	−1.734	−0.7542
50	+2.043	+0.8434	+0.2410	−0.8434	−0.6558	−0.1925	+0.0792	+0.006516	+0.001776	−2.892	−2.312	−0.8290
r/L = 0.2												
30	+17.38	+2.031	+0.0982	−2.031	−0.8742	−0.04288	+0.6950	+0.00726	+0.000353	−18.86	−8.233	−0.4893
35	+12.87	+3.538	+0.1259	−3.538	−1.794	−0.06530	+0.4733	+ 0.02065	+0.000678	−24.18	−12.54	−0.5564
40	+12.51	+5.544	+0.1524	−5.544	−3.282	−0.09256	+0.3715	+0.04590	+0.001145	−29.25	−17.77	−0.6159
45	+13.47	+7.956	+0.1748	−7.956	−5.447	−0.1233	+0.3402	+0.08862	+0.001757	−33.56	−23.67	−0.6651
50	+14.54	+10.50	+0.1899	−10.50	−8.270	−0.1546	+0.3610	+0.1526	+0.002478	−36.46	−29.70	−0.6986
r/L = 0.3												
30	+35.24	+9.474	+0.0891	−9.474	−4.010	−0.03866	+1.187	+0.06357	+0.000536	−86.64	−37.58	−0.4499
35	+37.31	+15.65	+0.1092	−15.65	−7.890	−0.05606	+0.9583	+0.1474	+0.000923	−106.2	−54.50	−0.4965
40	+42.38	+22.99	+0.1245	−22.99	−13.59	−0.07595	+0.9561	+0.2931	+0.001413	−121.0	−73.83	−0.5288
45	+46.06	+30.12	+0.1317	−30.12	−20.69	−0.09384	+1.073	+0.5020	+0.001927	−127.9	−91.21	−0.5434
50	+45.75	+35.11	+0.1285	35.11	−27.93	−0.1070	+1.209	+0.7422	+0.002334	−124.9	−104.0	−0.5402
r/L = 0.4												
30	+75.06	+26.10	+0.07760	−26.10	−11.06	−0.03483	+1.940	+0.2256	+0.000632	−239.2	−104.0	−0.4049
35	+87.57	+40.75	+0.09046	−40.75	−20.25	−0.04716	+1.924	+0.5080	+0.000994	−277.7	−145.2	−0.4320
40	+96.68	+55.00	+0.09596	−55.00	−32.68	−0.05921	+2.189	+0.9067	+0.001384	−293.8	−181.4	−0.4417
45	+96.10	+64.69	+0.09256	−64.69	−44.95	−0.06746	+2.500	+1.372	+0.001641	−284.1	−207.1	−0.4348
50	+85.15	+67.32	+0.08360	−67.32	−54.74	−0.07216	+2.591	+1.731	+0.001732	−255.7	−220.8	−0.4228
r/L = 0.5												
30	+142.2	+54.14	+0.06634	−54.14	−22.98	−0.02900	+3.218	+0.5995	+0.000644	−497.6	−216.6	−0.3602
35	+163.2	+78.77	+0.07239	−78.77	−39.97	−0.03811	+3.588	+1.202	+0.000945	−542.9	−285.9	−0.3706

(continued)

Table B1 (continued)

ϕ	Vertical edge load			Horizontal edge load			Shear edge load			Edge moment		
	Δ_V	Δ_H	ϕ	Δ_V	Δ_H	ϕ	Δ_V	Δ_H	ϕ	Δ_V	Δ_H	ϕ
40	+167.7	+97.43	+0.07124	−97.43	−58.47	−0.04484	+4.141	+1.938	+0.001160	−535.2	−336.5	−0.3668
45	+152.4	+104.9	+0.06505	−104.9	−74.31	−0.04883	+4.418	+2.566	+0.001242	−487.9	−366.2	−0.3568
50	+127.5	+103.9	+0.05747	−103.9	−87.21	−0.05166	+4.246	+2.945	+0.001199	−431.0	−387.5	−0.3493
r/L = 0.6												
30	−238.5	+88.47	+0.05559	−88.47	−39.78	−0.02434	+5.112	+1.218	+0.000620	−864.6	−378.6	−0.3247
35	−257.7	+126.7	+0.05714	−126.7	−64.74	−0.03049	+5.976	+2.242	+0.000834	−888.7	−474.2	−0.3185
40	−245.4	+144.9	+0.05366	−144.9	−88.10	−0.03396	+6.629	+3.250	+0.000932	−834.6	−528.2	−0.3171
45	−212.3	+149.5	+0.14783	−149.5	−108.3	−0.03700	+6.601	+3.942	+0.000918	−743.9	−575.4	−0.3042
50	−175.8	+143.7	+0.04304	−147.9	−127.9	−0.03967	+6.678	+4.324	+0.000865	−669.4	−618.7	−0.3086

Table B2 r/t = 200

φ	Vertical edge load			Horizontal edge load			Shear edge load			Edge moment		
	Δ_V	Δ_H	ϕ	Δ_V	Δ_H	ϕ	Δ_V	Δ_H	ϕ	Δ_V	Δ_H	ϕ
r/L = 0.1												
30	+13.59	+0.5598	+0.10563	−0.5598	−0.3112	−0.04555	−0.3446	+0.001423	+0.000252	−5.070	−2.186	−0.5143
35	+7.732	+0.9838	+0.13813	−0.9838	−0.4916	−0.07100	−0.2228	+0.003352	+0.000438	−6.630	−3.408	−0.5958
40	+5.711	+1.576	+0.1719	−1.576	−0.9269	−0.1036	−0.1579	+0.007139	+0.000718	−8.250	−4.974	−0.6749
45	+5.186	+2.354	+0.2050	−2.354	−1.462	−0.1434	−0.1237	+0.01391	+0.001117	−9.842	−6.883	−0.7503
50	+5.265	+3.292	+0.2352	−3.292	−2.576	−0.1895	−0.1083	+0.02506	+0.001649	−11.29	−9.097	−0.8195
r/L = 0.2												
30	+32.02	+8.267	+0.0986	−8.267	−3.492	−0.04257	+0.7716	+0.03561	+0.000385	−75.70	−32.70	−0.4866
35	+33.96	+14.07	+0.1245	−14.07	−7.090	−0.06423	+0.6124	+0.08739	+0.000700	−95.68	−49.33	−0.5497
40	+39.51	+21.38	+0.1468	−21.38	−12.62	−0.08898	+0.6071	+0.1815	+0.001123	−112.7	−68.34	−0.5990
45	+44.27	+28.96	+0.1600	−28.96	−19.84	−0.1132	+0.6944	+0.3237	+0.001601	−122.9	−86.91	−0.6273
50	+45.16	+34.63	+0.1600	−34.63	−27.44	−0.1321	+0.8010	+0.4950	+0.002014	−122.8	−101.0	−0.6299
r/L = 0.3												
30	+110.9	+37.12	+0.08756	−37.12	−15.70	−0.03796	+1.641	+0.2500	+0.000532	−340.4	−147.6	−0.4443
35	+123.0	+58.70	+0.10342	−58.70	−29.63	−0.05150	+1.782	+0.5520	+0.000868	−402.2	−200.2	−0.4792
40	+136.4	+78.62	+0.10899	−78.62	−46.73	−0.06707	+2.154	+0.9824	+0.001193	−423.7	−260.8	−0.4884
45	+131.1	+89.18	+0.10274	−80.18	−62.16	−0.07481	+2.432	+1.410	+0.001376	−399.4	−290.9	−0.4749
50	111.5	+89.27	+0.09013	−89.27	−73.18	−0.07807	+2.437	+1.695	+0.001382	−350.4	−303.5	−0.4550
r/L = 0.4												
30	+247.6	+98.82	+0.07423	−98.82	−41.89	−0.03237	+3.525	+0.9471	+0.000586	−912.9	−397.8	−0.3936
35	+283.3	+139.7	+0.07945	−139.7	−70.98	−0.04180	+4.354	+1.701	+0.000837	−976.3	−513.6	−0.4011
40	+273.8	+161.7	+0.07471	−161.7	−97.64	−0.04717	+4.977	+2.518	+0.000964	−918.0	−579.7	−0.3884
45	+232.0	+163.1	+0.06551	−163.1	−117.2	−0.04975	+4.925	+3.016	+0.000945	−805.0	−611.3	−0.3732
50	+187.9	+157.7	+0.05768	−157.7	−135.5	−0.05270	+4.456	+3.239	+0.000875	−708.8	−647.5	−0.3670
r/L = 0.5												
30	+473.4	+194.4	+0.06058	−194.4	−82.77	−0.02666	+6.899	+2.116	+0.000570	−1817.0	−800.0	−0.3421
35	+486.4	+244.3	+0.05905	−244.3	−125.5	−0.03167	+8.309	+3.527	+0.000704	−1772.0	−950.2	−0.3343

(continued)

Table B2 (continued)

φ	Vertical edge load			Horizontal edge load			Shear edge load			Edge moment		
	Δ_V	Δ_H	φ	Δ_V	Δ_H	φ	Δ_V	Δ_H	φ	Δ_V	Δ_H	φ
40	+424.2	+256.6	+0.05240	−256.6	−158.6	−0.03414	+8.519	+4.483 + 4.924	+0.000707	−1572.0	−1024.0	−0.3210
45	+349.1	+254.2	+0.04653	−254.2	−189.1	−0.03671	+7.792	+5.075	+0.000653	−1396.0	−1101.0	−0.3156
50	+295.1	+260.4	+0.04398	−260.4	−234.5	−0.04173	+6.701		+0.000616	−1319.0	−1252.0	−0.3226
r/L = 0.6												
30	+756.0	+315.3	+0.04838	−315.3	−135.1	−0.02155	+11.82	+3.979	+0.000510	−3009.0	−1341.0	−0.2961
35	+706.1	+360.8	+0.04430	−360.6	−188.1	−0.02429	+13.15	+5.790	+0.000555	−2756.0	−1511.0	−0.2847
40	+591.7	+367.3	+0.03930	−367.3	−232.8	−0.02637	+12.46	+6.714	+0.000519	−2445.0	−1640.0	−0.2778
45	+498.8	+374.8	+0.03615	−374.8	−287.2	−0.02931	+11.18	+7.265	+0.000486	−2249.0	−1823.0	−0.2782
50	+431.9	+389.3	+0.03392	−389.3	−357.3	−0.03284	+10.21	+7.983	+0.000481	−2110.0	−2043.0	−0.2807

Index

© The Author(s), under exclusive license to Springer Nature Switzerland AG 2022
M. Gohnert, *Shell Structures*, https://doi.org/10.1007/978-3-030-84807-1